STARBURSTS

TRIGGERS, NATURE, AND EVOLUTION

Les Houches School, September 17-27, 1996

Editors

Bruno GUIDERDONI
Ajit KEMBHAVI

EDP Sciences

7, avenue du Hoggar
Parc d'Activités de Courtabœuf
B.P. 112
91944 Les Ulis cedex A, France

Springer-Verlag
Berlin Heidelberg GmbH

EDP Sciences
875-81 Massachusetts Avenue
Cambridge, MA 02139, USA

Centre de Physique des Houches

Books already published in this series

1 Porous Silicon Science and Technology
J.-C. VIAL and J. DERRIEN, Eds. 1995

2 Nonlinear Excitations in Biomolecules
M. PEYRARD, Ed. 1995

3 Beyond Quasicrystals
F. AXEL and D. GRATIAS, Eds. 1995

4 Quantum Mechanical Simulation Methods
for Studying Biological Systems
D. BICOUT and M. FIELD, Eds. 1996

5 New Tools in Turbulence Modelling
O. MÉTAIS and J. FERZIGER, Eds. 1997

6 Catalysis by Metals
A. J. RENOUPREZ
and H. JOBIC, Eds. 1997

7 Scale Invariance and Beyond
B. DUBRULLE, F. GRANER
and D. SORNETTE, Eds. 1997

8 New Non-Perturbative Methods and
Quantization on the Light Cone
P. GRANGÉ, A. NEVEU, H.C. PAULI,
S. PINSKY and E. WERNER, Eds 1998

Book series coordinated by Michèle LEDUC

Editors of "Starbursts: Triggers, Nature, and Evolution" (N° 9)
Bruno Guiderdoni (Institut d'Astrophysique de Paris, CNRS, France)
Ajit Kembhavi (Inter-University Centre for Astronomy and Astrophysics, Pune, India)

ISBN 978-2-86883-334-1   ISBN 978-3-662-29742-1 (eBook)
DOI 10.1007/978-3-662-29742-1

© Springer-Verlag Berlin Heidelberg 1998
Originally published by Springer-Verlag Berlin Heidelberg New York in 1998
Softcover reprint of the hardcover 1st edition 1998

AUTHORS

Jean-Pierre Chièze, DSM/DAPNIA/Service d'Astrophysique, CEA Saclay,
91191 Gif-sur-Yvette cedex 01, France

Françoise Combes, Observatoire de Paris, DEMIRM, 61 avenue de l'Observatoire,
75014 Paris, France

Edith Falgarone, Ecole Normale Supérieure, Observatoire de Paris and CNRS,
24 rue Lhomond, 75005 Paris, France

Robert Kennicutt, Steward Observatory, University of Arizona, Tucson AZ 85721,
USA

Georges Meynet, Geneva Observatory, 1290 Sauverny, Switzerland

Francesco Palla, Osservatorio Astrofisico di Arcetri, Largo E. Fermi 5,
50125 Firenze, Italy

Joseph Silk, Departments of Astronomy and Physics, and Center for Particle
Astrophysics, University of California, Berkeley CA 94720, USA

FOREWORD

Starbursts are regions of unusually rapid star formation, often located in the central parts of galaxies. They differ from more normal regions of star formation in terms of the throughput of mass and the rapidity with which the gas is consumed. In the last twenty years, extensive observational data at most wavelengths have become available on starbursts, but many important issues remain to be addressed, observationally as well as theoretically.

How are strong episodes of star formation triggered? What is the quantity of gas converted into stars during bursts? What is the initial mass function of stars in these events? How does the feedback from stars influence the interstellar medium and self-regulate star formation? What is the subsequent chemical and photometric evolution? How do starbursts rule the formation and evolution of galaxies?

In recent years, many observational data at different wavelengths (optical, radio, infrared, X-ray) have become available. However, these observations are still fragmentary in the sense that different classes of objects have been observed in different ways, and the coverage is not consistently deep or complete. As a consequence, an overall observational picture of starburst galaxies is missing, and theoretical understanding and modelling have remained highly tentative. The purpose of the school *Starbursts: Triggers, Nature, and Evolution* was to gather theorists and observers with complementary approaches to the starburst phenomenon, in order to summarize the state-of-the-art of the observations and models, emphasizing the consistency of the various viewpoints.

Multiwavelength observations of starburst and "normal" galaxies now provide interesting clues to the parameters ruling large-scale star formation. On smaller scales, the hierarchy of gas clouds, from giant molecular clouds to compact cores, has been observed in great detail and has begun to be theoretically understood. It is now possible to address the formation of isolated and clustered stars, and to understand the mechanisms of feedback to the interstellar medium. These new ideas can also be studied through hydrodynamical simulations. The spectrophotometric and chemical evolution of starburst and post-starburst stellar populations can be followed, thanks to our knowledge of stellar evolution for the whole range of masses.

Extragalactic optical and infrared surveys have unveiled the increasing importance of galaxy interaction and merging for the triggering of starbursts. Numerical simulations now reproduce many original features of these galaxies, including their disturbed morphologies and gas inflows fuelling the starbursts. Finally, a consistent scenario of galaxy formation and evolution, which makes use of these ideas in the cosmological paradigm of hierarchical structure formation, is slowly emerging. In such an exciting context, we hope that the lecture notes gathered in this volume will provide a better understanding of these new insights into the physics of starbursts, and help in answering at least some of the questions raised above.

Acknowledgements

This session at the Centre de Physique des Houches and the present volume of lecture notes could not have been achieved without the financial support of the "Training and Mobility of Researchers" Programme of the European Commission (contract ERB-4064-PL-95-0251), the Division of "Permanent Training" of the Centre National de la Recherche Scientifique, the Inter-University Center for Astronomy and Astrophysics (Pune), the Institute of Astrophysics (Paris), the Division of Science and Technology of the Ministry of Foreign Affairs, the ACCES Programme of the Ministry of Higher Education and Research, the Department of Science and Technology of the Government of India, the French Embassy in Delhi, and the University Joseph Fourier of Grenoble.

We are grateful to Alain Omont and Jayant Narlikar for advice in the critical initial stages, to Michèle Leduc for her help and continuous support, to Ghislaine Chioso, Brigitte Rousset and Claire Simon for their patience and skill in the management of the session, to Mme J. Fichard for her help, to Jyotsna Apte, Manjiri Mahabal and Archana Kamnapure for secretarial assistance, and to Santosh Khadilkar for help with the poster. Finally, this volume would not have existed without the work and expertise of our lecturers, and without the presence of the students, who came from various countries. They attended the lectures with enthusiasm. We hope that these lecture notes, which hopefully will become a standard reference in the field, will aid them in their research work.

Bruno Guiderdoni and Ajit Kembhavi

CONTENTS

LECTURE 3

Elements of Hydrodynamics
Applied to the Interstellar Medium

by J.-P. Chièze

LECTURE 6

Starburst Triggering and Environmental Effects

by F. Combes

LECTURE 7
From Star to Galaxy Formation
by J. Silk

Fundamental Aspects of Star Formation in Galaxies

R. Kennicutt

Steward Observatory, University of Arizona
Tucson, AZ 85721, USA

1. INTRODUCTION

The starburst phenomenon is one of the most actively pursued problems in extragalactic astronomy, but it is just one aspect of the much broader subject of large-scale star formation in galaxies. This subject is important not only for understanding the physical context of star formation on the scale of individual stars and molecular clouds, but also for understanding the formation and evolution of galaxies in a broader context.

Systematic studies of star formation in external galaxies traces back to the seminal papers of Hubble and Shapley in the 1920's and 1930's. Roberts (1963) was one of the first to systematize what was known about the global star formation properties and histories of galaxies. This approach was quantified in the 1970's with the first applications of evolutionary synthesis models (Tinsley 1972; Searle et al. 1973; Huchra 1977; Larson & Tinsley 1978). The latter papers introduced the concept of the star formation burst and applied rudimentary burst models to examples of interacting and peculiar galaxies. This was followed in the 1980's by the development and calibration of a number of more precise methods for measuring star formation rates (SFRs), and their application to the systematic properties of nearby galaxies (e.g., Kennicutt 1983; Gallagher et al. 1984; Donas et al. 1987).

The discovery of the starburst phenomenon stimulated the interest of a much broader community of extragalactic astronomers and expanded the field almost overnight. Optically-selected starburst galaxies were revealed already in the 1970's by the Byurakan surveys and several emission-line surveys, and the term "starburst" itself was coined to describe an optically-selected emission-line galaxy (Weedman et al. 1981). However it was the discovery of a new class of ultraluminous starbursts in the infrared (Rieke & Low 1972), and the subsequent discovery of hundreds of such objects in the IRAS survey that stimulated much of the subsequent interest and progress in this field.

Starbursts represent the extreme of what is actually a continuous spectrum of star formation scales, which span more than a million fold range in mass and luminosity from familiar regions such as the Orion nebula to giant HII regions, normal galaxies, emission-line galaxies, and finally the ultraluminous infrared galaxies. The main goal of my lectures is to introduce the context of starbursts, by reviewing the star formation properties of the normal, quiescent galaxies that are the hosts to starbursts. I begin by describing the techniques used to measure SFRs in galaxies, and then apply them to the global star formation properties of normal galaxies. The next lecture is a brief review of the initial mass function (IMF) in star forming galaxies. The final lecture reviews our current understanding of the physical processes that determine the large-scale SFR.

2. STAR FORMATION PROPERTIES OF NORMAL GALAXIES

Before we consider the properties of starbursts in later lectures, it is useful to review the star formation properties of normal galaxies. I begin with a brief description of the methods used to measure current star formation rates (SFRs) in galaxies, followed by a summary of the star formation properties of the Hubble sequence, and the underlying evolutionary interpretation of this sequence. More complete tutorials on this subject can be found in Kennicutt (1990; 1992b; 1997).

2.1. Quantitative Diagnostics of Star Formation Rates

Accurately determining the large-scale SFR of a galaxy is difficult business, because the massive young stars are unresolved in all but the closest systems, even with HST, and the low-mass stellar component of the stellar population is not directly observable in any case. Several techniques have been developed to measure SFRs from the integrated light of galaxies, and I describe the most widely used methods, and their particular advantages and weaknesses for different types of galaxies.

2.1.1. Resolved Stellar Populations

For the nearest galaxies it is possible to construct HR diagrams of young regions, and take a direct census of the massive stellar populations. This technique is time consuming and obviously limited to a handful of systems, but

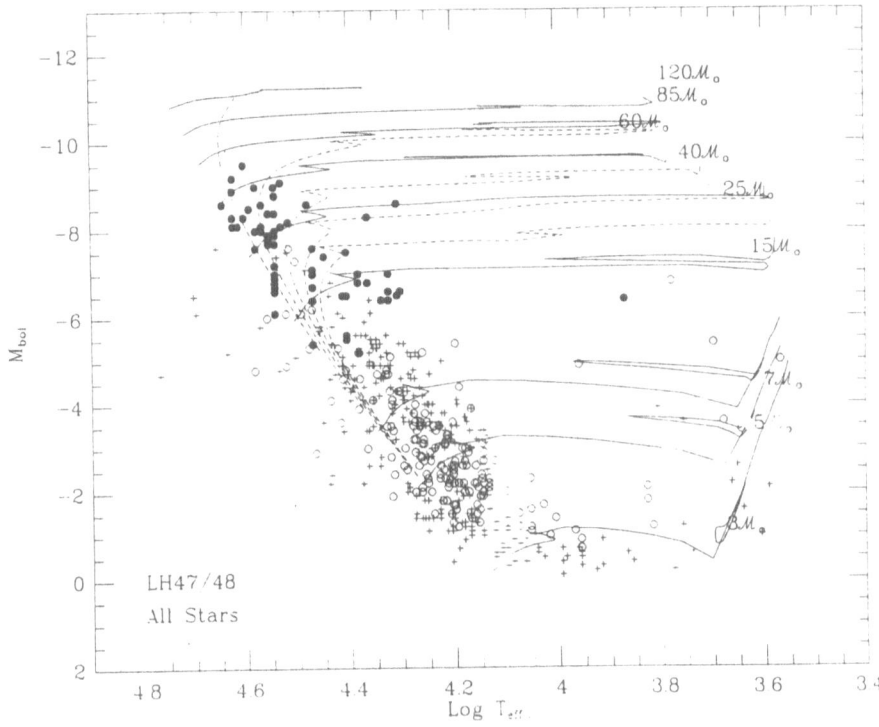

Fig. 1. — HR diagram for the OB association LH47/48 in the LMC, with evolutionary tracks superposed, from Oey & Massey (1995).

such data provide a fundamental foundation for calibrating the other methods. The most extensively studied galaxy to date is the LMC, where complete stellar surveys have been carried out for ~20 OB associations as well as the LMC field (e.g., Parker & Garmany 1993; Massey et al. 1995a; Oey 1996). Accurate measurements of the massive stellar contents require a combination of broadband photometry and stellar spectroscopy for the hottest stars, because colors alone are insensitive to stellar type for stars above ~10 $M_\odot$. Figure 1 shows an example of this technique applied to the LMC OB association LH47/48 (Oey & Massey 1995). By superimposing the HR diagram on theoretical evolutionary tracks it is possible to measure the mass in young stars directly. A large database of photometry alone (i.e., without corresponding stellar spectra) exists for many other galaxies, but the degenerate visible colors of the massive stars make it impossible to obtain complete samples and accurate masses and SFRs. Systematic errors of an order of magnitude are not uncommon in those studies.

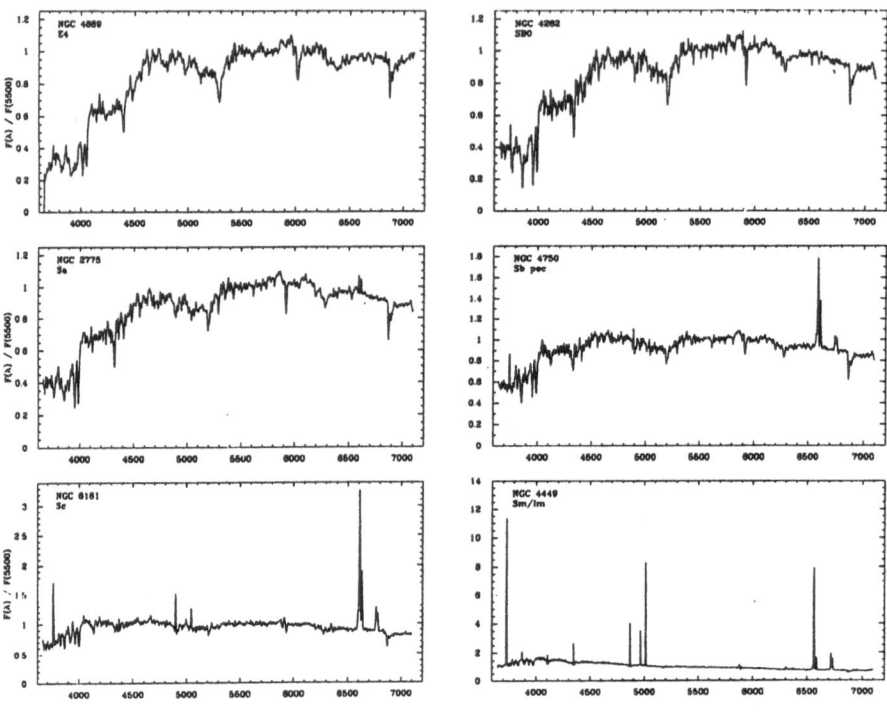

Fig. 2. — Integrated spectra of elliptical, spiral, and irregular galaxies. The fluxes
have been normalized to unity at 5500 Å (Kennicutt 1992a).

2.1.2. Integrated Colors and Spectra

Stellar crowding, combined with the difficulty of obtaining spectra for large
numbers of individual stars, makes it nearly impossible to obtain complete
samples of young massive stars in galaxies outside of the Local Group. Fortunately, the combined effects of a young stellar population on the integrated
spectra of galaxies are readily apparent, and several techniques can be used
to quantitatively measure the aggregate luminosity and mass in young stars.
Figure 2, taken from Kennicutt (1992a), shows the progression of spectral properties of 6 nearby galaxies spanning the range in Hubble types from elliptical
to Magellanic irregular. As one progresses to more active star forming systems
several changes in the spectrum are apparent: an increase in the blue con-

tinuum (mainly dominated by A-type stars), a change in the character of the stellar absorption spectrum, and most pronounced of all, a dramatic increase in the strengths of the nebular emission lines, especially Hα.

Although the integrated spectra contain contributions from the full range of stellar spectral types and luminosities, it is easy to show that at visible wavelengths the dominant contributors are intermediate-type main sequence stars (mainly A to early F) and bright red (G–K) giants. As a result the integrated colors and spectra of normal galaxies fall on a relatively tight sequence, with the spectrum of any given object dominated by the ratio of early to late-type stars, or alternatively by the ratio of young (< 1 Gyr) to old (3–15 Gyr) stars. This allows one to extract the stellar age ratio (and an estimate of the SFR) by modelling the spectrum with a relatively small number of colors or spectral indices.

This most widely applied version of this method is known as evolutionary synthesis modelling, and it is described in depth in the lectures by G. Meynet. Consequently I only briefly summarize the method as applied to the broadband colors of galaxies. Stellar evolution models are used to derive the effective temperatures and bolometric luminosities for various stellar masses as a function of time, and these are converted into broadband luminosities vs time using stellar atmosphere models. Those luminosities in turn are weighted by an IMF and summed to synthesize the integrated luminosities and colors of model star clusters as a function of time. These "single-burst models" can then be added in linear combination to synthesize the spectrum or colors of a galaxy with an arbitrary star formation history, usually parameterized as an exponential function of time. The models are often parameterized in terms of the ratio of the current SFR to the average past SFR over the age of the system, denoted here as b.

Although these models contain several parameters, including the star formation history, age, metallicity, and IMF, the colors of normal galaxies are remarkably well represented by a single parameter sequence of models with fixed age, composition and IMF, varying only in the time dependence of the SFR (Searle et al. 1973; Larson & Tinsley 1978). This is illustrated in Figure 3, which shows the colors of normal galaxies from the RC2, along with a set of models from Kennicutt (1983). Each line corresponds to a set of coeval populations with fixed IMF, and exponential star formation histories with b ranging from 0 to 2. The best fitting model (middle line) is for roughly Salpeter IMF and age of 15 Gyr (Kennicutt 1983).

The models allow one to infer b directly for a given galaxy, simply by matching its colors to the closest point on a model sequence (e.g., Kennicutt et al. 1994). The models also predict the SFR per unit luminosity in a given bandpass as a function of b (Searle et al. 1973; Larson & Tinsley 1978; Tinsley & Danly 1980), and this can used to crudely estimate the absolute SFR. The method is no longer widely applied to nearby galaxies, because the derived values are relatively inaccurate (colors vary logarithmically with the SFR), and are subject to large systematic errors from reddening, metallicity, age, and IMF variations.

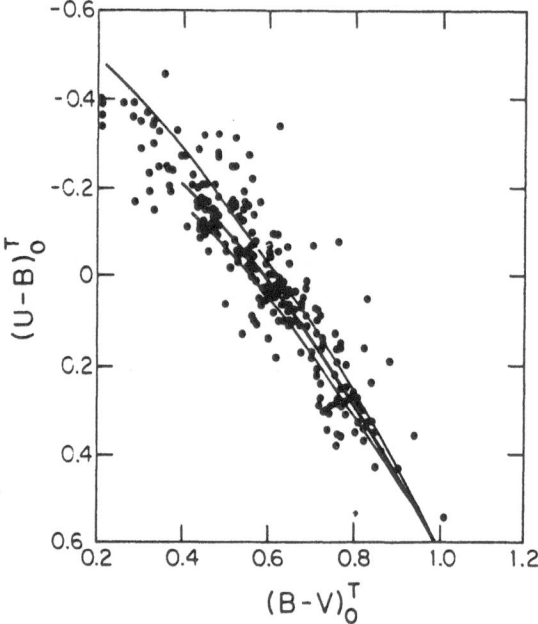

Fig. 3. — *UBV* colors of spiral galaxies and simple synthesis models with constant age and IMF, and varying only in the star formation history (Kennicutt 1983).

Nevertheless the technique is enjoying a modest revival as applied to samples of high-redshift galaxies, where often colors are the only available information. The same technique, applied to full spectral syntheses of galaxies, are also a primary observational constraint on starburst properties, as described in later lectures.

2.1.3.   UV Continuum Fluxes

Many of the limitations encountered in modelling of the stellar continuum can be avoided if one observes at wavelengths where the continuum is dominated by

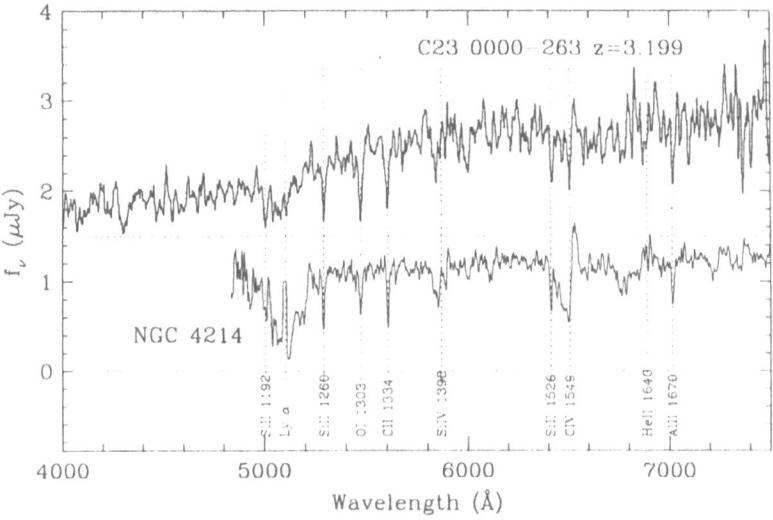

Fig. 4. — Ultraviolet spectra of star forming galaxies near and far. The upper spectra are for $z = 3$ objects identified with the Keck telescope, while the lower spectra are of an HII/OB association in the nearby galaxy NGC 4214, observed with HST (Leitherer et al. 1996). Figure from Steidel et al. (1996).

young stars; in that case the SFR will scale linearly with the monochromatic flux. For active star-forming galaxies, the best continuum wavelengths are in the range 1300–2500 Å, longward of the Lyman absorption lines but short enough to minimize spectral contamination from older stellar populations. This wavelength range has the additional advantage that it is accessible from the ground in highly redshifted systems ($z \simeq 2 - 5$). A beautiful example is shown in Figure 4, which displays the near-UV spectrum of an OB association in the Magellanic irregular NGC 4214 observed with HST (Leitherer et al. 1996) next to the redshifted spectra of two $z > 3$ starburst galaxies observed with the Keck telescope (Steidel et al. 1996).

The inaccessibility of this part of the spectrum for low-redshift galaxies has limited the application of this technique, but the situation is improving rapidly, thanks in large part to HST and the Ultraviolet Imaging Telescope (e.g., Meurer et al. 1995; Maoz et al. 1996; Smith et al. 1996). A growing database of UV spectra are also becoming available from HST (e.g., Fig. 4) and from IUE (e.g., McQuade et al. 1995). The most complete work on global SFRs is based on a series of balloon and space experiments from a group at Marseille (e.g., Buat 1992; Deharveng et al. 1994).

Calculating the conversion between the flux at an arbitrary UV wavelength and the SFR follows the same procedures as outlined earlier. SFR calibrations in terms of monochromatic UV fluxes are available from Buat et al. (1987) and Leitherer & Heckman (1995). Since the UV luminosity is directly coupled to the youngest population ($<$ 20 Myr) the conversion is much more straight-forward, but only massive star formation is directly measured, and total SFRs require that the form of the IMF be assumed. In systems with very low SFRs, contamination from older post-AGB populations can become important, and the UV continuum only yields upper limits on the SFR. Extinction is the most serious systematic error with this technique, and the best SFRs are based on extinction corrections which take into account the patchy structure of the dust and the radiative transfer of the UV radiation (Calzetti et al. 1994). Nevertheless the UV spectrum remains a very powerful tool for characterizing the SFRs of galaxies, especially at high redshift where it is readily observed with ground-based telescopes.

2.1.4. Emission-Line Fluxes

Figure 2 demonstrates that the most dramatic change in the integrated spectrum with galaxy type is the rapid increase in the strengths of the nebular emission lines. Since the nebular lines effectively re-emit the integrated luminosity of galaxies shortward of the Lyman limit, they provide a direct and sensitive measure of the SFR for the massive hot stars that dominate the ionizing continuum. The most extensive surveys are based on Hα measurements (Kennicutt & Kent 1983; Romanishin 1990), though any hydrogen recombination line can be used in principle.

The conversion factor between the Balmer flux and the SFR is computed by constructing model star clusters as described earlier, except that the stellar atmospheres are integrated shortward of the Lyman limit to derive the ionizing photon flux. These are then converted to Balmer emission-line fluxes using standard recombination theory. Widely used conversions are published by Kennicutt (1983), Gallagher et al. (1984), Kennicutt et al. (1994), and Leitherer & Heckman (1995). The conversions are sensitive to the assumed IMF slope and stellar mass limits, and different authors adopt different conventions, so I refer the reader directly to the primary papers for details. In the blue, higher-order Balmer lines can be used, but underlying stellar absorption often is difficult or impossible to remove. The [OII] doublet can serve as a substitute for the Balmer lines, and is especially useful for high-redshift galaxies, where Hα becomes unobservable. Kennicutt (1992a) provides a rough empirical calibration of the [OII] SFR scale.

The primary advantages of this technique are its sensitivity and direct coupling between the nebular emission and the massive SFR. With CCD detectors and narrow band filters, the Hα emission can be mapped at high resolution with even a small telescope, so this technique has provided our most detailed information about the distribution and structure of the star forming regions in galaxies. Observations of the redshifted Hα line have even been applied to starburst galaxies at $z = 2.72$ (Bechtold et al. 1997). The chief disadvantages of the method are its sensitivity to extinction, the IMF, and the escape of ionizing photons from the galaxy. As with the UV continuum fluxes, the emission-line fluxes only trace the most massive stars ($M \geq 10\ M_\odot$), and the SFR for lower-mass stars must be derived in some other way. Kennicutt (1983) used measurements of the visible continuum together with Hα to constrain the IMF down to $\sim 1\ M_\odot$, but it is not clear whether this IMF is valid in extreme starburst environments (next lecture). Extinction is the other problem, especially in luminous starburst galaxies. The extinction in normal galaxies can be constrained by measuring the reddening of HII regions, or by comparing the integrated Hα and thermal radio fluxes of galaxies (e.g., Kennicutt 1983; van der Hulst et al. 1988). These results show that the typical extinction correction is about 1 mag. However it can be much higher in the nuclear regions of galaxies, and in the strongly concentrated starbursts found in many mergers. For those objects, the SFR is best determined instead from infrared or submillimeter recombination lines or from modeling of the far-infrared emission from dust (below).

2.1.5. Far-Infrared Continuum

A significant fraction of the bolometric luminosity of a galaxy is absorbed by interstellar dust, and this radiation is re-emitted in the thermal infrared, at wavelengths of roughly 10–300 μm. Since the absorption cross section of the dust is strongly peaked in the ultraviolet, this far-infrared (FIR) emission can be a sensitive tracer of the SFR. The IRAS survey has produced a database of over 30,000 galaxies (Moshir et al. 1992), offering a rich reward to those who can calibrate an accurate SFR scale from the 10–100 μm FIR emission.

Calibration of this SFR scale for normal galaxies has proven to be difficult, however. The dust grains absorb radiation with a broad range of wavelengths, and hence the heating function contains contributions from young and old stars alike. Modelling this heating, and separating the component associated with star formation is complicated by the different spatial distributions of young stars, old stars, and dust. Young stars appear to dominate the FIR emission of many galaxies, especially in a warm component peaking near 60 μm, but the contribution of heating from cooler, older stars may be significant and is a subject of much debate (e.g., Devereux & Young 1990; Walterbos & Greenawalt 1996; Buat & Xu 1996). Uncertainty over the nature of this cool component has stymied efforts to construct a linear SFR calibration based on some combination of the IRAS spectral indices.

The physical situation is often much simpler in dusty starburst regions, however. There the stellar radiation field is dominated by a young stellar population, and the optical depth of the dust often is very high. In such objects the interpretation of the FIR emission is simple; it effectively measures the bolometric luminosity of the starburst. This in turn makes the FIR flux a very powerful probe of the total SFR, and often the only reliable probe available.

2.2. SFRs of Normal Galaxies

Applications of these methods reveal an enormous diversity in the star formation properties of normal galaxies, and strong correlations with the Hubble
types and other properties of the parent galaxies. Here I present a very brief
synopsis, emphasizing those results which are the most useful for interpreting
the properties of interacting galaxies. For simplicity I will focus mainly on results from $H\alpha$ surveys, but most other studies lead to similar conclusions (with
exceptions discussed below).

Absolute SFRs measured for nearby (normal) galaxies range over $0 - 10$
$M_\odot\,yr^{-1}$, with the large range reflecting differences in both galaxy mass and
morphological type (Kennicutt 1983; Gallagher et al. 1984; Caldwell et al.
1991, 1994; Kennicutt et al. 1994). This is quantified in Figure 5, which
shows the distribution of $H\alpha$ + [NII] equivalent widths in a sample of 150
nearby spiral galaxies, subdivided by Hubble type. This index is defined as the
emission-line luminosity normalized to the adjacent continuum flux, and hence
is proportional to the SFR per unit (red) luminosity. In all but the strongest
emission-line galaxies, the red continuum is dominated by old stars, so this
provides an approximate index of the SFR per unit galaxy mass.

The SFR per unit luminosity is a strongly increasing function of Hubble
type, increasing from zero in E/S0 galaxies (within the observational errors)
to several solar masses per year for an Sc–Irr L^* galaxy (roughly comparable
to the Milky Way). Also apparent in Figure 5 is a very large dispersion in
relative SFR among galaxies of the same type. The scatter, roughly an order
of magnitude for most types, is much larger than would be expected from
photometric errors or extinction effects, so most of it must reflect real variations
in the SFR. Several mechanisms appear to be responsible, including variations
in gas content (Kennicutt 1989), nuclear emission, interactions, and temporal
variation in the SFR within individual objects (Kennicutt et al. 1994).

Similar qualitative trends in the SFR are seen in UV continuum properties,
broadband colors, and integrated spectra of galaxies (see references earlier).
However the integrated FIR emission tends to show a considerably weaker dependence on Hubble type (e.g., Devereux & Young 1996; Tomita et al. 1996;
Devereux & Hameed 1997). Several factors probably contribute to this inconsistency. The FIR observations sometimes reveal very intense nuclear emission in
early-type galaxies, that often appears to be associated with very dense, dusty
central starbursts. This activity tends to be strongly associated with strongly
barred spirals, and it reveals that some early-type spirals can have substantial
SFRs that are not always apparent in the visible or ultraviolet data. It is also
likely that some of the elevated FIR emission in early-type systems is associated with dust heated by older stars, even in systems with low or negligible
SFRs. Nevertheless this work has revealed an important bias in the optical/UV
based studies, and the importance of circumnuclear star formation, especially
in early-type systems.

High-resolution imaging of individual galaxies in $H\alpha$ reveals that the system-

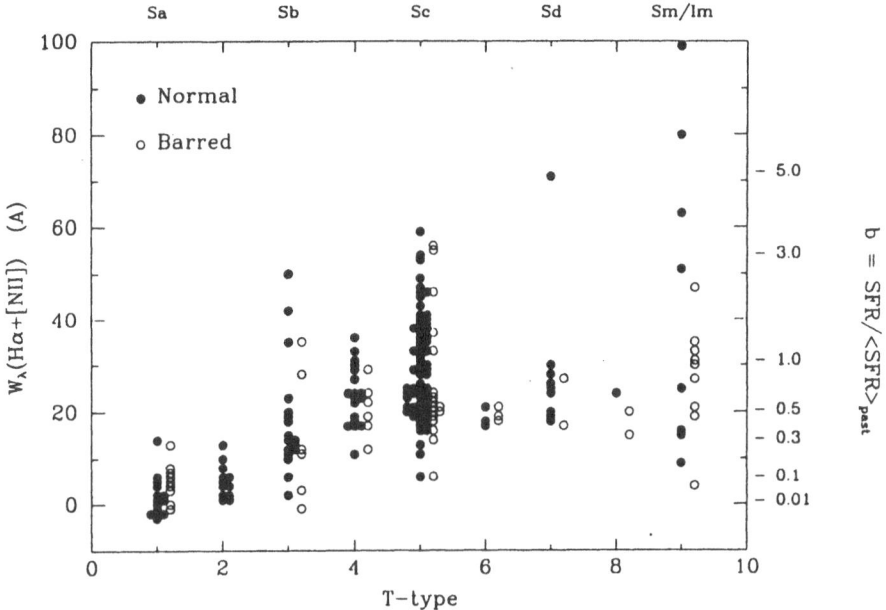

Fig. 5. — Distribution of Hα + [NII] emission-line equivalent widths for a large sample of nearby spiral galaxies, subdivided by type and into normal and barred systems. The scale on the right shows the approximate values of the birthrate parameter b.

atic change in the total SFR along the Hubble sequence is produced in roughly equal parts by an increase in the total number of star forming regions per unit mass or area, and an increase in the characteristic masses of individual regions (Kennicutt et al. 1989). The changes in the mass spectra of the dominant star forming regions may have important implications for the structure and evolution of their respective interstellar media. For example, in an Sa or early Sb spiral the average OB star forms in an HII region of roughly the size of the Orion nebula, in a cluster containing only a handful of massive stars (or singly). By contrast, the average massive star in a late-type Sc or Irr galaxy forms in a giant HII/OB association containing hundreds or thousands of OB stars, and often in dense clusters analogous to the "blue globular" clusters in the Large Magellanic Cloud. The physical or dynamical conditions in these gas-rich, late-type galaxies have altered not only the amounts but also the character of the star formation, and we should not be surprised to find even more dramatic differences in the star formation environments in the most active interacting systems.

What other large-scale properties of a galaxy influence its global SFR? As Figure 5 illustrates, the presence of a bar does not seem to significantly influence the total SFR, though it does appear to trigger nuclear star formation in many galaxies, especially early-type systems (references above). Likewise the presence of a grand-design spiral structure does not appear to significantly affect the total SFR, but rather simply changes its spatial distribution (e.g., Elmegreen & Elmegreen 1986; McCall & Schmidt 1986). The strongest dynamical influence appears to be the presence of a tidally interacting companion, as quantified in dozens of studies (Barnes et al. 1997 and references therein).

2.3. Interpretation: Star Formation Histories

The strong trends observed in the relative SFRs of galaxies along the Hubble sequence mirror underlying trends in their past star formation histories (Roberts 1963; Kennicutt 1983; Gallagher et al. 1984; Sandage 1986; Kennicutt et al. 1994). This is shown directly in Figure 5, which shows the distribution of the birthrate parameter b (right margin scale) as derived in this case from the Hα equivalent widths (Kennicutt et al. 1994). The figure shows that the typical late-type spiral has formed stars at a roughly constant rate ($b \sim 1$), consistent with direct measurements of the stellar age distribution in the Galactic disk (e.g., Scalo 1986). By contrast, early-type disks are characterized by rapidly declining SFRs, with $b \sim 0.01 - 0.1$. These changes in disk properties are much larger than those caused by changes in bulge/disk ratio, though differences in bulge structure may be important for driving the disk evolution. Unfortunately it is not possible with these kinds of measurements to characterize the detailed time dependence of the star formation histories across the Hubble sequence, but a schematic illustration of one such interpretation is shown in Figure 6, taken from Sandage (1986). Note that other time dependences could fit the observed b constraints, for example a high initial rate followed by a rapid decline, with the duration of the peak SFR increasing with later Hubble type. As more direct observations of distant galaxies become available, it should be possible to constrain the possibilities.

3. THE INITIAL MASS FUNCTION

3.1. Introduction

It should be clear from the preceding section that interpreting the star formation properties of galaxies depends critically on knowing the relative distribution of stellar masses, or the initial mass function (IMF). The tracers of the massive SFR in integrated light (UV and FIR continua, emission lines) are produced predominantly by stars more massive than 5–10 $M_\odot$, a population that typically accounts for only 10–20% of the total mass of the population. Consequently when these techniques are used to estimate the total SFR in a

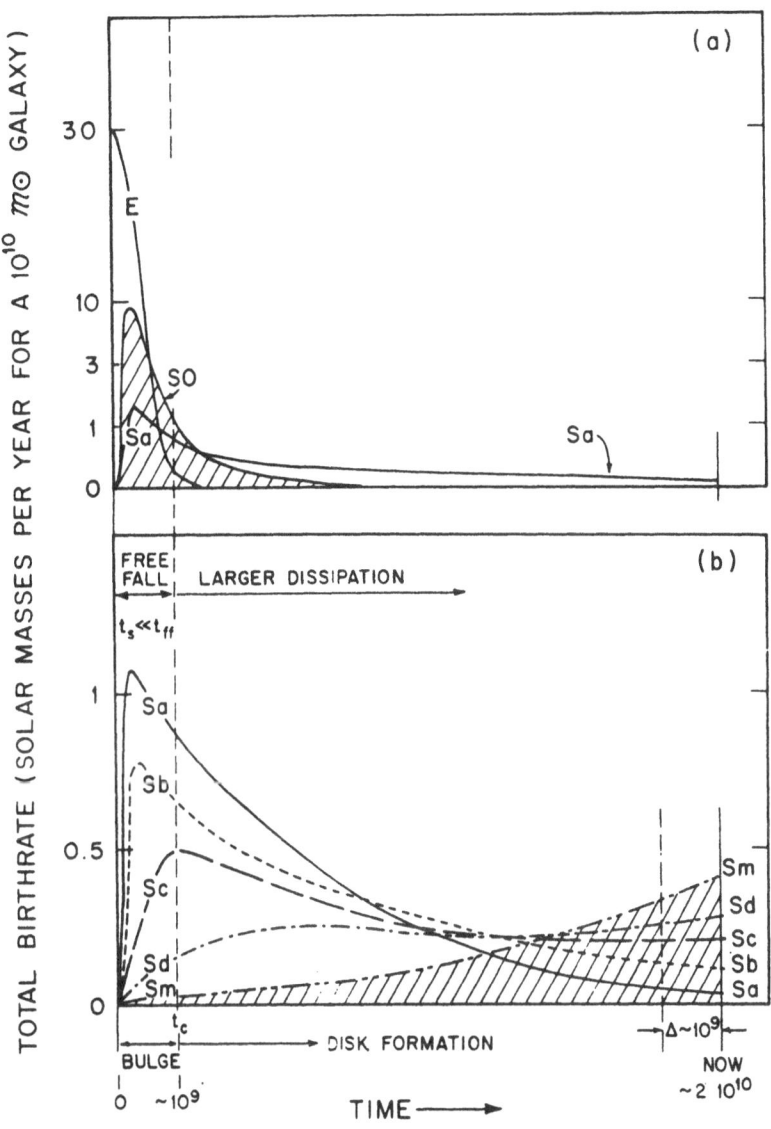

Fig. 6. — A schematic illustration of how the stellar birthrate may have changed with galaxy type, based on data of the type shown in Fig. 5. Figure taken from Sandage (1986).

galaxy, they involve extrapolations of 5–10 times the directly observed stellar mass. It goes without saying that an accurate determination of the IMF is a prerequisite to being able to accurately measure the SFRs and star formation histories of galaxies and starbursts. Unfortunately measuring the IMF in regions beyond the solar neighborhood is notoriously difficult, and we are only beginning to place meaningful constraints on the form and variation of the IMF in external galaxies. This uncertainty is one of the largest remaining obstacles to understanding and modelling galaxy evolution, and this subject will doubtless be a major focus of observational and theoretical work in coming years.

There are few subjects in modern astronomy that have been as profoundly influenced by a few seminal papers as the IMF. Salpeter (1955) made one of the first systematic measurements of the IMF in the solar neighborhood, and the resulting Salpeter function remains widely used today. A major attack on the problem was initiated by Scalo in the 1970's, the culmination of which was a major review article that fills an entire journal issue (Scalo 1986). Scalo's review remains the fundamental reference in this subject, and it is especially useful as a pedagogical text, because Scalo describes the methods used to measure the IMF in great detail. It is impossible in a few pages to summarize fully the state of this subject or cite even a fraction of the relevant literature. My goals instead are to briefly review the steps taken to determine the IMF, and summarize the most important changes that have taken place in the field since Scalo's article was written.

3.2.   Nomenclature

One of the difficulties in this field is the confusing and often inconsistent nomenclature that is used to describe and parameterize the IMF. Mistakes abound even in the published literature, so to avoid confusion I begin with a brief set of definitions and nomenclature that I will apply uniformly in this lecture.

The *Initial Mass Function (IMF)* is the distribution function that describes the number of stars formed per unit time in a fixed volume, per unit mass interval or logarithmic mass interval (more on that later). It is usually normalized to unity when integrated over the entire stellar mass spectrum, and hence the IMF can be thought of as a probability distribution.

The *Present Day Mass Function (PDMF)* is the function describing the presently observed distribution of stellar masses in a fixed volume of space. In a star cluster the PDMF is identical to the IMF for stars below the main sequence turnoff (assuming that stars have not escaped), but of course it is entirely different above the turnoff mass. For a field population the PDMF/IMF conversion may also be affected by variations in the local SFR as a function of stellar age.

The *Luminosity Function (LF)* is the distribution of stellar luminosities, usually expressed as a function of absolute magnitude, per fixed absolute magnitude interval. It is related to the PDMF by the stellar mass-luminosity relation.

The *cluster* IMF is self-explanatory, the distribution of stellar masses at birth in star clusters or OB associations. The definition of the *field star* IMF is the subject of much confusion, however. Many authors apply the term to the integrated stellar population, including clusters, associations, and isolated stars. However some authors apply the term to isolated stars only, excluding stars in clusters and associations. There is little theoretical justification for supposing that the cluster and field IMFs need be the same at all, though as shown later, observations suggest that the IMFs are indeed quite similar.

The final bit of nomenclature concerns the functional form of the IMF. Salpeter (1955) originally approximated the solar neighborhood IMF as a power law truncated at an upper and lower stellar mass limit:

$$f(m) \, dm = Am^{-\gamma} \qquad \gamma = 2.35 \tag{1}$$

and this power-law form for the IMF has been adopted by almost all authors since that time. There choice of a power-law form is quite arbitrary, and some evidence suggests that a log normal form may be more realistic. Nevertheless the power-law form is very convenient, and I will use it extensively here. We shall see later that a single power-law cannot describe the observed IMF over the entire stellar mass spectrum, but this can be accommodated to first order by combining two or more power-law segments over successive mass intervals.

There is a considerable variation in nomenclature for the power-law formulation as well, and readers must be very careful to check which formulation is being adopted. Many authors express the IMF in terms of the number of stars per unit *logarithmic* mass interval:

$$F(\log m) d(\log m) = Am^{\Gamma} \tag{2}$$

It is easy to show that $\Gamma = \gamma - 1$ for a given power-law IMF, hence $\Gamma = 1.35$ for the Salpeter IMF. To complicate matters further, many authors use a sign convention that is reversed from that given above, or interchange the use of γ, Γ and other abbreviations. The only sure way to avoid problems is to carefully check each paper to ascertain which convention is being used. To minimize confusion here I will use the index γ $(df(m)/dm)$ throughout the remainder of this lecture.

The power-law form for the IMF only has physical meaning when upper and lower stellar mass limits are applied. Again, there is little uniformity in the literature. Most investigators truncate the lower end of the IMF in the range 0.07–$0.1 \, M_{\odot}$, close to the hydrogen burning limit, though it is an open question whether a mass spectrum of substellar objects (brown dwarfs) persists to lower masses. Upper mass cutoffs usually are in the range 30–$120 \, M_{\odot}$, though again there is no uniform convention.

Consequently even a single power-law IMF really contains three free parameters, the slope γ, and the upper and lower mass limits. SFRs determined using the Pop I tracers described in the preceding lecture are sensitive to all of them. For example, a change in the slope γ by unity changes the inferred

SFR for a fixed Hα luminosity by roughly an order of magnitude (Kennicutt & Chu 1988), and changing the upper mass limit can have comparable influence. Perhaps the most famous example of how the presumed IMF can affect the interpretation of an astronomical observation is the measurement of the cosmological deceleration parameter q_0 from the classical brightest cluster galaxy test (Tinsley 1980); changing the assumed power-law slope by unity changes the inferred value of q_0 by approximately 0.5! These examples illustrate the broad relevance of the IMF for the full gamut of extragalactic astrophysical applications, and provide ample motivation for measuring the IMF.

3.3.   The Solar Neighborhood IMF

Scalo (1986) and Kroupa et al. (1993) provide detailed discussions of the steps required to measure the field IMF in the solar neighborhood, and I only briefly review them here. The first step is to construct a stellar luminosity function. Ideally this would be based on a large sample of stars with measured trigonometric parallaxes, and here the upcoming *Hipparcos* satellite data release should have a major impact. Otherwise the luminosity function is derived from a combination of direct parallax measurements, statistical, and spectroscopic parallaxes.

The luminosity function is best determined for stars in the absolute magnitude range $-4 < M_V < 9$, corresponding to stellar masses ranging from a few tenths to several solar masses. The LF for the most massive stars is much less well determined. Such stars are very rare (they lie in the tail of the IMF and are very short lived), and are highly clustered in the Galactic plane. Consequently a very large volume of the disk, kiloparsec in radius, must be analyzed to ensure a representative sample, and problems of extinction and incompleteness are severe. Incompleteness is also a severe problem at the low end of the IMF, because the stars are so faint they can only be detected to small distances (<100 pc). Fortunately this is an area where application of near-infrared imaging is having a major impact, and the statistics are improving rapidly (e.g., Tinney 1993; Kroupa et al. 1993).

When these surveys are combined the result is a composite luminosity function which traces the entire main sequence mass spectrum. The astute reader will note that this composite LF will be highly composite in character, having been averaged over kiloparsec volumes for the most massive stars, down to volumes of 25–100 pc in radius for the lowest stellar masses. This inhomogeneity may be important for the upper and lower mass extremes, where statistics are poor, but it is probably unimportant in a physical context, because orbital mixing of the Galactic disk ensures that even a local volume of space includes stars with formation sites spread over a region of order 1 kpc or larger.

The next step is to convert this observed luminosity function to a present day mass function, using either an empirical or theoretical stellar mass-luminosity relation. This is another area in which considerable progress has been made in the past decade, and error in the mass-luminosity relation no longer is a major source of uncertainty, except again for the upper and lower mass extremes.

The final step is to convert the field star PDMF to an IMF. For stars below the turnoff mass the two functions are identical; every star that ever formed remains on the main sequence, and one needs only to divide the observed number of stars by the disk age to determine the average birthrate. For stars with masses well above the turnoff one makes a similar computation, except in this case the observed numbers of stars need to be corrected by an incompleteness factor which includes the ratio of the main sequence stellar lifetime to the disk age, and the ratio of the present-day SFR to the SFR averaged over the age of the disk. For stars with masses near the turnoff range for the disk, the conversion is more problematic, because the population of stars observed today is a function of the star formation history of the disk. The latter can be obtained from direct measurements of stellar age distribution, or by forcing that the IMF be continuous through the turnoff mass range. Fortunately both prescriptions are consistent, and imply a roughly constant stellar birthrate over the past 10 Gyr in the solar neighborhood (Scalo 1986; Kennicutt 1992b).

The local IMF has been determined in this way by several authors over the past decade, with the most widely applied results including those of Scalo (1986), Basu & Rana (1992), and Kroupa et al. (1993). Figure 7 shows one of the most recent examples, from Kroupa et al. (1993). Averaged over the entire stellar mass spectrum, the average IMF slope is close to Salpeter's original estimate of $\gamma = 2.35$ (a power-law function in this plot would be a straight line). However the data show a clear curvature, with the IMF slope becoming increasingly shallow at lower stellar masses, a behavior seen already by Miller & Scalo (1979) and Scalo (1986). When parameterized in terms of multiple power law, the slope γ increases from 1.3 below 0.5 $M_\odot$ to 2.7 for $M > 1\ M_\odot$. The turnover below 1 $M_\odot$ is very important for the interpretation of observations of external galaxies and starbursts, because it greatly reduces the amount of mass locked up in unevolving (and usually unobserved) stars below the turnoff (Kennicutt et al. 1994). Once again I refer the reader to Scalo (1986) and Kroupa et al. (1993) for details.

3.4.   Galactic Clusters and Associations

Star clusters offer obvious advantages for measuring the IMF, because they provide clean distance-limited samples of stars and a coeval population. Constructing the IMF for clusters is a nontrivial task, however, because the limited numbers of stars in clusters provide sparse sampling of the IMF at high masses (if such stars have not evolved away), and the large angular sizes of most nearby clusters and associations require wide-field digital detectors. This is another area in which great progress has been made in the past few years, and I will cite only a few examples of recent studies.

The most comprehensive study to date of a young star cluster/association is a survey of the Orion cluster by Hillenbrand (1997). She combined visible wavelength photometry with infrared photometry and spectroscopy of the red-

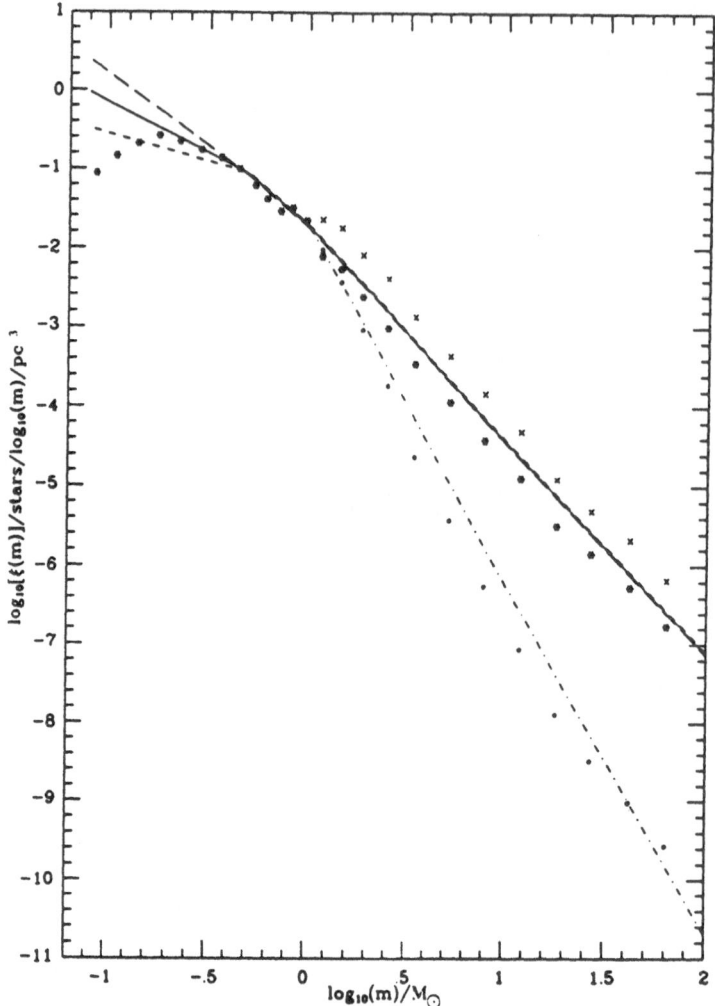

Fig. 7. — The local field IMF (points and solid line) and PDMF (dashed line), from
Kroupa et al. 1993.

der stars to compile for the first time a data set that spans the full range from
OB stars to late M dwarfs near the stellar low-mass limit. Figure 8 shows
a composite HR diagram from this study. The IMF derived from these data
is consistent with a Scalo function at most radii, and evidence of radial mass
segregation is seen. This study provides some of the strongest evidence that
the local field IMF may apply to clusters as well.

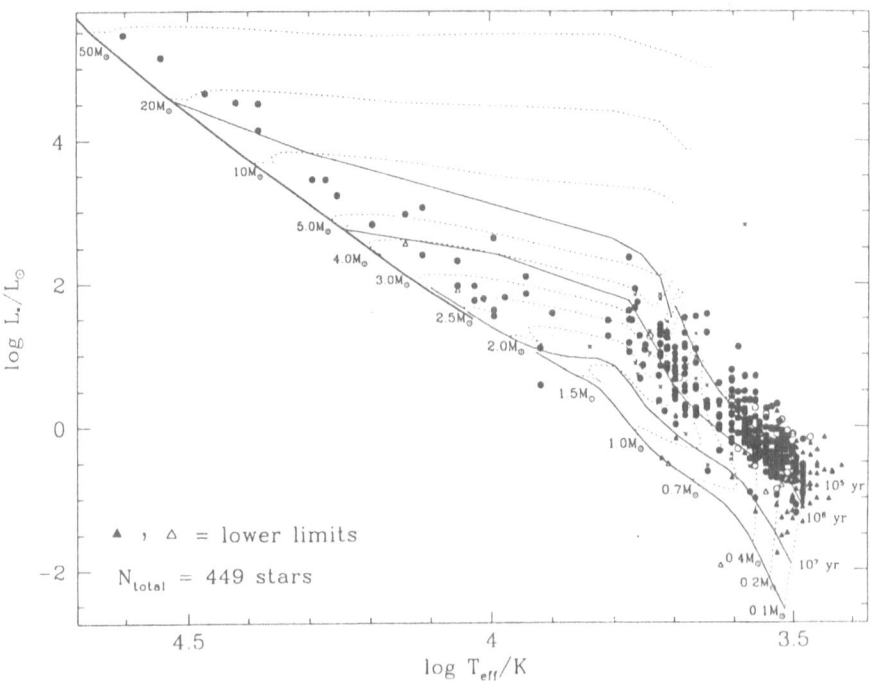

Fig. 8. — HR diagram of the Orion nebula cluster, with isochrones and mass contours superimposed, from Hillenbrand (1997).

Numerous studies have been published for other Galactic clusters and associations, and they provide more accurate information on the shape of the IMF at high masses, where large samples are required. Phelps & Janes (1993) measured color-magnitude diagrams, and combined the data for 8 well measured clusters to derive a composite IMF. Their data are consistent with a single power law slope $\gamma = 2.40$ over the mass range 1.4–7.9 $M_\odot$, nearly identical to the Salpeter slope of 2.35. Massey et al. (1995b) have carried out a survey of 14 Galactic OB associations, and find values of γ in the range 1.8–2.6, with a considerable uncertainty in individual objects. When their data are combined they find a mean slope $\gamma = 2.1 \pm 0.1$ over the mass range 4–100 $M_\odot$. All of the associations have IMFs that are consistent with this function, given the statistical uncertainties. Note that this IMF is considerably shallower than the slope $\gamma \sim 2.7$ derived from field studies over the same mass range. I return to this point later.

Clusters should also provide invaluable information on the IMF at very low masses, though progress here has been slower due to the limited field sizes of infrared cameras. Aside from the Orion study discussed earlier, extensive studies have been made of Praesepe and the ρ Ophiucus clusters (Williams et al. 1995a, b). The data are consistent with a mean index $\gamma = 1.1$-1.3 over the range 0.1-1 $M_\odot$, and is roughly consistent with the field IMF over the same mass range. Several attempts have also been made to measure the IMF in globular clusters, as a test for population-dependent IMF changes, but these are outside the scope of this lecture.

3.5. The Magellanic Clouds

Observations of the IMF in other galaxies offer the exciting possibility of testing for systematic variations as functions of metallicity, galaxy type, or star forming environment. Such studies are generally restricted to the most massive stars, and extreme care needs to be taken to account for the effects of incompleteness and stellar crowding in the analysis. The Magellanic Clouds are an exception, however, and they offer perhaps the best laboratories for these studies, because they offer a wide range of metallicity and environments, yet are close enough to provide data on resolved stellar populations down to 1 $M_\odot$ or lower.

Several studies of OB associations in the LMC have been carried out (e.g., Hill et al. 1994; Massey et al. 1995a; Oey 1996), using techniques similar to those applied to the Galactic OB associations, and described in the first lecture. Figure 1 shows an example of one such association. The richest associations are those studied by Massey et al. (1995a). They show an average IMF slope $\gamma = 2.3 \pm 0.3$, similar again to the Salpeter value. They also attempted to measure the IMF for field OB stars in the LMC and SMC, with "field" defined in this case as stars located outside of a cluster or OB association. They find much steeper slopes for these samples, with $\gamma = 5.1$ for the LMC and 4.7 for the SMC, for stars with masses above ~ 10 $M_\odot$. It is not clear how to interpret these results; OB stars are predominantly formed in associations, and all of the studies of the field IMF in the Galaxy (previous sections) included these association members in their statistics. The Magellanic Cloud results serve to quantify this strong clustering of the massive stars, but without comparable data for lower mass stars it is not clear whether this segregation applies uniquely to the most massive stars.

Several investigators have derived the upper IMF in the extraordinary cluster/association NGC 2070 in the 30 Doradus nebula (Parker & Garmany 1993; Hunter et al. 1995, 1996). Over 2000 massive stars have been identified in the central 50 pc radius of this region, with over 200 located in the central R136 compact cluster. Once again however the IMF slope is consistent with a Salpeter function, with $\gamma = 2.22 \pm 0.06$ over the mass range 3-100 $M_\odot$. HST observations of the core show evidence for the same kind of mass segregation that is observed in the Orion nebula cluster.

The Magellanic Clouds contain a population of young to intermediate age populous clusters, often referred to as "blue globular clusters," that is unlike any population in the Milky Way, and these also have been used to measure the IMF (e.g., Mateo 1988; Vallenari et al. 1994; Will et al. 1995). Most of these measurements are sensitive to the mass range between 1–10 $M_\odot$, above the incompleteness limit and below the turnoff mass of the youngest objects. Again most of the measurements are consistent with $\gamma \sim 2$–2.4, though a few objects may show slightly steeper IMFs.

Taken together the measurements of the Magellanic Clouds suggest an IMF in the 1–100 $M_\odot$ range which is consistent with a roughly Salpeter slope, and is similar to that seen in Galactic clusters and associations, but somewhat shallower than the IMF typically derived in the solar neighborhood field studies. Studies of the field IMF in the Magellanic Clouds are under way, based mainly on HST measurements (e.g., Holtzman et al. 1997).

3.6.   Other Galaxies

Beyond the Local Group one can only infer the IMF indirectly, by modelling the integrated spectra or emission-line properties of the population. The techniques are similar to those described for determining the SFR, and the results share the same limitations and systematics that apply to the SFRs, mainly in that they are limited to characterizing the massive stellar populations that dominate the integrated light of the population. However it is possible to impose at least rough limits on the variation of the upper IMFs in galaxies that span a much wider range of physical environments than can be probed in the Galaxy or the Magellanic Clouds. This is a rapidly developing field and I will cite only two examples of what can be learned.

The advent of high resolution, high S/N ultraviolet spectroscopy with HST makes it possible to directly infer the massive stellar contents of giant HII regions and starbursts using what amount to classical spectral classification criteria, with the addition of spectral synthesis modelling of the composite OB/W-R stellar populations. The methodology is described in detail by Robert et al. (1993), and makes use of not only the equivalent widths of the stellar absorption lines but also the widths and P-Cygni profiles of the principal resonance lines. An example of the technique is illustrated in Figure 3, which shows the ultraviolet spectrum of a star forming knot in NGC 4214, along with two synthesis models with different assumed star formation histories (Leitherer et al. 1996). The best fits are obtained for a continuous burst with $\gamma = 2.35$ or an instantaneous burst with $\gamma = 3$. The degeneracies between age and IMF do not allow for a unique characterization of the IMF for individual objects, but this is a very promising technique for application to larger samples of star forming regions.

More traditional visible-wavelength spectral synthesis modelling has been applied to evolved starbursts in many interacting and merging galaxies (e.g.,

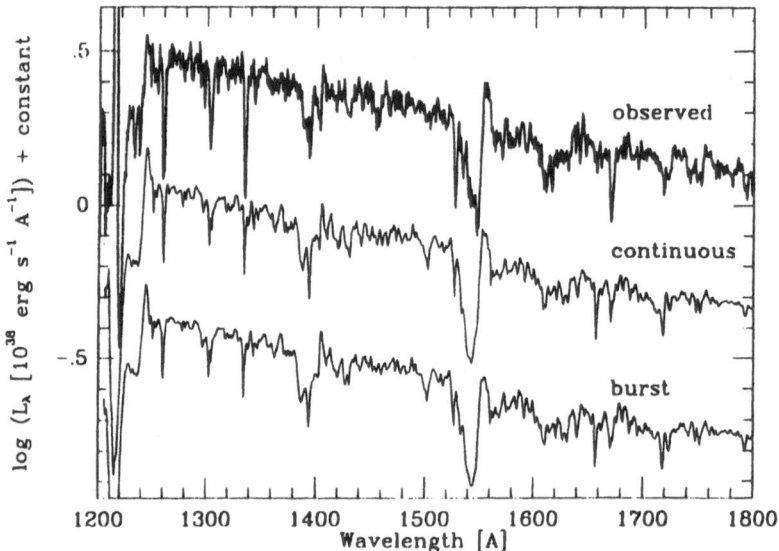

Fig. 9. — Observed UV spectra and synthesis models for a star forming cluster in the nearby irregular galaxy NGC 4214, from Leitherer et al. (1996).

Bernlöhr 1993; Fritze v. Alvensleben & Gerhardt 1994; Leonardi & Rose 1996; Turner 1997). These provide information primarily on the SFRs and ages of the starbursts, but can provide some rough constraints on the IMFs in the bursts. Since this topic is covered in depth by Meynet elsewhere in this volume, I refer the reader to his lectures.

The most comprehensive analysis of the *global* IMF in star forming galaxies has come from combined Hα and multicolor measurements (Kennicutt 1983; Kennicutt et al. 1994). As discussed in the first lecture, galaxies follow a well-defined trajectory in the UBV and Hα–color planes, and these provide interesting constraints on the average IMF in the 1–100 $M_\odot$ range. This is illustrated in Figure 10, taken from Kennicutt et al. (1994). Shown is the correlation of the integrated equivalent widths of the Hα emission line with $B - V$ color, with several synthesis models superimposed. The left plot shows models with an IMF slope $\gamma = 2.5$ above 1 $M_\odot$, while the left plot shows the same models but using the considerably steeper Scalo (1986) IMF for the solar neighborhood. The Hα equivalent width is very sensitive to the IMF slope in the range 1–30 $M_\odot$, because it essentially measures the ratio of ionizing flux (produced by stars in the range 10–100 $M_\odot$) to the red continuum, which is dominated by giants (typical masses 0.7–3 $M_\odot$). It is clear that the Scalo IMF simply produces too few massive stars to account for the observed line emission, especially when extinction is taken into account, whereas a roughly Salpeter

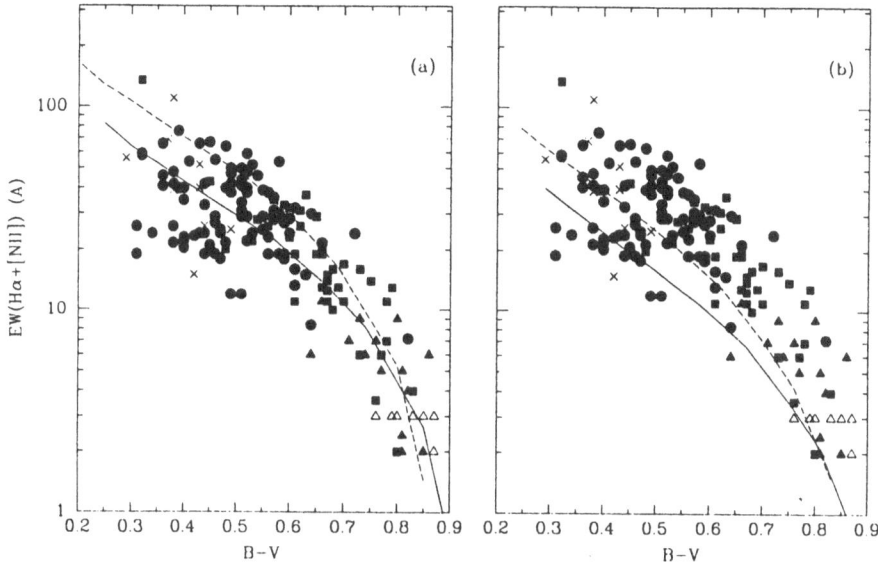

Fig. 10. — Hα vs color relation for nearby spiral galaxies, with synthesis models superimposed, as described in the text. Left panel shows models with a roughly Salpeter IMF, the right panel is with a Scalo IMF. Figure taken from Kennicutt et al. (1994).

slope is consistent with observations. This suggests that the Scalo IMF under represents the massive stellar population by factors of ∼2–3. Similar techniques can be applied to individual HII regions, to use the distribution of Hα equivalent widths to place limits on the systematic variation of IMF as a function of metal abundance or galaxy type. This has been carried out recently by Bresolin & Kennicutt (1997), who find no evidence for systematic changes in the IMF slope or upper mass limit in a diverse sample of galaxies.

3.7. Conclusions

By now I hope that a pattern is evident in this wide range of results. Although the existing data fall far short of being able to determine whether there exists a "universal" IMF, the data present a remarkably uniform picture, with most observations consistent with the solar neighborhood IMF. The largest inconsistency appears to be for the upper IMF (above a few solar masses), where

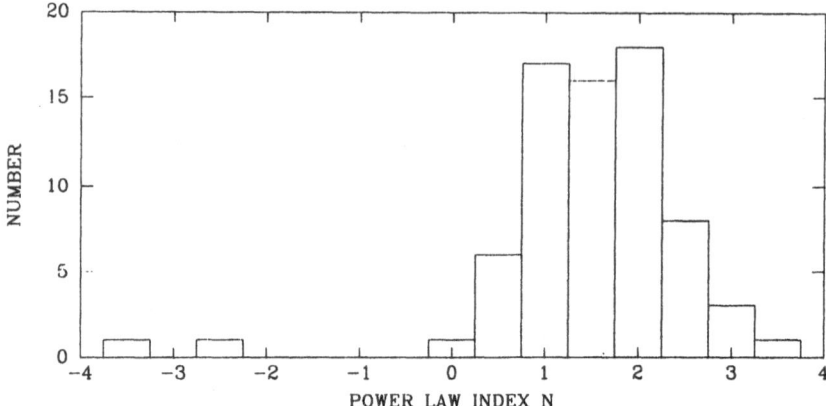

Fig. 11. — Best fitting values of Schmidt law index N, from a compilation of the literature.

a wide range of direct and indirect measurements of star forming regions and galaxies suggest a slightly shallower slope than is typically derived for the solar neighborhood. Chemical evolution arguments also favor a somewhat more top-heavy IMF than the Scalo function (Peimbert et al. 1994). Although the source of this discrepancy with the solar neighborhood studies remains unexplained, I am inclined not to attach any deep physical significance to the disagreement, and suspect instead that the solar neighborhood simply contains too few massive stars to reliably determine the IMF, or that those studies suffer from incompleteness at the high-mass end of the IMF.

With these qualifications in mind, I would maintain that *all* existing observations of nearby normal galaxies are consistent with a common IMF, with approximately the Salpeter slope for $M > 1$ M$_\odot$, and a turnover consistent with the Scalo (1986) or Kroupa et al. (1993) IMF at lower masses. Very roughly this would correspond to a two-segment power law with $\gamma = 2.4 \pm 0.1$ above 1 M$_\odot$, and $\gamma \sim 1.3 \pm 0.3$ at lower masses. I seriously doubt whether such an IMF actually is universal, but I would challenge observers to identify environments where this function does not fit their observations within the empirical uncertainties.

Are there *any* extragalactic environments where a radically different IMF is required to explain the observations? The only strong candidate I am aware of are the dusty starbursts in infrared-luminous galaxies, as typified by the starburst galaxy M82. Several studies of M82 and other infrared-luminous galaxies have shown that the extinction-corrected ionizing fluxes of these objects require enormous masses in young stars if a solar neighborhood IMF is applied (e.g., Rieke et al. 1980, 1994). In some cases the inferred masses exceed the dynamical masses of the regions, and the simplest explanation is that the IMF is truncated or severely suppressed below a few solar masses. The same observations suggest an *upper* truncation of the IMF above 15–30 $M_\odot$, which implies a very peculiar mass function, with Scalo slope between $\sim$3–20 $M_\odot$, and few if any stars formed above or below these limits (Rieke et al. 1994). Before such a radical IMF is adopted I believe that alternatives need to be more thoroughly explored, e.g., shallower (e.g., Salpeter) IMF with a higher stellar mass limit, before one can unambiguously rule out a normal lower stellar mass limit. However the analysis of M82 by Rieke et al. (1994) suggests that such IMFs are problematic as well, and it is abundantly clear that a Scalo IMF is completely ruled out. Given the extraordinary physical conditions in starbursts (see other lectures) it would not be surprising if the IMF were abnormal. I refer the reader to other lectures in this volume, particularly those by Terlevich, Meynet, and Joseph, for more discussion of this and related issues.

4. PHYSICAL REGULATION OF SFRs IN GALAXIES

4.1. Introduction

Understanding the physical processes that govern the conversion of interstellar gas into stars is crucial for a host of fundamental problems: galactic structure, the nature of the Hubble sequence, galaxy formation and evolution, and chemical evolution. Although our understanding of large-scale star formation and its relationship to the interstellar medium (ISM) is at a primitive state, a recent flurry of observations and theoretical modelling has provided important new insights into the star formation law and its physical basis. The goals of this work are twofold, to define the empirical form of the SFR vs gas density law, and beyond that to understand the physical processes that underlie the observed star formation law. The empirical relations by themselves are invaluable as "recipes" for numerical models and simulations of galaxy formation and evolution, and understanding the physical basis of these recipes will solve one of the major theoretical impediments to understanding galaxy evolution itself.

4.2. The Schmidt Law

The roots of this field trace to the seminal paper of Schmidt (1959), which introduced the star formation rate (SFR) vs gas density power law that is still used today. Nearly 40 years later it remains the most useful parameterization of the SFR for many applications. For extragalactic applications it is usually parameterized in terms of the projected surface densities of gas and stars:

$$\Sigma_{sf} = A\Sigma_{gas}^{N} \qquad\qquad (3)$$

Since 1959 roughly 50 papers have been published which attempt to measure the value of N, by correlating the surface densities of a young Pop I stellar tracer with gas column densities. I have compiled the literature and the distribution of measured values of N is shown in Figure 11. The values cover a wide range, with a broad peak between $N = 0.8$ and 2.5 and a full range of $N = \pm 3.5$. This large dispersion can be attributed partly to observational factors, such as beam smearing effects or fitting to only the atomic or molecular gas densities. However much of the scatter is real, being caused by such factors as deviations from a power law or spatial variations in N across a galaxy. Furthermore, the point-by-point correlations that form the basis of most of the Schmidt law fits in Figure 11 often exhibit an enormous scatter, which suggests that the large-scale law represents a statistical average of a much more stochastic relation on small scales. An excellent illustration is shown in Figure 12, which shows the correlation of Hα surface brightness on H I column density in cells of 220 pc in the LMC (Kennicutt et al. 1995). When the data are binned in column density a monotonic Schmidt law is apparent, with $N = 1.75 \pm 0.3$, consistent with previous studies. However the distribution of individual points is a virtual scatter diagram, with the correlation evident only when the data are binned.

Nevertheless there are applications for which a Schmidt law *does* provide a useful description of the SFR, especially when applied to the integrated SFRs of galaxies. As an illustration, Figure 13 shows the correlation between average disk surface brightness in Hα, proportional to the SFR per unit area, and average gas surface density. The plot is an updated version of one shown in Kennicutt (1989), and is based on a compilation of galaxies with spatially resolved Hα, HI, and CO distributions, so that the various surface densities can be measured in a self-consistent manner.

The lines shown in Figure 13 correspond to a slope $N = 1$, and correspond to global star formation efficiencies of 1%, 10%, and 100% per 10^8 years, using the SFR calibration and extinction corrections in Kennicutt (1983). The observed relation is clearly nonlinear, with a best-fitting slope $N \simeq 1.5$. Similar results have been derived by other authors using a variety of star formation tracers (e.g., Buat et al. 1989; Buat 1992; Boselli 1994; Deharveng et al. 1994). The average zero point of the relation in Figure 13 corresponds to a SFR of ~ 5% per 10^8 yr; in other words the average star forming spiral converts 5% of its interstellar gas to stars every 0.1 Gyr (within the optical radius). The reciprocal of this rate is the average gas depletion timescale; in the absence of stellar recycling, the timescale is $\sim$ 2–2.5 Gyr. Recycling actually increases these times by a factor of 1.5–2 (Kennicutt et al. 1994).

Several of the studies above have analyzed the dependence of the disk-averaged SFR on the molecular and atomic surface densities separately. In most cases the SFR is well correlated with the HI surface density, but the coupling with the molecular surface density is surprisingly weak, probably due to variations in the CO/H_2 conversion factor.

Although Figure 13 shows some evidence for a nonlinear Schmidt law, there is considerable scatter which weakens the strength of the correlation. Such a

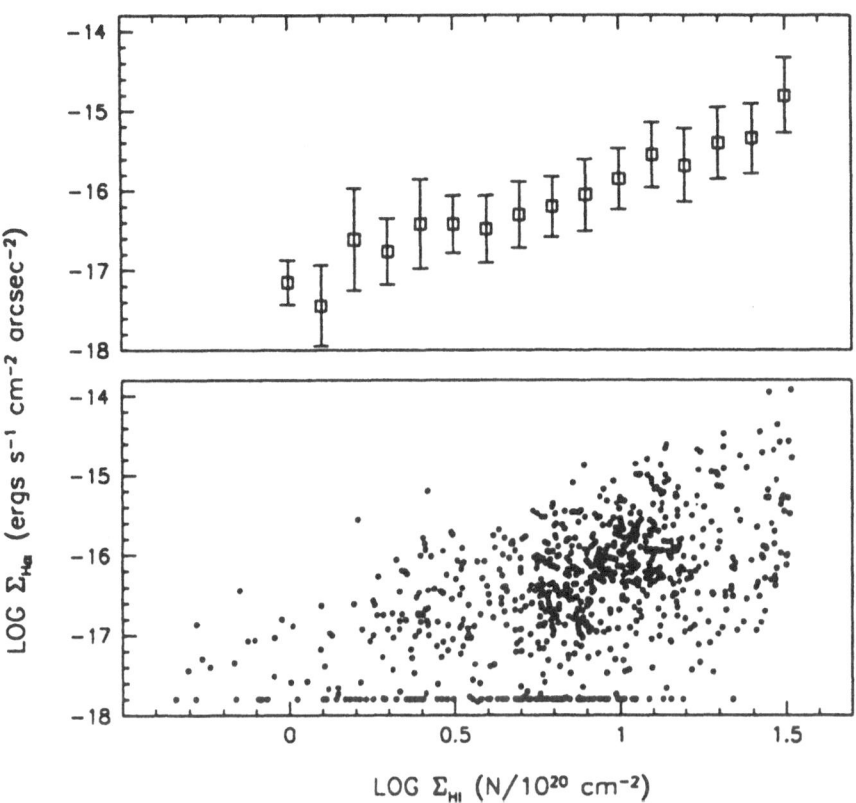

Fig. 12. — Hα vs H I correlation for the LMC. Points represent individual 220 pc cells, while the bars show mean trends when the points are binned in column density. From Kennicutt et al. (1995).

scatter should not be surprising. Variations in extinction produce a scatter at the ∼50% level (a constant average correction is applied to all of the points), but most the dispersion is probably real, reflecting the fact that we have averaged the SFRs and gas densities over the entire star forming disks, and therefore are averaging a nonlinear local relation over regions spanning orders of magnitude in local SFR and density. Furthermore the disk-averaged SFRs and densities span a relatively limited dynamic range. We can expand the range of densities probed by plotting instead the spatially resolved Hα, HI, and CO data, and this is shown in Figure 14 for a representative sample of spirals. Each line shows the radial run of gas density and SFR in a single galaxy; each galaxy

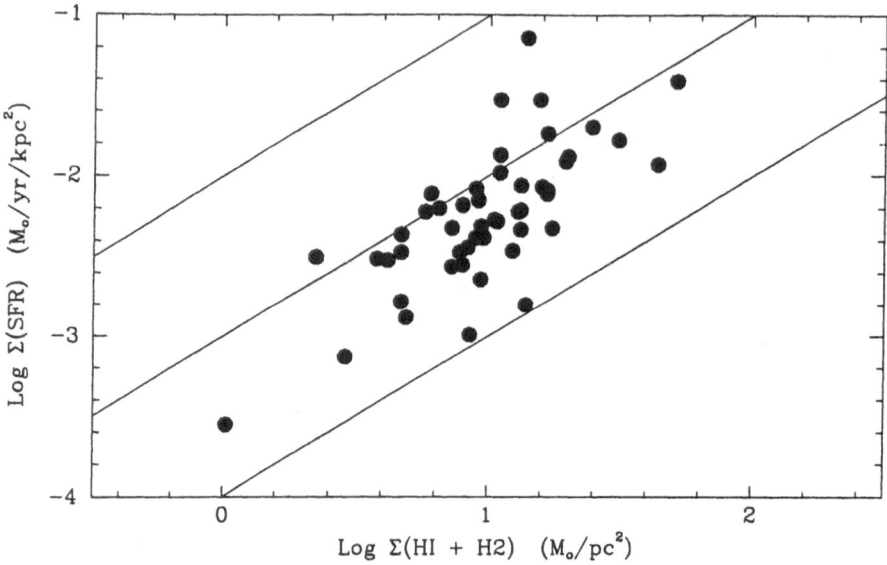

Fig. 13. — Correlation between average Hα surface brightness of optical disks and the average surface density of atomic plus molecular gas. Lines indicate constant efficiencies or timescales for gas consumption, as described in the text.

is typically measured at 8-15 points, and the individual points have been left off the plot for clarity. I have superimposed the relations for each galaxy to illustrate the similarities in the behaviors of the star formation laws seen over the sample.

The most obvious features in these relations are the presence of abrupt star formation thresholds, which will be discussed in the next section. At high gas densities, however, the SFRs are well represented by a Schmidt law, again with an average index $N \sim 1.5$, and in this case extending to densities 5–10 times higher than seen in the disk-averaged relation in Fig. 13. Also note the overlap of the Schmidt laws for the individual galaxies. Except for the regions below the star formation thresholds, the SFR at a fixed gas density is roughly the same in all 22 galaxies. Apparently there is some common physical mechanism that regulates the SFR at a given density, largely independent of the other properties of the host galaxy.

Is it possible that this large-scale Schmidt law extends to even higher gas densities? This question cannot be addressed using SFRs determined from Hα emission, because for normal gas/dust ratios a region with $\Sigma_H > 100$ M$_\odot$ pc^{-2}

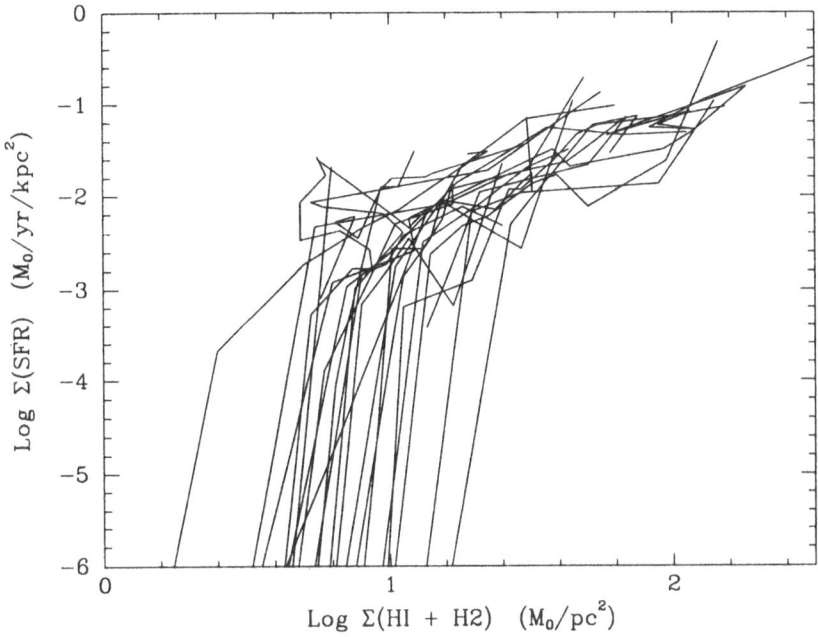

Fig. 14. — Azimuthally averaged correlation between the SFR per unit area (as derived from $H\alpha$) and total atomic + molecular gas density for 22 nearby galaxies, from Kennicutt & Martin (1997).

is expected to suffer an extinction at $H\alpha$ of 1–2 magnitudes. Hence it is not surprising that the observations in Figures 13–14 do not extend beyond this limit. To probe the Schmidt law at higher densities we must rely on SFRs determined either from infrared recombination lines or from far-infrared emission. I recently investigated this question by compiling data on CO distributions, FIR fluxes, and (when available) Brackett line fluxes for a sample of infrared-luminous circumnuclear starbursts. These were used to investigate the SFR vs gas density correlation, on the assumption that the FIR emission was produced by a starburst of age $10^7 - 10^8$ yr, and that the gas in the central region was predominantly molecular, reasonable assumptions for these objects. The nuclear starbursts are characterized by much higher densities than are found in normal disks, and hence this provided a sample of regions with gas densities in the range $10^2 - 10^5$ $M_\odot$ pc^{-2}.

To my surprise these objects also are well fitted by a Schmidt law, but with an average SFR/gas ratio (the coefficient A in eq. [3]) that is typically 10-30 times higher than for the disks plotted in Fig. 13. This simply is a reflection

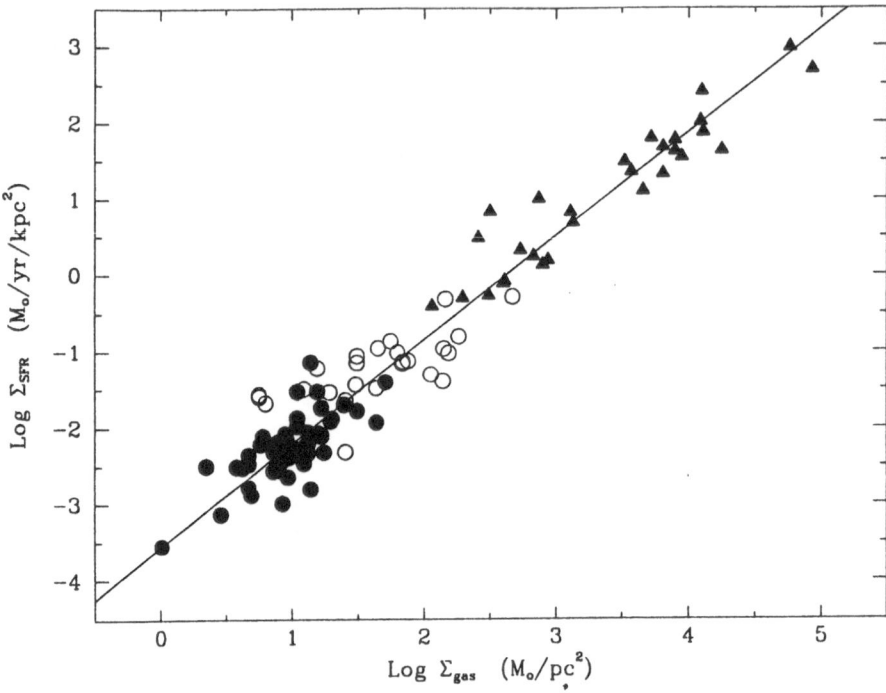

Fig. 15. — Correlation between SFR per unit area and gas surface density, for normal disks (circles) and circumnuclear starbursts (triangles), as described in text. Solid circles are disk-averaged SFRs as in Fig. 13, open circles are values for the central regions of the normal spirals. Adapted from Kennicutt & Martin (1997).

of the high SFRs and short gas consumption timescales in starbursts. However when one plots the data for the starbursts and the normal disks together it is clear that two sets of objects define a common Schmidt law! This is illustrated in Figure 15. The line shows the best fitting Schmidt law index $N = 1.36$. The largest uncertainties in this fit are introduced by systematic uncertainty in the CO/H_2 conversion factor and the conversion from FIR flux to SFR in the starbursts, however this introduces an uncertainty of only ~ 0.1–0.15 in the derived index N. Remarkably, a single Schmidt law appears to provide a excellent "recipe" for the large-scale star formation law over ·5 orders of magnitude in gas density or 7 orders of magnitude in SFR per unit area.

4.3. Physical Interpretation of the Schmidt Law

It is tempting to ascribe the tight correlation in Figure 15 to a simple physical scaling law. There is a rich literature on this subject, as reviewed in Larson (1987). For example, a self-gravitating disk with constant scale height might be expected to exhibit a Schmidt law with $N \sim 1.5$, on the assumption that the SFR scales with the gas mass divided by the gravitational free-fall timescale. However other interpretations can also account for the correlation in Fig. 15. For example Wyse & Silk (1987) and Silk (this volume) have proposed a model in which the SFR scales as $\Sigma\Omega$, where Σ is the gas surface density and Ω is the angular rotation speed of the disk; this is tantamount to postulating that a constant fraction of the gas is converted to stars per disk orbit. Figure 16 shows the correlation between the SFR and $\Sigma\Omega$, for the same data as plotted in Fig. 15, with a line of slope unity superposed. The fit is nearly as good as for the Schmidt law, and if anything is tighter for the normal disks. Hence we have found two physical "recipes" which reproduce the observed global SFRs with comparable success. This illustrates the perils of inferring a physical relation from observed scaling laws alone. A complete discussion of these results is given in Kennicutt & Martin (1997).

4.4. Star Formation Thresholds

As shown in Figure 14, the Schmidt law breaks down at low gas densities, below which the SFR declines abruptly. The thresholds in Fig. 14 correspond to abrupt outer edges to the star forming disks, well inside the gaseous HI disks. The critical column density varies between about 10^{21} and 10^{22} H cm^{-2} ($\sim$1–20 M$_\odot$ pc^{-2}), and there is evidence for systematic variation in this threshold density between different types of galaxies and within the same galaxy, as discussed below. Similar thresholds are observed in Magellanic irregular galaxies (e.g., Skillman 1987). Although star formation is not completely suppressed outside the threshold region, there is an abrupt transition radius in most late-type disks. Other thresholds are observed in gas-rich S0 galaxies (e.g., Eder 1990) and in low surface brightness galaxies (e.g., van der Hulst et al. 1993), where galaxies with gas masses of up to 10^{11} M$_\odot$ possess abnormally low SFRs, far below the extrapolation of a linear Schmidt law. Another nonlinearity is often observed in the disks of grand-design spiral galaxies, where relatively small arm vs interarm contrasts in gas column density induce order-of-magnitude contrasts in the SFR (e.g., Knapen et al. 1992; Rand 1993).

4.5. Gravitational Stability Thresholds

Kennicutt (1989) proposed that these observed thresholds could be physically associated with a large-scale gravitational threshold for the growth of density perturbations in the gas disk (cf. Quirk 1972). He estimated the critical densities for a set of star forming spirals in terms of the Toomre (1964) criterion

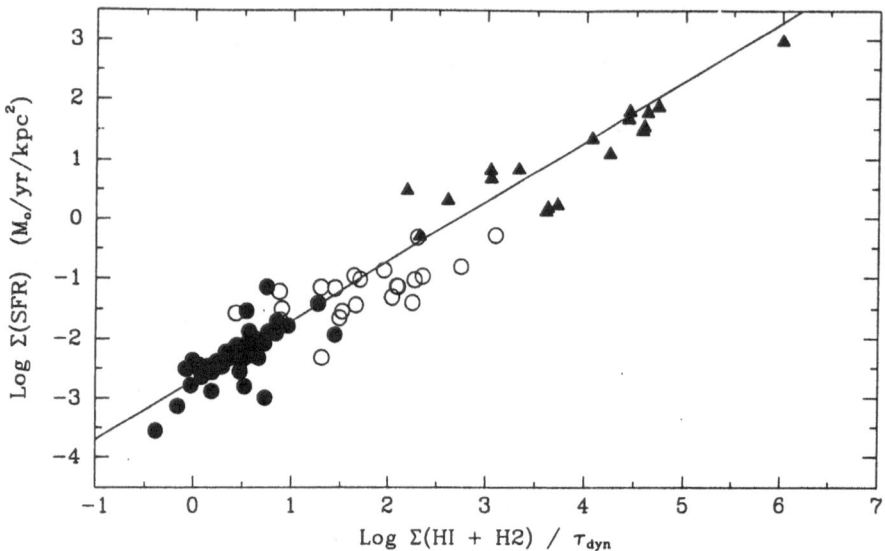

Fig. 16. — Correlation between SFR per unit area and the ratio of the gas surface density to the local disk orbital time. The line is the best fit with a slope of unity. Adapted from Kennicutt & Martin (1997).

for the growth of large-scale perturbations in a self-gravitating gas disk:

$$\Sigma_c = \alpha \frac{\kappa c}{\pi G} \, , \qquad (4)$$

$$Q \equiv \frac{\Sigma_c}{\Sigma_g} \, , \qquad (5)$$

where the critical surface density Σ_c is dictated by the epicyclic frequency of rotation κ, the gas velocity dispersion c, and a dimensionless constant α which takes into account the deviation of real disks from the idealized Toomre model. The critical parameter for determining whether cloud growth and star formation is suppressed is the famous Toomre Q parameter, the ratio of the critical density at a given radius to the actual gas density. In this picture, star formation should be strongly suppressed in regions with $Q \gg 1$, and it should proceed unimpeded in regions with $Q \ll 1$. In intermediate regimes, where $Q \sim 1$, the disk would be globally stable against cloud growth but could be easily perturbed locally. Since star formation tends to deplete the disk until Q approaches unity, one would expect evolved disks to hover in the metastable

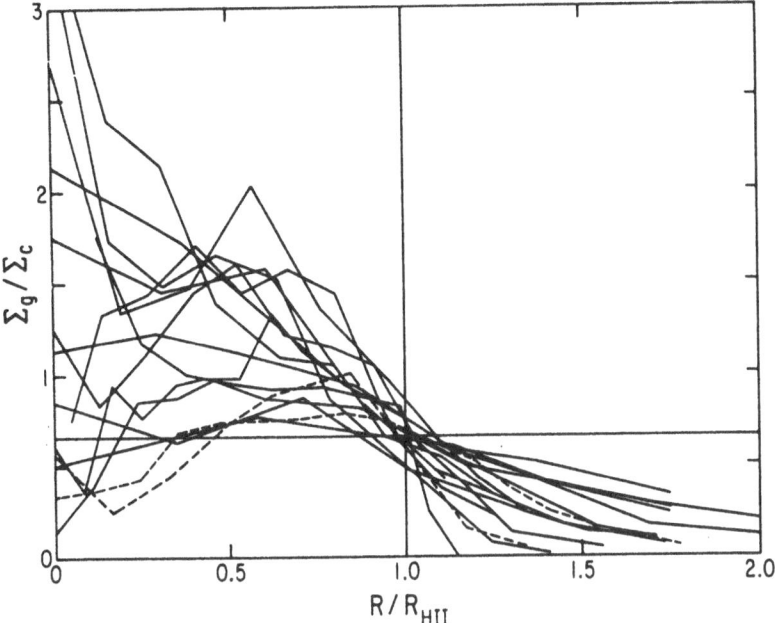

Fig. 17. — Radial behavior of the Toomre stability parameter Q for active star forming disks, from Kennicutt (1989).

regime near $Q \sim 1$. For typical values of κ and c the critical density is of order
10% of the total disk surface density. This potential supply of fresh gas for star
formation is the key to understanding the fueling of starbursts in interacting
early-type galaxies.

This admittedly over simplistic gravitational model is surprisingly successful at reproducing many of the observed star formation properties of normal
galaxies. Figure 17, taken from Kennicutt (1989), shows the typical radial variation in Q for active star forming spirals. It is interesting that even though
the absolute gas density drops by 1–2 orders of magnitude over the radius of a
typical disk, the stability index Q varies by factors of a few or less, suggesting
that the thresholds partly regulate the radial behavior of the gas distribution.
The thresholds also predict with surprising precision the threshold radii and
column densities of the star forming disks; see Kennicutt (1989) for details.
The same criterion successfully accounts for the observed star formation properties of low surface brightness galaxies (van der Hulst et al. 1993), gas-rich
S0 galaxies (Kennicutt 1989; Eder 1990), and star forming irregular galaxies
(Taylor et al. 1994).

Despite these successes, the Q model is flatly contradicted by one class of galaxies— low-mass spirals. Two galaxies of this type, M33 and NGC 2403, are indicated by the dashed lines in Figure 17. Both exhibit subcritical surface densities over much of their inner disks, even though vigorous star formation is taking place in these regions. Another nearby spiral, NGC 300, appears to show similar behavior (A. Ferguson, private communication). These spirals differ from their more massive counterparts in two significant respects; they have lower disk metal abundances, which might affect the CO/H_2 conversion ratio, and they exhibit slowly rising rotation curves over most of the inner disks. The former effect might lead to an underestimate of the gas densities in the inner disks, and explain the apparently subcritical densities in the active star forming regions. The latter effect might lead to a breakdown of the Toomre criterion for predicting the critical densities for cloud formation and star formation.

The physical motivation for applying the Safranov/Toomre criterion in Kennicutt (1989) was that the onset of large-scale star formation might be dictated primarily by the growth of gravitational perturbations leading to cloud formation. However the Toomre criterion— which governs long-wavelength disk instabilities— may not describe these cloud growth instabilities in all conditions. Elmegreen (1993b) and Kenney et al. (1993) have proposed alternative critical density conditions in which the epicyclic frequency κ is replaced with the shear rate or tidal criteria. These critical densities converge to the Toomre value in the limit of constant rotation velocity (with a small scaling factor), but tend toward zero in the limit of solid body rotation (vs a constant critical density for the Q criterion). Consequently they produce much lower thresholds in the central parts of galaxies, or in low-mass spiral and irregular galaxies where there is relatively little shear. This could explain the high SFRs in spirals like M33, NGC 2403, or NGC 300, where there is an apparent failure of the Q model. However it does not explain the observation of star formation thresholds in Magellanic irregular galaxies (Skillman 1987, Taylor et al. 1994), which are accounted for quite readily by the Q model. Perhaps the best way to distinguish between these respective models is to measure the SFRs and gas densities near the centers of massive spiral galaxies, where the Q model predicts a much higher threshold density in most cases.

Further complications are introduced by the presence of both stellar and gaseous disks in real galaxies. It is well known that the interaction between the stellar and gaseous disks can act to destabilize the gas disk (Jog & Solomon 1984). The consequences of adopting a more realistic two-fluid stability condition have been explored by numerous authors (e.g., Elmegreen 1991, 1994b; Gammie 1994). While the threshold conditions predicted by these models are in qualitative accord with the simpler Toomre condition, Elmegreen (1994b) concludes that the differences in predicted threshold radii and densities should be measurable. The paucity of data on the velocity dispersions of the stellar *and* gas disks of nearby galaxies hampers such a test at present. Elmegreen (1991) and Gammie (1994) also consider a wide range of combined instability mechanisms for galactic gas layers, including gravitational, shear, magnetic Rayleigh-Taylor, and thermal instabilities. The interplay between these different instabilities may regulate star formation so that $Q \sim 1$ (Elmegreen 1991).

4.6. Other Threshold Mechanisms

Gravitational instabilities are not the only types of physical mechanisms that might lead to an abruptly nonlinear star formation law. If spiral density waves are the primary triggering agent for large-scale star formation, they could produce threshold features as well (Wyse 1986; Wyse & Silk 1987). These authors have proposed a modified Schmidt law of form:

$$R = A\Sigma^n(\Omega - \Omega_p) \qquad (6)$$

where Ω and Ω_p are the disk and spiral pattern angular frequencies. Such a model will produce a cutoff in star formation at the corotation radius, and a roughly $1/r$ term from $\Omega - \Omega_p$ inside this radius. Kennicutt (1989) has argued that the range of star formation properties in grand-design and flocculent spirals suggests a decoupling between the stellar disk instabilities (which lead to spiral density waves) and the gas instabilities (which lead to large-scale star formation), but this conjecture has not been tested observationally.

The close association of star formation with molecular cloud complexes implies that the formation of gravitationally bound clouds in itself is not a sufficient condition for large-scale star formation, and it raises the question of whether it is the formation of molecular gas, rather than the formation of gravitationally bound clouds, that is the crucial regulator of star formation. Are the formation of bound clouds and molecular clouds manifestations of the same threshold process, or are gravitational and molecular formation thresholds distinct processes which are decoupled from each other (and from star formation) in some environments?

An insightful exploration of the interplay between the formation of gravitational bound clouds and molecular clouds has been published by Elmegreen (1993a). In the solar neighborhood the formation of gravitationally bound clouds and molecular clouds are closely associated, because the threshold column densities for forming both objects are nearly the same. However the dependence of these thresholds on the interstellar pressure and metallicity are quite different, so in other environments the formation of bound clouds and molecular gas may be decoupled. For example, in low-mass irregular galaxies, where low ISM pressures and low metallicities lead to threshold densities for molecular cloud formation which are much higher than the corresponding threshold for the formation of gravitationally bound clouds. In those environments gravitational instabilities will produce bound *atomic* cloud complexes, which later form molecular cores and stars within the larger atomic clouds (Elmegreen 1993a). At the other extreme, in metal-rich and/or high pressure interstellar environments, such as that found in the inner Galaxy or in massive gas-rich spirals such as M51, conditions will favor the formation of molecular gas at column densities which are too low to form gravitationally bound clouds. There we would expect the observed star formation thresholds to occur in a largely molecular medium, consisting of bound molecular clouds within a substrate of diffuse molecular and atomic gas. In such galaxies it is the gravitational threshold, not the molecular formation (self-shielding) threshold, which is critical for initiating large-scale star formation. However in low-metallicity or low-pressure environments, such as in dwarf galaxies or the outer disks of spirals, gravitational stabilities may be secondary to molecule formation in leading to efficient large-scale star formation (*cf.* Lin & Murray 1992).

There is some observational evidence to support this basic picture. CO observations of the inner Galaxy and some starburst galaxies show evidence for a diffuse molecular medium surrounding the molecular clouds (e.g., Stark et al. 1991; Aalto et al. 1994). Observations of the CO and cold HI in the Magellanic Clouds (e.g., Israel et al. 1993) are consistent with the cloud structure described by Elmegreen (1993a), though the interpretation of these data is complicated by possible changes in the CO/H_2 coupling in these metal-poor environments. It should be possible to separate these gravitational and molecular thresholds by obtaining sensitive maps of the CO, HI, and $H\alpha$ in nearby galaxies which span a broad range of interstellar pressures, metal abundances, and rotation curves.

Acknowledgments

I wish to express special thanks to my collaborators in the research presented here, especially my current and former graduate students Audra Baleisis, Fabio Bresolin, Paul Harding, Crystal Martin, Sally Oey, and Anne Turner. I am grateful to Lynn Hillenbrand and Chuck Steidel for providing me with postscript copies of Figs. 4 and 8, and to the editors A. Kembhavi and B. Guiderdoni for providing postscript versions of several other figures. My research is supported by the National Science Foundation though grant AST-9421145.

References

[1] Aalto, S., Booth, R.S., Black, J.H., Koribalski, B., & Wielebinski, R. 1994, A&A, 286, 365
[2] Barnes, J., Kennicutt, R., & Schweizer, F. 1997, Galaxies: Interactions and Induced Star Formation, 26th Saas-Fee Advanced Course, (Berlin: Springer-Verlag), in press.
[3] Basu, S., & Rana, N.C. 1992, ApJ, 393, 373
[4] Bechtold, J., Yee, H.K.C., Elston, R., & Ellingson, E. 1997, ApJ, 477, L29
[5] Bernlöhr, K. 1993, A&A, 270, 20
[6] Boselli, A. 1994, A&A, 292, 1
[7] Buat, V. 1992, A&A, 264, 444
[8] Buat, V., Deharveng, J.M., & Donas, J. 1989, A&A, 223, 42
[9] Buat, V., Donas, J., & Deharveng, J.M. 1987, A&A, 185, 33

[10] Buat, V., & Xu, C. 1996, A&A, 306. 61

[11] Caldwell, N., Kennicutt, R., Phillips, A.C., & Schommer, R.A. 1991, ApJ, 370, 526

[12] Caldwell, N., Kennicutt, R., & Schommer, R. 1994, AJ, 108, 1186

[13] Calzetti, D., Kinney, A.L., & Storchi-Bergmann, T. 1994, ApJ, 429, 582

[14] Deharveng, J.M., Sasseen, T.P., Buat, V., Bowyer, S., Lampton, M., & Wu, X. 1994, A&A, 289, 715

[15] Devereux, N.A., & Hameed, S. 1997, AJ, 113, 599

[16] Devereux, N.A., & Young, J.S. 1990, ApJ, 350, L25

[17] Donas, J., Deharveng, J.M., Laget, M., Milliard, B., & Huguenin, D. 1987, A&A, 180, 12

[18] Eder, J.A. 1990, PhD thesis, Yale University

[19] Elmegreen, B.G. 1991, ApJ, 378, 139

[20] Elmegreen, B.G. 1993a, ApJ, 411, 170

[21] Elmegreen, B.G. 1993b, Proc. 4th EIPC Conference, p335

[22] Elmegreen, B.G. 1994a, in Violent Star Formation, ed G. Tenorio-Tagle, Cambridge Univ Press, p220

[23] Elmegreen, B.G. 1994b, ApJ, 427, 384

[24] Elmegreen, B.G. 1994c, ApJ, 433, 35

[25] Elmegreen, B.G., & Elmegreen, D.M. 1986, ApJ, 311, 554

[26] Fritze-v. Alvensleben, U., & Gerhard, O.E. 1994, A&A, 285, 775

[27] Gallagher, J.S., Hunter, D.A., & Tutukov, A.V. 1984, ApJ, 284, 544.

[28] Gammie, C.F. 1994, ApJ, in press

[29] Guiderdoni, B. 1987, A&A, 172, 27

[30] Hill, R.J., Madore, B.F., & Freedman, W.L. 1994, ApJ, 429, 204

[31] Hillenbrand, L.A. 1997, AJ, in press

[32] Holtzman, J. et al. 1997, AJ, 113, 656

[33] Huchra, J.P. 1977, ApJ, 217, 928

[34] Hunter, D.A. et al. 1995, ApJ, 448, 179

[35] Hunter, D.A. et al. 1996, ApJ, 459, L27

[36] Israel, F.P. et al. 1993, A&A, 276, 25

[37] Jog, C.J., & Solomon, P.M. 1984, ApJ, 276, 127

[38] Kenney, J.D.P., Carlstrom, J.E., & Young, J.S. 1993, ApJ, 418, 687

[39] Kennicutt, R.C. 1983, ApJ, 272, 54

[40] Kennicutt, R. C. 1989, ApJ, 344, 685

[41] Kennicutt, R.C. 1990, in The Interstellar Medium in Galaxies, ed. H.A. Thronson & J.M. Shull, (Dordrecht: Kluwer), 405

[42] Kennicutt, R.C. 1992a, ApJ, 388, 310

[43] Kennicutt, R.C. 1992b, in Star Formation in Stellar Systems, ed. G. Tenorio-Tagle, M. Prieto, & F. Sánchez (Cambridge: Cambridge Univ. Press), 191

[44] Kennicutt, R.C., & Chu, Y.-H. 1988, AJ, 95, 720

[45] Kennicutt, R.C., Edgar, B.K., & Hodge, P.W. 1989, ApJ, 337, 761

[46] Kennicutt, R.C., Keel, W.C., van der Hulst, J.M., Hummel, E., & Roettiger, K.A. 1987, AJ, 93, 1011

[47] Kennicutt, R.C., & Kent, S.M. 1983, AJ, 88, 1094

[48] Kennicutt, R.C., & Martin, C.L. 1997, in preparation

[49] Kennicutt, R.C., Tamblyn, P., & Congdon, C.W. 1994, ApJ, 435, 22

[50] Köppen, J., Theis, Ch., & Hensler. G. 1994, A&A, in press

[51] Kroupa, P., Tout, C.A., & Gilmore, G. 1993, MNRAS, 262, 545

[52] Larson, R.B. 1987, in Starbursts and Galaxy Evolution, eds T.X. Thuan,
      T. Montmerle, & J. Tran Tranh Van, Editions Frontierières, p467

[53] Larson, R.B., & Tinsley, B.M. 1978, ApJ, 219, 46

[54] Leonardi, A.J., & Rose, J.A. 1996, AJ, 111, 182

[55] Leitherer, C., & Heckman, T.M. 1995, ApJS, 96, 9

[56] Leitherer, C., Vacca, W.D., Conti, P.S., Filippenko, A.V., Robert, C., &
      Sargent, W.L.W. 1996, ApJ, 465, 717

[57] Lin, D. N. C., & Murray, S. D. 1992, ApJ, 394, 523

[58] Massey, P., Lang, C. C., DeGioia-Eastwood, K., & Garmany, C.D. 1995a,
      ApJ, 438, 188

[59] Massey, P., Johnson, K.E., & DeGioia-Eastwood, K. 1995b, ApJ, 454, 151

[60] Maoz, D., Filippenko, A.V., Ho, L.C., Macchetto, D., Rix, H.-W., &
      Schneider, D.P. 1996, ApJS, 107, 215

[61] Mateo, M. 1988, ApJ, 331, 261

[62] McCall, M.L., & Schmidt, F.H. 1986, ApJ, 311, 548

[63] McQuade, K., Calzetti, D., & Kinney, A.L. 1995, ApJS, 97, 331

[64] Meurer, G.R., Heckman, T.M., Leitherer, C., Kinney, A., Robert, C., &
      Garnett, D.R. 1995, AJ, 110, 2665

[65] Miller, G.E., & Scalo, J.M. 1979, ApJS, 41, 513200

[66] Moshir, M. et al. 1992, Explanatory Supplement to the IRAS Faint Source
      Survey, Version 2, JPL D-10015 8/92 (Pasadena: JPL)

[67] Oey, M.S. 1996, ApJ, 465, 231

[68] Oey, M.S., & Massey, P. 1995, ApJ, 452, 210

[69] Parker, J.W., & Garmany, C.D. 1993, AJ, 106, 1471

[70] Peimbert, M., Colin, P., & Sarmiento, A. 1994, in Violent Star Formation,
      ed G. Tenorio-Tagle. Cambridge Univ Press, p79

[71] Phelps, R.L, & Janes, K.A. 1993, AJ, 106, 1870

[72] Quirk, W.J. 1972, ApJ, 176, L9

[73] Rand, R.J. 1993, ApJ, 410, 68

[74] Rieke, G.H., Lebofsky, M.J., Thompson, R.I., Low, F.J., & Tokunaga,
      A.T. 1980, ApJ, 238, 24

[75] Rieke, G.H., Loken, K., Rieke, M.J., & Tamblyn, P. 1993, ApJ, 412, 99

[76] Rieke, G.H., & Low, F.J. 1972, ApJ, 176, L95

[77] Robert, C., Leitherer, C., & Heckman, T.M. 1993, ApJ, 418, 749

[78] Roberts, M.S. 1963, ARAA, 1, 149

[79] Romanishin, W. 1990, AJ, 100, 373

[80] Salpeter, E.E. 1955, ApJ, 121, 161

[81] Sandage, A. 1986, A&A, 161, 89

[82] Scalo, J.M. 1986, Fund Cos Phys, 11, 1

[83] Schmidt, M. 1959, ApJ, 129, 243

[84] Searle, L., Sargent, W.L.W., & Bagnuolo, W.G. 1973, ApJ, 179, 427

[85] Skillman, E.D. 1987, in Star Formation in Galaxies, ed C. J. Lonsdale Persson, NASA Conf Publ 2466, p263

[86] Smith, E.P. et al. 1996, ApJS, 104, 287

[87] Stark, A.A., Gerhard, O.E., Binney, J., & Bally, J. 1991, MNRAS, 248, 14P

[88] Steidel, C.C., Giavalisco, M., Pettini, M., Dickinson, M., & Adelberger, K.L. 1996, ApJ, 462, L17

[89] Taylor, C.L., Brinks, E., Pogge, R.W., & Skillman, E.D. 1994, AJ, 107, 971

[90] Tinney, C.G. 1993, ApJ, 414, 279

[91] Tinsley, B.M. 1972, A&A, 20, 383

[92] Tinsley, B.M. 1980, Fund Cos Phys, 5, 287

[93] Tinsley, B.M., & Danly, L. 1980, ApJ, 242, 435

[94] Tomita, A., Tomita, Y, & Saito, M. 1996, PASJ, 48, 285

[95] Toomre, A. 1964, ApJ, 139, 1217

[96] Vallenari, A., Aparicio, A., Fogotto, F., & Chiosi, C. 1994, A&A, 284, 424

[97] van der Hulst, J.M., Kennicutt, R.C., Crane, P.C., & Rots, A.H. 1988, A&A, 195, 38

[98] van der Hulst, J.M., Skillman, E.D., Smith, T.R., Bothun, G.D., McGaugh, S.S., & de Blok, W.J.G. 1993, AJ, 106, 548

[99] Walterbos, R.A.M., & Greenawalt, B. 1996, ApJ, 460, 696

[100] Weedman, D.W., Feldman, F.R., Balzano, V.A., Ramsey, L.W., Sramek, R. A., & Wu, C.-C. 1981, ApJ, 248, 105

[101] Will, J.-M., Bomans, D.J., & de Boer, K.S. 1995, A&A, 295, 54

[102] Williams, D.A., Rieke, G.H., & Stauffer, J.R. 1995a, ApJ, 445, 359

[103] Williams, D.A., Comerón, F., Rieke, G.H., & Rieke, M.J. 1995b, ApJ, 454, 144

[104] Wyse, R.F.G. 1986, ApJ, 311, L41

[105] Wyse, R.F.G., & Silk, J. 1987, ApJ, 339, 700

LECTURE 2

From Giant Molecular Clouds to Compact Cores

E. Falgarone

Ecole Normale Supérieure, Observatoire de Paris and CNRS
24 rue Lhomond, 75005 Paris, France

1. INTRODUCTION

A classical scenario for the process of star formation begins with an exponentially growing instability triggered by a perturbation of an unstable equilibrium configuration. In an isothermal gas of uniform density, the largest masses are the most unstable because the long wavelength modes are those which grow the fastest [1, 2]. Fragmentation is controlled by the growth of density fluctuations of smaller and smaller size as the initial mass of gas, set in free–fall collapse, contracts and cools down. The main difficulty with this description is that it is not in accord with some observational facts. First, stars do not form within structures in free–fall collapse. They are observed to form within dense cores, themselves parts of larger scale structures, the molecular clouds and complexes, none of them being in free–fall collapse. In addition, over the last twenty years, observations of molecular clouds have progressively revealed that thermal support in molecular clouds is negligible compared to the non-thermal support provided by supersonic internal motions and magnetic field, that giant molecular clouds (hereafter GMCs) are extremely heterogeneous entities, and that the bulk of their mass is contained in small structures of large local densities. The picture of molecular clouds which emerges now is that of a hierarchy of dynamically connected scales, extending down to about the size of the Solar System. The question of the initial conditions for collapse and triggering of star formation has therefore to be raised again, together with the role of the environment of the densest cores, taken as a whole up to the scale of the giant molecular clouds (GMCs) and possibly even further.

The complexity of the star formation process and its link with the structure and dynamics of the placental gas is such that we must often rely on observations to get key elements. To provide the reader with some insight into the difficulties met in describing the actual mass distribution of the dense gas and the potential flaws inherent in all the observational deductions, this lecture therefore starts with a full section (Section 2) devoted to the tracers of dense molecular gas. The main challenge is indeed to rely on the observations to derive the mass distribution and the density contrasts at each scale, and the mechanisms of dynamical coupling and energy exchanges between scales. Observational results relevant to the nearest GMCs and the comparison of their characteristics are given in Section 3. Then, a section is devoted to the observational evidence we have for the connection between scales (Section 4), and physical processes other than self–gravity are put in perspective as presumably at the origin of the connection between scales, *i.e.* the fact that the medium is turbulent and that it is threaded by magnetic field (Section 5). Last, a few scenarios of the time-dependent evolution of the star formation activity in galaxies are presented (Section 6). Note that this lecture refers to observations in the Galactic disk. Unlike in the case of star bursts, the gravitational potential well is mostly due to the gas itself; the contribution of preexisting stars is dominant only in the case of young dense clusters.

2.   THE TRACERS OF COLD MOLECULAR MATERIAL

2.1.   Molecular Hydrogen: a Tracer of the Interface between Dense Molecular Gas and Atomic Gas in a UV-rich Environment

The bulk of the mass of molecular clouds is sufficiently shielded from the UV photons so that hydrogen there is almost entirely molecular. The critical densities of the lowest rotational transitions of H_2 (Table I) show that these levels are easily thermalized for the densities in molecular clouds, but the upper level energies are high compared to the gas kinetic temperature in dense clouds, so that the lowest rotational lines of H_2 are not collisonally excited in cold molecular gas and the pure rotational line emission may be used as tracers of molecular gas only in hot cores. Further, the Einstein coefficients for spontaneous emission are extremely low because the H_2 molecule has no permanent dipole moment and the transitions are quadrupole electric transitions. The critical density of a transition is the density of collisional partners (H, He or H_2) required to maintain a thermal population in the rotational levels, $n_{cr} = A_{ul}/\langle \sigma v \rangle$, where $\langle \sigma v \rangle$ is the excitation collision rate, the brackets standing for an average of the relative velocities v over the partner velocity distribution.

The rotational-vibration transitions of the molecule are used to diagnose the thin interface between shielded and unshielded regions in molecular clouds.

Table I. — Characteristics of the first v=0 pure rotational levels of H_2

	J_u-J_l	λ (μm)	E_{ul}/k_B (K)	A_{ul}(s^{-1})	n_{cr}(cm^{-3})
S(0)	2-0	28.2	510	2.9×10^{-11}	2.9
S(1)	3-1	17.0	1020	4.7×10^{-10}	47
S(2)	4-2	12.3	1699	2.7×10^{-9}	270
S(3)	5-3	9.66	2549	9.8×10^{-9}	980
S(4)	6-4	8.03	3569	2.6×10^{-8}	2.6×10^3

Those regions are called PDRs, for Photon Dominated Regions. This atomic to molecular interface in dense gas is extremely thin because molecular hydrogen is dissociated by absorption of photons in its own lines and therefore self–shields efficiently from the dissociative radiation. An estimate of the thickness of the interface is $l \sim 1/n_{tot}\sigma(1000\text{Å})$, where $\sigma(1000\text{Å}) = 6 \times 10^{-22}$ cm^2 is the cross section for H_2 photodissociation, so that $l \sim 5 \times 10^{-3}$pc in a gas of total density $n_{tot} = 10^5$ cm^{-3} [3]. In these interface regions, the vibrational levels are populated by fluorescence. The molecule is brought into an excited electronic level by absorption of a far UV photon (912Å–1100Å). The outcome of the radiative de-excitation to the electronic ground state is molecule dissociation in 10% of the cases, and in all the others radiative decay from highly excited vibrational levels of the electronic ground state to eventually the vibrational ground state. The transitions which are the most commonly observed are the v=1-0 S(1) (J=3-1) line at $\lambda = 2.1\mu$m and the S(2) (J=4-2) line at $\lambda = 2.2\mu$m. Their Einstein coefficients for spontaneous emission are large, $A_{ul} \sim 3 \times 10^{-7}$s^{-1}. The position of the HI/H_2 interface depends only on the ratio n_{tot}/G_0 where the radiation field in the range 912Å–1100Å is $\Phi = G_0\Phi_0$. $\Phi_0 = 1.6 \times 10^{-3}$ erg s^{-1} cm^{-2} is the average interstellar radiation field in the Solar Neighborhood [4]. The emission of these lines is therefore able to probe the physical conditions at the illuminated surfaces of molecular clouds being photo–dissociated [5].

In addition, the size scales accessible by direct imaging of PDRs in the near IR H_2 lines are significantly smaller than those reachable by any other technique. Recent observations of the H_2 fluorescent emission, carried out at 2.2 μm with the technique of adaptive optics on large telescopes, have shown for the first time that the interface between dense cores and the reflection nebula NGC2023 [6] or an HII region in the Orion bar [7] is structured down to scales as small as ~ 100AU, never observed directly before. This structure is very likely the intrinsic small–scale structure of the dense core, made fluorescent in the H_2 lines by the UV photons of the young illuminating stars, but it might also result from instabilities in the dense interface.

Table II. — Characteristics of a few rotational and of the (v=1,J=0–v=0,J=1) vibrational transitions

molecule	rotation					vibration	
	μ (D)	J_u-J_l	E_u/k_B (K)	A_{ul} (s^{-1})	n_{cr} (cm^{-3})	A (s^{-1})	$h\nu/k$ (K)
CO	0.11	1-0	5.53	7.2×10^{-8}	3.0×10^3	30.6	3000
CS	1.95	2-1	7.05	1.6×10^{-5}	3.0×10^5	17	1830
CS		7-6	65.8	8.3×10^{-4}	2.5×10^7		
HCN	2.98	1-0	4.25	2.4×10^{-5}	3.6×10^5	0.04	2900
HCN		4-3	42.5	2.0×10^{-3}	1.7×10^8		
HCO$^+$	4.07	1-0	4.28	4.5×10^{-5}	2.0×10^5		

2.2. Transitions of Polar Molecules: Tracers of Density and Temperature

Most molecular species present in molecular clouds have permanent dipole moments and therefore electric dipole transitions with large Einstein coefficients. Although their abundances are low, usually in the range 10^{-12}–10^{-4}, the combination of the large range of frequencies and therefore of rotational levels now accessible to observations allows rotational and vibrational transitions to probe a large range of physical conditions in molecular clouds. This is illustrated in Table II which, for a set of molecules and rotational levels, gives the energy of the upper level J of a rotational transition $E(J) = h\, B_0\, J(J+1)$, the Einstein coefficient for spontaneous emission,

$$A_{J,J-1} = 9.3 \times 10^{-20} B_0^3 \mu^2 \frac{J^4}{(2J+1)}\ \text{s}^{-1}$$

where B_0, expressed in MHz, is the rotational constant of the molecule and μ, in Debye, its dipole moment. As the density of collisional partners increases, the transition eventually thermalizes. The critical density n_{cr} at which it occurs, is given in Table II, for a temperature $T_k = 20$K, characteristic of cold molecular clouds.

The rotational levels of a molecule are populated by collisions and radiation and, assuming statistical equilibrium, multitransition observations of the same molecule allow a simultaneous determination of the density and column density of the emitters, under assumptions on the radiative transfer of the millimeter line photons. The method is illustrated for instance by CS measurements in [8, 9]. These determinations are plagued by the fact that a given pair of detected transitions seems to always provide a density value close to the largest of the critical densities of these transitions. Higher densities cannot be derived because in that case both transitions are thermalized and their ratio is essentially independent of density. Neither can lower densities be inferred, because the intensities of the lines drop exponentially below the sensitivity limit. Observations of a given pair, in practice, just provide a lower limit to the density, close to the largest critical density of the pair.

Symmetric top molecules like methylacetylene CH_3CCH, CH_3 or NH_3 are used as thermometers because the effects of temperature and density in the excitation can be separated. Their rotational energy is a function of the two quantum numbers (J, K), the total angular momentum and its projection along the axis of symmetry of the molecule. As long as the molecular dipole moment is along this axis, the selection rule for radiative transitions $\Delta K = 0, \Delta J = 0, \pm 1$ prevents radiative transitions between different K–ladders. Transitions between the same pair of J numbers in different K–ladders have almost the same frequency but their energy levels are different in each K–ladder. The relative populations of these levels in the different K–ladders are therefore a function of the temperature only. The additional advantage of selecting these transitions is that they can be observed simultaneously with a single receiver, therefore reducing the problems inherent in the relative calibration of lines observed with different telescopes (see references in [10]). The rotational temperatures derived from the observations of CH_3CCH in the dense cores in Orion range between 20K and 60K.

A special mention of the transition $^{12}CO(J=1-0)$ should be made because it is currently used to trace the mass of giant molecular clouds and scales below. The conversion factor X between the integrated $^{12}CO(J=1-0)$ emission $W(CO)$ and H_2 column density $N(H_2)$ has been calibrated in the Milky Way only and its use in external galaxies is a matter of ongoing debate. At large scale, the total amount of gas may be determined by several independent methods, γ-ray emission (a cosmic ray interaction with a nucleon produces an unstable π^0 which disintegrates by emitting two γ photons), near IR extinction of the stellar bulge of the Galaxy, and far infrared emission at $100\mu m$ outside star forming regions. At small scale, the column density of H_2 is determined by H_2 line absorption in the UV against O stars. The most recent determination of X based on the data produced by the high–energy detectors aboard the Compton Gamma–Ray Observatory [11, 12], is

$$X = \frac{N(H_2)}{W(CO)} = (1.1 \pm 0.2) \times 10^{20} \, cm^{-2} \, (K \, km \, s^{-1})^{-1}$$

In general, larger beam calibration methods which give more weight to extended lower density gas tend to give larger values for the X factor. A CO to H_2 conversion factor $X = 2.6 \times 10^{20} \, cm^{-2} \, (K \, km \, s^{-1})^{-1}$ has been used to determine the mass of H_2 in several large–scale ($\sim 200 \, pc$) regions in Orion and the total masses of gas (HI and H_2) are in remarkable agreement with the mass determinations based on the submillimeter emission of dust observed by the COBE satellite [13]. Calibrations carried out in nearby high latitude clouds, on the basis of the comparison of the 100μm dust emission and the HI and ^{12}CO(J=1–0) emissions, assuming the same 100μm emissivity per H nucleon, whether it is in atomic gas or in the molecular component, tend to find smaller X values, $X \sim 0.5$ to $1\times10^{20}\,\mathrm{cm}^{-2}$ ($\mathrm{K\,km\,s^{-1}})^{-1}$. Variations in the values of the X factor at small scale may be due to the existence of gas rich in H_2 and poor in CO.

2.3. The Shapes and Properties of the Rotational Line Profiles: Tracers of the Velocity Field

The ^{12}CO(J=1–0) transition is considered to be a good tracer of the molecular H_2 mass in the Milky Way, in spite of the large optical depths $\tau(\nu)$ of the lines and the low photodissociation energy (D=11.1 eV) of the molecule. This unexpected behavior, also true to a lesser extent for the CS molecule, is quite general. The ^{12}CO emission is in general very optically thick, as indicated by the observed isotopic line ratios. The line temperature is defined as $T_L(\nu) = (1 - e^{-\tau(\nu)})[B(\nu, T_{ex}) - B(\nu, T_{BG})]$ where $T_{BG} = 2.7$K is the temperature of the cosmic background. Under a few assumptions (optically thin ^{13}CO line, optically thick ^{12}CO line and similar excitation temperatures of the two transitions), the observed line ratio provides an estimate of the optical depth of the ^{12}CO line:

$$\frac{T_L(^{12}\mathrm{CO})}{T_L(^{13}\mathrm{CO})} \sim \frac{1}{\tau_{13\mathrm{CO}}} = \frac{[^{12}\mathrm{CO}]}{[^{13}\mathrm{CO}]} \frac{1}{\tau_{12\mathrm{CO}}}.$$

The observed ^{12}CO to ^{13}CO line ratio is in general much smaller than the isotopic abundance ratio $[^{12}\mathrm{CO}]/[^{13}\mathrm{CO}] = 75$, indicating ^{12}CO optical depths much larger than 1.

However, CO molecular lines behave as optically thin tracers, since a correct mass estimate of the clouds is derived from this tracer. The small *beam-averaged* optical depth of the ^{12}CO lines is indeed due to the combination of the velocity field and spatial distribution of the emitting gas in the medium *i.e.* the large escape probability of the photons is presumably due to a very lacunar structure of the gas in space *and* velocity.

The radiative transfer along a line of sight is governed by the optical depth $\tau(x, y, v_z)$, x and y being the two coordinates in the sky of the line of sight and v_z the gas velocity projected on the line of sight direction. The optical depth over a velocity interval δv is the integration of the absorption coefficient κ over a length fixed by δv *i.e.* by the velocity field, or by the gas density distribution. This length may be the size of a clump of matter at the velocity of interest, but it may also be defined as $l \sim \delta v/(dv/dz)$ in a region where the velocity gradient dv/dz is large. The emergent intensity in a beam is governed by the average of $1 - e^{-\tau(x,y,v_z)}$ over the beam. If the medium is extremely clumped in space and velocity, at scales much smaller than the beam size, the line of sight optical depths exhibit large fluctuations from one line to the next within the beam, and $\langle 1 - e^{-\tau(x,y,v_z)}\rangle_{beam} \ll 1 - e^{-\bar{\tau}}$ where $\bar{\tau} = \langle\tau(x,y,v_z)\rangle_{beam}$. For a detailed demonstration, the reader is referred to [14]. At the opposite end, if there is little or no fluctuation of $\tau(x, y, v_z)$ within the beam, the beam-averaged opacity is close to the opacity along individual lines of sight.

The line widths all being much larger than the gas thermal velocity at the temperatures of a few 10K, Doppler line broadening is attributed to non-thermal motions present in the clouds, either due to systematic motions like rotation, or to turbulence or to waves pervading the medium. The properties of the CO lines of molecular clouds, and in particular the large escape probability of the CO photons are well understood in the framework of macroturbulence, as opposed to microturbulence. In the microturbulent case, there is no structure in the velocity field, it is random at small scale, like in the thermal case. There is a null velocity correlation length. The photons which contribute to the emergent profile at any frequency (*i.e.* velocity in the line profile) may have been emitted at many different locations in the cloud because on each line of sight (LOS), the entire velocity field of total velocity dispersion σ_{tot} is realized. The velocity dispersion along a line of sight is therefore $\sigma_{LOS} \sim \sigma_{tot}$. The LOS optical depth is $\tau \propto N_{tot}/\sigma_{tot}$ and has little variation from one line of sight to another within the beam. All positions in a beam are possibly radiatively coupled which explains why in this case computed line profiles of optically thick lines are often self-reversed. In the macroturbulent case, the velocity field is characterized by a large velocity correlation length l compared to the photon mean free path λ, *i.e.* on each LOS the projected velocity stays about the same. The internal velocity dispersion along any LOS is $\sigma_{LOS} \sim \sigma_l$ and the optical depth on individual LOS, $\tau_{LOS} \propto N_l/\sigma_l$ is large because σ_l is small, and fluctuates from one LOS to the next. There is little radiative coupling between different parts of the cloud.

We illustrate these differences in Fig 1 drawn from [15]. The emergent CO profiles have been computed in a uniform density medium. The structure only originates in the velocity field described by a correlation length l and a random term $f(s)$ so that the variation of any component of the velocity in a LOS direction s is:

$$\frac{\partial v_i}{\partial s} = -\frac{v_i}{l} + f(s)$$

Fig 1 reveals the profound changes in the line shapes and line ratios as only the properties of the velocity field are changed [15, 16].

This point has not always been fully appreciated. It is emphasized here to illustrate the fact that the observed properties of the CO line profiles and line ratios between various isotopes or different J transitions are in principle powerful tools in the analysis of the sub-beam structure of the gas in space and velocity.

2.4.   The Thermal Dust Emission: Tracer of Gas Column Density

Although only a tiny fraction of the gas mass of the interstellar medium (m_g) is in the form of dust particles $m_D/m_g \sim 10^{-2}$, the thermal radiation of these

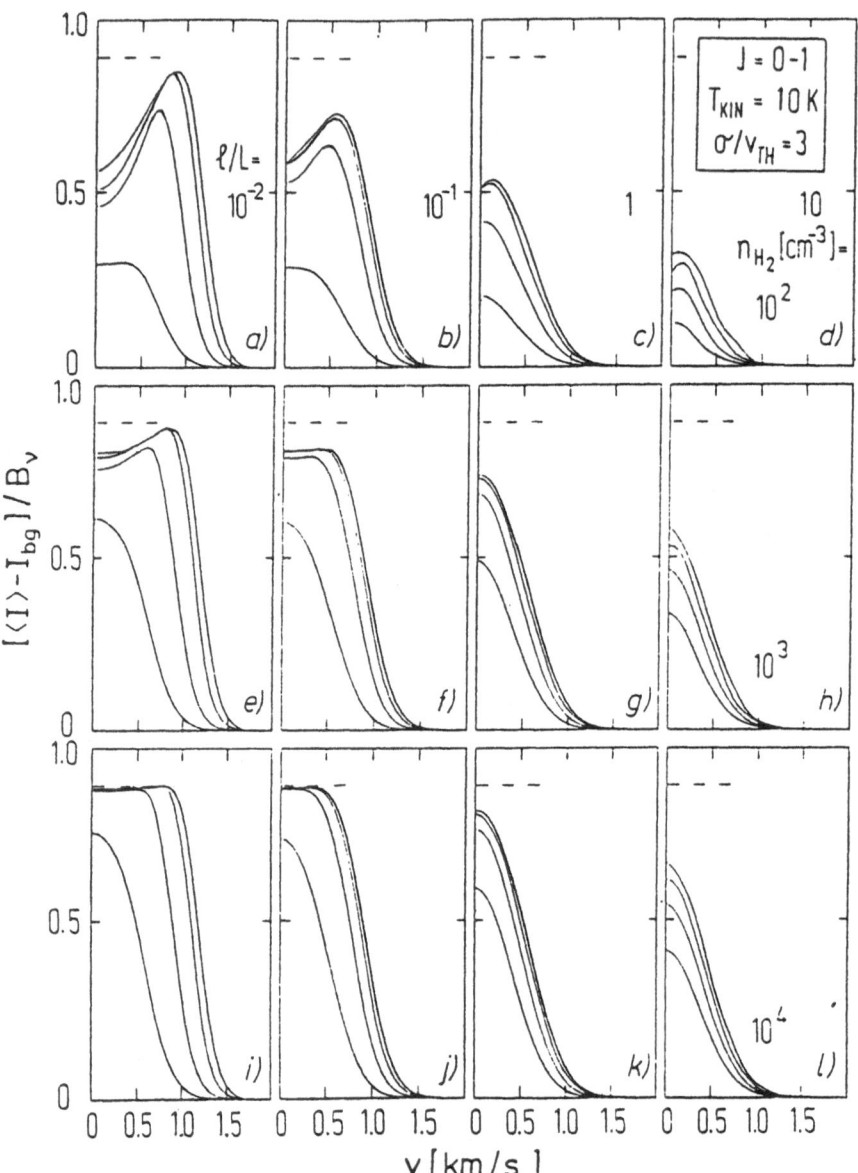

Fig. 1. — CO(J=1-0) line profiles computed for a turbulent velocity dispersion $\sigma_{tot}/c_s = 3$, ratios of the velocity correlation length to cloud size l/L ranging between 10^{-2} and 10, and H_2 densities $n_{H_2} = 10^2, 10^3, 10^4 \, cm^{-3}$. The different profiles in one panel correspond to CO column densities $N_{CO} = 10^{16}, 10^{17}, 10^{18}, 10^{19} \, cm^{-2}$.

particles heated by stellar photons dominates the emission of the interstellar medium in the range $\lambda = 10\,\mu\text{m}$ to 1mm by orders of magnitude. This emission is a powerful and sensitive tracer of the distribution of interstellar matter, and much effort has been devoted in the last 10 years to model it.

In the standard model [18], dust grains are dielectric particles made of silicates and graphite. An illuminating review of the behavior of dust grains in a radiation field and a discussion of all the assumptions made in deriving their optical properties is given in [19]. The size distribution, $n(a) \propto a^{-3.5}$ extending from $a = 0.01\,\mu\text{m}$ to $0.025\,\mu\text{m}$, necessarily steeper than a^{-3} at the large size end to keep the dust mass finite, has been inferred from the observed scattering and absorption properties of the dust in the diffuse interstellar medium [20]. The optical properties used are the following. At wavelengths $\lambda \gg 2\pi a$, a dust particle of size a has an absorption cross–section $\sigma_{abs}(\lambda) = \pi a^2 Q_{abs}(\lambda)$ with, to a good approximation,

$$Q_{abs}(\lambda) \sim \frac{2\pi a}{\lambda} q_0 (\lambda/\lambda_0)^{-\alpha}$$

while for $\lambda \leq 2\pi a$, $Q_{abs}(\lambda) \sim 1$.

The parameters q_0 and α have been determined empirically in several environments because in shielded and dense regions molecular ices deposited at the grain surfaces modify their optical properties. In [21], the dust absorption cross–section per H atom on the line of sight is estimated to be

$$\sigma_H(\lambda) = 7 \times 10^{-22} b \frac{Z}{Z_\odot} \lambda_{\mu\text{m}}^{-1.5} \text{cm}^2$$

for $40\,\mu\text{m} < \lambda < 100\,\mu\text{m}$, and as steep as

$$\sigma_H(\lambda) - 7 \times 10^{-21} b \frac{Z}{Z_\odot} \lambda_{\mu\text{m}}^{-2} \text{cm}^2$$

for $\lambda > 100\,\mu\text{m}$. The parameter b is not the same in all media, and takes values $b = 1$ in a diffuse medium, $b = 1.5$ in gas at densities $n_{H_2} < 10^5\,\text{cm}^{-3}$ and $b = 3$ to 4 in dense cores with $n_{H_2} > 10^6\,\text{cm}^{-3}$. Z is the metallicity in Solar units.

In the standard model, dust grains are assumed to be in thermal balance between the power they absorb from the ambient UV and visible radiation field of energy density $u(\lambda)$ and their own thermal radiation in the infrared. Since the thermal emission of a dust grain is is optically thin in the infrared, the equilibrium temperature T_D is derived from the thermal balance equation of a single grain:

$$c \int u(\lambda)\sigma_{abs}(\lambda)d\lambda = 4\pi \int B(\lambda, T_D)\sigma_{abs}(\lambda)d\lambda,$$

where $B(\lambda, T_D)$ is the Planck function. In the integral on the left hand side, the relevant wavelengths cover the visible and UV ranges for which $Q_{abs}(\lambda) \sim$

1. The LHS integral is therefore proportional to the total energy radiation density in this range, or $U_{vis,UV} = \int_{vis,UV} u(\lambda)d\lambda \sim 0.4\text{eV cm}^{-3}$ in the Solar Neighborhood. The right hand side integral is dominated by the IR range for which $\sigma_{abs}(\lambda) \propto \lambda^{-2}$ so that $\int B(\lambda,T_D)\lambda^{-2}d\lambda \propto T_D^6$. The equilibrium dust temperature is therefore a very weakly dependent function of the energy density and size, since $T_D \propto (U_{vis,UV}/a)^{1/6}$, the small grains being hotter. For diffuse interstellar dust in the Solar Neighborhood, the integral of the power radiated by dust grains between $\sim 10\,\mu$m and $100\,\mu$m, expressed per H nucleon, is indeed equal to

$$L_{IR,H} = 5.7 \times 10^{-24}\text{erg s}^{-1}\text{H}^{-1}$$

The far infrared emission, in these conditions, may be regarded as a tracer of the total gas column density with a scaling between the 100μm emission and total column density of gas, $I_\nu(100\mu\text{m}) \sim 10\text{MJy sr}^{-1}\text{mag}^{-1}$ valid up to $A_v \sim 1$mag on average in the local interstellar radiation field. The IR emission of dust in the Galaxy (far from star forming regions) has an average color ratio $I_\nu(60\mu\text{m})/I_\nu(100\mu\text{m}) = 0.15$ to 0.2, determined from IRAS images of the Galactic plane.

At wavelengths $\lambda < 60\,\mu$m the standard model fails by orders of magnitude to reproduce the observed dust emission spectrum. Recent models extend the size distribution to a few Å and introduce small dust grains out of thermal balance, heated to temperature $\sim 10^3$K with single photon absorption [22] (see Fig 2). The small grains are either three-dimensional particles or two-dimensional polyaromatic hydrocarbons (PAHs). The small grain population is responsible for the bulk of the energy radiated in the short wavelengths range and is thought to be the carrier of the prominent, infrared bands at 3.3, 6.2, 7.7 and 11.3 μm due to vibrational modes of C-C and C-H in an aromatic lattice (see the review [23]).

An illustration of the existence of gas components with various dust temperatures is provided in Fig 3. The total 100μm emission of the Taurus–Auriga complex is compared to that of the cold dust [17]. The latter is the map of $I_\nu(100\mu\text{m})$-$I_\nu(60\mu\text{m})/0.15$ and is intended to trace the distribution of the cold dust component. It reveals much more clearly than in the 100μm map the long filaments which cross the complex. These filaments are also seen in the ^{13}CO(J=1–0) maps [17], while they are not easily identified in the ^{12}CO(J=1–0) map [24].

2.5.   The Tracers of Magnetic Field

2.5.1.   Tracer of its Direction: the Polarization of Dust Emission and Absorption

Dust grains have a spin thought to be due to the angular momentum received from molecules forming on the grain surface, when they leave this surface at velocities much larger than the gas thermal velocities (they have to release a fraction of their formation energy which is often a fraction of eV). This

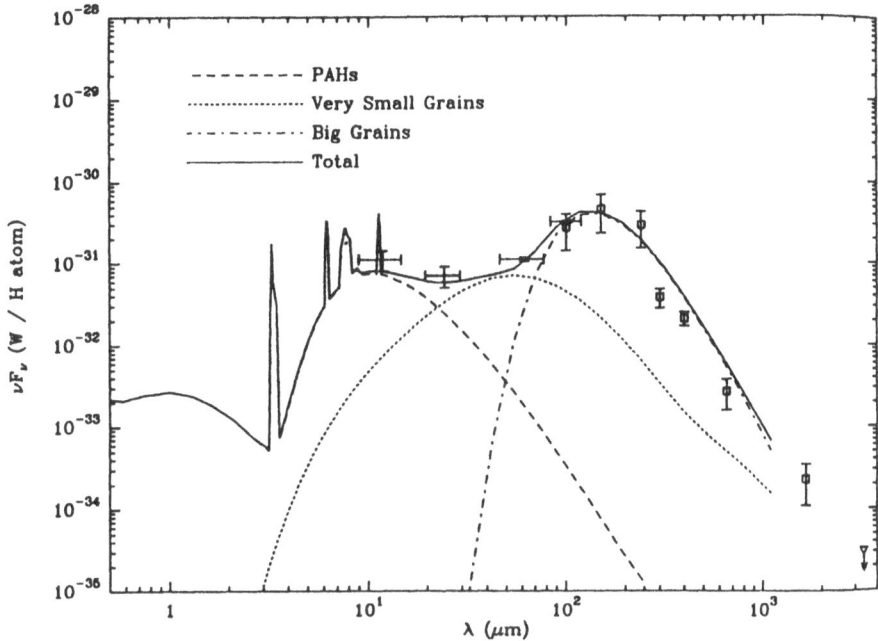

Fig. 2. — Dust emission spectrum. Observations (crosses) pertain to the cirrus interstellar diffuse medium. The model resulting spectrum (continuous line) is the sum of the three components: PAHs, very small grains and big grains.

angular momentum is much larger than that resulting from collisions with gas at velocities close to thermal. The induced rotation velocities are as large as $\omega \sim 10^9$ rad s^{-1}. The dust grains also have unpaired electrons and therefore a permanent magnetic moment $\mathcal{M}$ parallel to their angular momentum $\mathbf{J}$. The adopted line of argument to explain the grain alignment in a magnetic field is the following. Equipartition of the rotational kinetic energy $\frac{1}{2} I \omega^2$ about the three rotational axis of the grains is assumed so that ω is smaller about the axis of larger moment of inertia, and therefore the dominant component $I\omega \propto \omega^{-1}$ is that about the axis of larger inertia. Thus, under this assumption of energy equipartition, needle-like grains rotate preferentially about an axis perpendicular to their own axis. They precess about the magnetic field under the action of the torque $\mathcal{M} \times \mathbf{B}$. Paramagnetic relaxation causes the component of $\mathbf{J}$ perpendicular to the magnetic field to disappear within $\tau \sim 2 \times 10^4$ to 10^6 years in a field B=20μG. In their equilibrium configuration, needle-like grains therefore have their axis perpendicular to the magnetic field. This preferential orientation is responsible for the polarization observed. First, if a star located

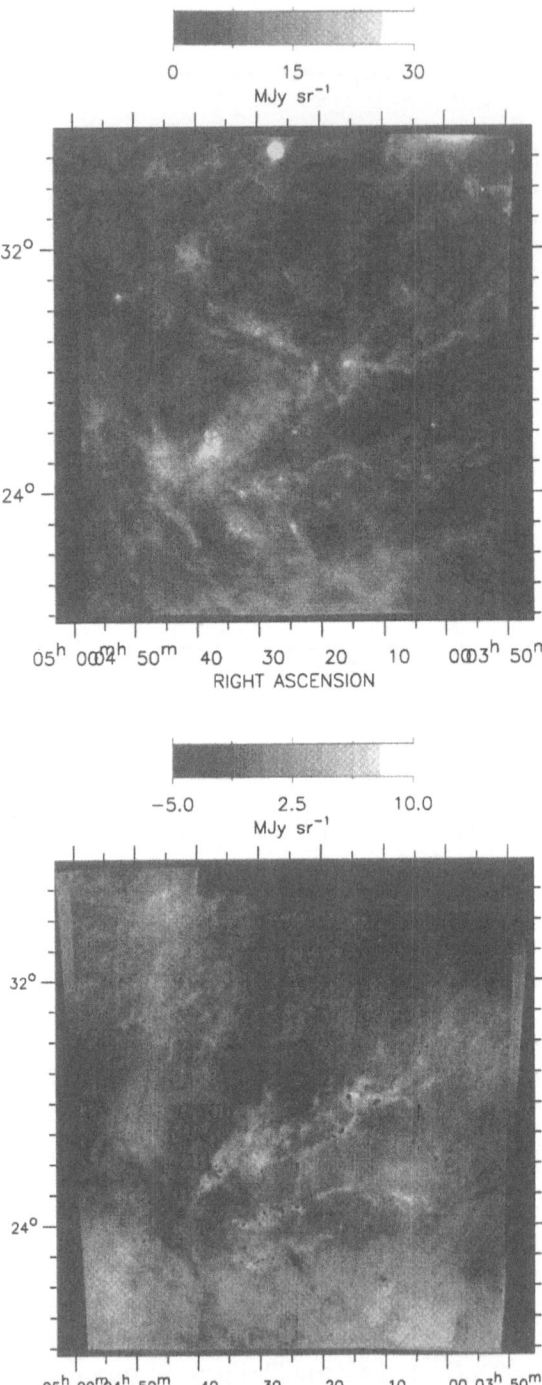

Fig. 3. — Large-scale IRAS emission of the Taurus–Auriga complex. (a) map of the total 100μm emission, $I_\nu(100\mu$m) (b) the cold dust emission traced by $I_\nu(100\mu$m)-$I_\nu(60\mu$m)/0.15.

behind a cloud is observed, the transmitted star light (*i.e.* the fraction which has not been absorbed by dust) has its electric vector perpendicular to the dust grains axis (because the grains have preferentially absorbed the component of **E** parallel to their axis) and is therefore polarized perpendicular to the dust grain axis and parallel to **B**. On the other hand, if the dust emission is observed, the polarization is different because **E** is preferentially aligned with the dust grain axis and the emitted radiation has a polarization parallel to the grain axis and perpendicular to **B**. This polarization should not be mistaken with polarization of stellar emission in the near IR due to scattering by large grains, the efficiency of this process, $Q_{sca} \propto (2\pi a/\lambda)^4$, being a sharp function of grain size a. This polarization mechanism is often used at $2\,\mu$m to determine the position of embedded stellar objects, otherwise hidden at all wavelengths.

An illustration of the results obtained from optical polarization measurements of the radiation of background stars after their passage through molecular clouds is given in Fig 4 from [25]. Note that the magnetic field is aligned along the direction of the main filaments in some of the cases, but is also observed perpendicular to the filaments at some other locations. Similar results have been derived from polarization measurements in the $2\,\mu$m range.

2.5.2. Tracer of its Intensity: Zeeman Splitting

For a detailed understanding of this effect, the reader should refer to [26] for the theory and to the very exhaustive review [27] for the astrophysical observations. The magnetic moment of molecules may have several origins: orbital motions of unpaired electrons, as in O_2 which has two unpaired electrons with parallel spins but has never been detected in interstellar clouds, or localized charges in rotating radicals such as CN, CCS and SO. The intrinsic difficulty of any Zeeman splitting measurement lies in the fact that the frequency separation of the displaced Zeeman components of any line is extremely small compared to the line width. With the magnetic field intensity expressed in μG, the displacement in Hz is $2\nu_Z = a \mid \mathbf{B} \mid$. The factor a is only 2.8 Hz μG^{-1} for the 1420 MHz HI line, 3.27 and 1.96 Hz μG^{-1} for the 1665 and 1667MHz OH lines and is equally small for other molecular radicals. Firm detections of the Zeeman splitting have been obtained with the HI line and therefore refer to the interstellar atomic medium. Values of the magnetic field intensity projected on the line of sight $B_{LOS} \sim 20\mu$G have been detected in the HI line. In star forming dense cores, only upper limits of the order of 100μG have been determined so far from molecular radicals [28], except for a value as large as $B_{LOS} \sim 400\mu$G derived from OH line splitting at high angular resolution [29].

3. THE LOCAL GIANT MOLECULAR CLOUDS: COMPARISON OF THEIR PROPERTIES

It is only recently that we have begun to have access simultaneously to the global and local properties of GMCs. This is mostly due to the developments

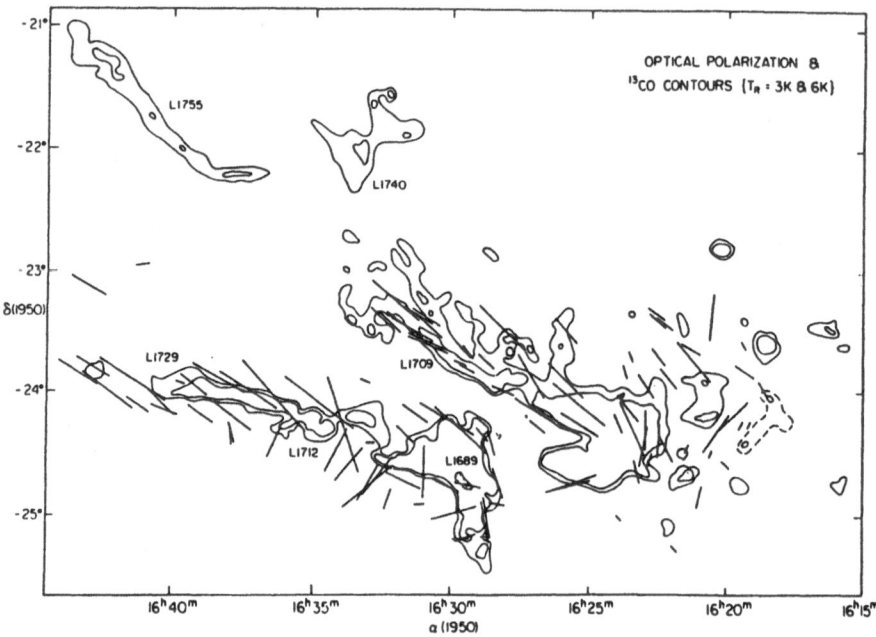

Fig. 4. — Orientation and relative magnitude of optical polarization vectors toward field stars at the cloud edges in the ρ Ophiuchus complex. Solid contours are $T_L=3$ and 6K in the ^{13}CO(J=1–0) line.

of arrays of heterodyne receivers in the millimeter range, and to their increased sensitivity which allows fast mapping. The heterodyne detection is the only one to provide information on the gas velocity field at a high enough resolution ~ 0.05 km s^{-1} or even better. Several GMCs exist in the Solar Neighborhood and although none of them is forming stars at the rate of the brightest star forming regions in our Galaxy such as W49A or NGC3603 (see the lectures of R. Kennicutt), nor of course of the starbursts, their proximity provides us with a unique opportunity to search for a connection between large–scale properties and the mass distribution at protostellar scales, one stepping stone on the way to understanding star formation.

3.1. Large–scale Properties

The large–scale properties of a few nearby GMCs, Orion A and B, MonR2 [30], Taurus–Auriga–Perseus [24], ρ Ophiuchus [31] and Chamaeleon [32], all

Table III. — Large–scale properties of a few GMCs derived from ^{12}CO(J=1–0) data

	Δv km s^{-1}	L pc	M_{gas} $10^5 M_\odot$	$\bar{n}$ cm^{-3}	τ_{ff} Myr	τ_{dyn} Myr	$\mid \Omega \mid /2T$
M31-B292	7	37	3.6	284	2.6	5.2	0.74
OriA-OriB	5	42	1.8	100	4.5	8.4	0.63
Mon R2	4	55	0.9	20	9.8	14	0.38
Taurus	4	66	0.5	7	16	16	0.17
ρ Oph	3.4	57	0.1	2	32	17	0.05
Chamaeleon	3.5	20	0.04	20	10	5.7	0.06

derived from ^{12}CO(J=1–0) observations, are displayed in (Table III) in addition to those of a GMC in M31, an external galaxy similar to the Milky Way, M31-B292 [33]. These properties are the velocity coverage of the complex, Δv, provided by the half-power width of the line integrated emission, its size L and H$_2$ mass, called M_{gas}, derived from the ^{12}CO(J=1–0) integrated emission. The average density is derived from the mass estimate and size, assuming spherical geometry. The one-dimensional internal velocity dispersion is $\sigma = \Delta v/2(2\ln 2)^{1/2}$.

Table III gives the free-fall time τ_{ff} at $T_k = 10$K defined as

$$\tau_{ff} = \frac{44}{\bar{n}^{1/2}}\text{Myr}$$

where $\bar{n}$ is the average density of the gas on the large scale, the dynamical time $\tau_{dyn} = L/\Delta v$ is defined as the large–scale structure crossing time and $\mid \Omega \mid /2T$ is the ratio of the gravitational potential energy (assuming spherical symmetry) to the non–thermal kinetic energy expressed in the units of Table III:

$$\frac{\cdot\mid \Omega \mid}{2T} = \frac{1}{270}\frac{M_{gas}}{L\Delta v^2}$$

with $\mid \Omega \mid = \frac{3}{5}\frac{GM^2}{R}$ and $2T = M\sigma^2$. Note that τ_{ff} depends only on $\bar{n}$ and T_k, τ_d only on Δv, and $\mid \Omega \mid /2T$ is a combination of all these quantities.

The features to be recognized in Table III are the fact that all these GMCs have comparable sizes ~ 50pc, however their average density and mass cover two orders of magnitude, their free fall times vary by a factor larger than 10 and the ratio $\mid \Omega \mid /2T$ by a factor 14. In Table III, the GMCs are ordered by decreasing gas mass which happens to make them also ordered by decreasing values of $\mid \Omega \mid /2T$ and it is remarkable that several OB associations exist in M31-B292 and OriA and OriB while no stars more massive than B stars are found in ρ Ophiuchus and only low mass stars exist in Chamaeleon, [34]. These results suggest that the mass of the most massive stars formed in a complex scales with the mass of gas accumulated within ~ 50 pc and with the ratio of gravitational to kinetic energy.

3.2.   Filamentary Structures: Large and Small–scale Structures

A striking and recently unveiled property of the morphology of GMCs is the existence of massive filaments of length $\sim$100 times their thickness. These filaments are not clearly delineated in the ^{12}CO maps but are very conspicuous either in the ^{13}CO maps or in the maps of cold dust (see Fig 3). From the available ^{13}CO(J=1–0) observations on large and small scale, we infer the properties of the most opaque regions in the GMCs and the small scales. Fig 3 shows that the coldest dust (therefore the most opaque parts) in the Taurus–Auriga–Perseus complex are filamentary structures of extremely large aspect ratio. Young stars (visible as black dots in figure 3b) appear to be nested in various spots within these filaments. The situation is the same in OriA where most of the present star formation activity is taking place in the conspicuous S–shape filament visible in the ^{13}CO(J=1–0) map (Fig 5 from [35]). Filamentary structure is also conspicuous in ρ Ophiuchus (see Fig 4 from [25]).

Table IV gathers the dynamical characteristics of the dense filaments observed in three nearby complexes. They share the following properties:

- their aspect ratios are close to 10 or larger. These structures are at the same time large scale (their length is comparable to that of the GMC size) and small scale (their thickness is barely larger than that of the dense cores) and their recognition requires high angular resolution maps over large areas,

- given their length, width and density, the mass in the dense filaments is estimated to be about 30% of the mass seen in ^{13}CO,

- the spatial transition between the filaments seen in ^{13}CO and the surrounding medium is extremely sharp and occurs within a few 0.01 pc. This is not due to an artifact of the photodissociation of ^{13}CO because the same sharpness is seen in the cold FIR emission (see Fig 3). This raises the issue of their confinement,

- these filaments harbor numerous young stellar objects,

- although they have comparable widths and lengths, they have masses per unit length which differ by a factor 10, and the filament with the largest mass per unit length is found in the most massive and active GMC, Orion.

The scaling of the filaments properties with those of the parent molecular complex and star formation activity suggests that understanding star formation implicitly means understanding the formation mechanisms of such massive filaments.

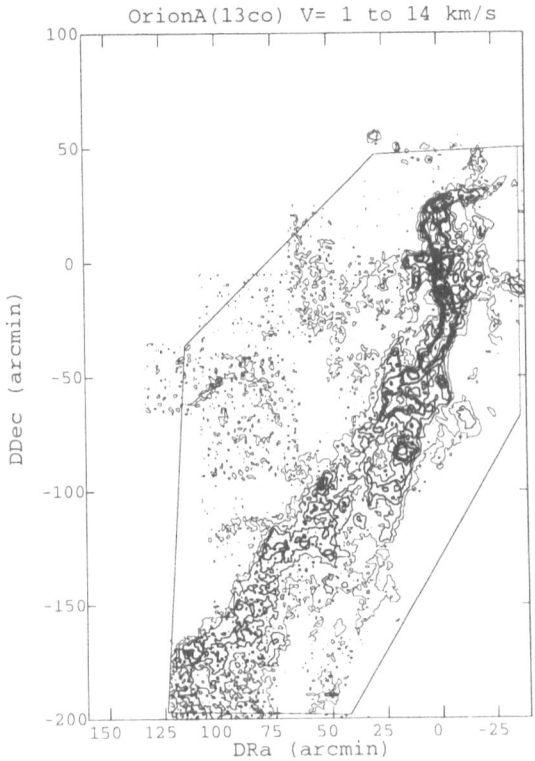

Fig. 5. — ^{13}CO(J=1-0) integrated emission of the OriA complex

Table IV. — Properties of the filamentary structures in a few GMCs from ^{13}CO(J=1-0) data

	W(^{13}CO) $K\,km\,s^{-1}$	mass/length $M_\odot\,pc^{-1}$	width pc	length pc
OriA	54	390	0.5	15
ρ Oph	6–12	30–50	0.4–0.8	~ 6
Taurus	8	40	0.6	18

3.3. The Dense Cores and the Unseen Scales

Dense cores in molecular clouds have been defined observationally, either as regions of large visual extinction ($A_v > 4^m$) in nearby clouds, or as regions in which high J line emission of large dipole moment molecules, like CS(J=5-4), is detected, or as peaks of millimeter and submillimeter continuum thermal emission from cold dust. In the Solar Neighborhood, they have often been identified with the largest thermally supported self–gravitating structures because their mass and size are consistent with those of a self–gravitating isothermal sphere of gas at $T_k = 10$K bounded by an external medium of pressure $P_0/k_B = 3 \times 10^4$ cm^{-3} K [36]. This pressure is that estimated to pervade molecular clouds in the Solar Neighborhood (see Section 4). The size scale of the dense cores is a few 0.1 pc.

3.3.1. The Internal Structure of Dense Cores

Several recent observational results seem to impair the description of dense cores as self–gravitating isothermal structures. The first is the fact that non–thermal but subsonic motions are observed within dense non–star-forming cores. Line widths of heavy molecules like CCS are far above thermal. For instance in TMC1, the CCS line widths observed with an unprecedented velocity resolution of 80m s^{-1} are $\sim$ 0.2km s^{-1} (the thermal line width of CCS would be 0.08 km s^{-1} at $T_k = 10$K), and as large as those of the CS lines in the same core [37]. These line widths are presumably due to subsonic turbulence within the cores. The second is that gravitationally unbound structures exist within dense cores and have been observed down to $\sim$ 1500AU=7.5$\times 10^{-3}$ pc [37, 38]. These results suggest that processes other than self–gravity are at work within dense cores to generate substructure. The third is the dependence, often different from a r^{-2} law, of the average density of the core on its radius (assuming that a *center* can be defined observationally which is not always the case *e.g.* the elongated dense core in Polaris [40, 38]).

The average density, defined by the H$_2$ column density divided by the length of the line of sight through the core, supposed spherical, may be derived from several tracers, as discussed in Section 2. Fig 6 from [41] displays the radial distribution of gas density derived from dust continuum observations in the millimeter range. In a pre–stellar core, the density, assuming an isothermal structure for the dust, seems to be almost constant in the inner core ($n \propto r^{-0.4}$ for $r <$4000AU) and then steeply declines ($n \propto r^{-2}$ for 4000 AU$< r <$1.6 $\times 10^4$AU=0.08 pc) while in a dense core containing a young stellar object, for a decreasing dust temperature with radius ($T_k \propto r^{-0.4}$), the density fall-off is a single power law $n \propto r^{-1.5}$ across the whole core, consistent with infall.

In ρ Ophiuchus, dense cores identified as bright sources in a DCO$^+$(J=3-2) survey of the central area of the complex [42], appear in the ISOCAM images at 6.7μm and 15μm as dark regions against the bright background (see Fig 7 from [43]). The background at these wavelengths is due to emission in the broad band features attributed to very small grains or PAHs, (see Section

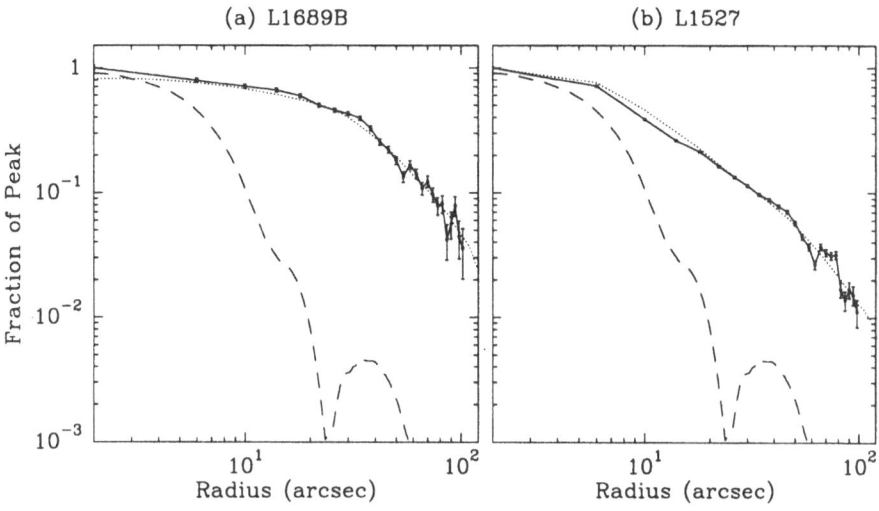

Fig. 6. — Comparison of the radial intensity profile of the dust continuum emission at 1.3mm (a) in the prestellar core L1689B, and (b) in a core containing a protostar L1527. Profiles that fit the data are shown as dotted lines. The beam profile is shown as dashed lines.

2), excited by the UV radiation of young stars associated with the complex. The angular resolution of the ISOCAM observations is 6" or 4.8×10^{-3} pc at the distance of the cloud, and the dependence of the average density on the distance to the center has been determined at an unprecedented resolution within the cores. The absorption profiles are consistent with a core of uniform density $n_{H_2} = 6.7 \times 10^5$ cm^{-3} up to $r = 0.035$ pc and an extremely steep density gradient at larger distances, $n \propto r^{-\beta}$ with $\beta \sim 3$ for 0.035 pc $< r < 0.05$ pc. The steep density gradients are not consistent with the structure predicted by an isothermal self-gravitating sphere, and an additional confinement mechanism is implied. It might be a magnetic pressure gradient or a steep gas temperature gradient toward the outside producing a sharp density drop [44], but there is no evidence for such a sharp rise of the temperature (more than a factor 10 over a few hundredths of pc). The confinement by magnetic field pressure gradient is more plausible and hydrostatic equilibrium would lead to plausible values of the magnetic field, $B \sim 150\mu$G, at the boundary $r = 0.035$ pc.

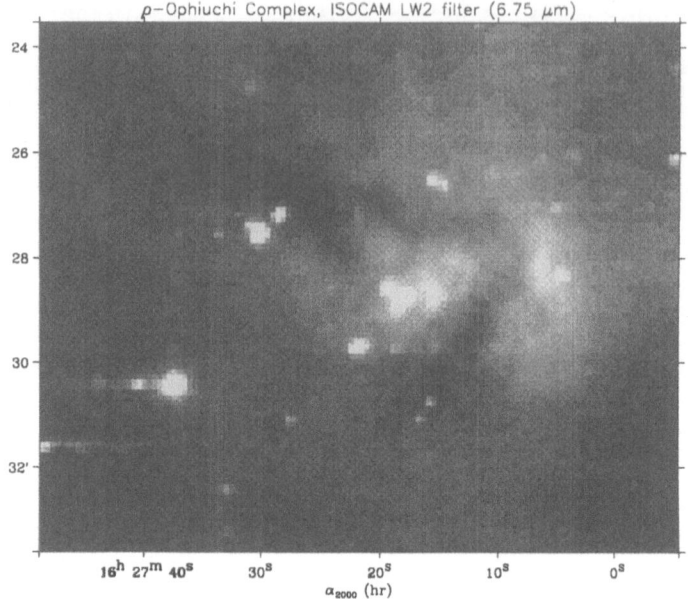

Fig. 7. — ISOCAM image of OphB2: the dense core appears in absorption against the bright background at 6.7 μm.

3.3.2. The Unseen Scales

In addition to the structure directly observed below 0.1 pc, substructure down to a few hundreds AU or even less is inferred from line profile properties. In particular, in massive ($> 100\,M_\odot$) as in low–mass ($\sim\,M_\odot$) dense cores, line shapes of the most opaque transitions of molecular species (for instance CS and CCS) are the same as those of the most transparent transitions ($C^{34}S$ and $CC^{34}S$), although their optical depths differ by at least the Solar isotopic abundance ratio $[^{32}S]/[^{34}S]=22$ [37, 9]. One interpretation is that both sets of lines are optically thin and that the line profiles trace the velocity distribution of the gas in the dense core. But it is not consistent with the line intensity ratios, which are < 22, unless the isotopic abundance ratio for sulfur in the interstellar medium is much lower than Solar. The other possibility is that, following the arguments developed for the CO isotopic lines, CS lines form in cells much smaller than the beam (and than the cores) in macroturbulent conditions.

This conclusion has also been reached following another line of argument, in several multitransition studies of dense cores with CS lines or other high dipole moment molecules. In a study of HC_3N in massive dense cores, [8], the values of the density $3 \times 10^5 \, \text{cm}^{-3} < n_{\text{H}_2} < 5 \times 10^6 \, \text{cm}^{-3}$ derived from a multitransition analysis show no evidence for large–scale variations across the dense cores and the authors conclude that the densest gas is confined to small unresolved structures which fill only a small fraction ($\leq 5\%$) of the core volume. This result has also been observed in lower mass cores [38] and with other tracers.

In conclusion, it seems that substructure exists within dense cores, with large density contrasts between the densest substructures and the surrounding medium (> 100 in the case of HC_3N in M17, Orion and Cepheus, [8]) and that the mass contained in the smallest structures is orders of magnitude below a stellar mass. In a few dense cores, the mass of the inferred substructures is $\sim$ a few $10^{-6} \, M_\odot$ from ^{13}CO, $C^{18}O$ and CS multitransition observations while the smallest observed structures in the same cores have masses $\sim$ a few $10^{-4} \, M_\odot$, [39].

3.3.3. Cores Forming Clusters of Stars and Cores Forming Just a Few Stars

Recent near infrared imaging of molecular clouds have led to the discovery of a large population of rich embedded clusters containing hundreds of newly formed stars [46, 47]. These young clusters are so rich and numerous that they might account for most of the present star formation in our Galaxy. This is the reason why dedicated efforts have been devoted in the last few years to the understanding of cluster formation in dense cores. Maps of dense cores associated with stellar clusters have been performed in several molecular transitions to characterize their spatial and kinematic structure. For comparison, the same has been done for dense cores forming just a few stars. Results are still scarce and the observed properties are plagued by the fact that they invariably bear the signatures of the interaction of the already formed stars with the remnant gas. Despite the bias due to evolutionary effects and the scarcity of the results, trends have already emerged, like in the Rosette molecular clouds, where the seven young embedded clusters recently discovered are associated with some of the most massive clumps in the complex [45]. The authors suggest that large mass is a necessary condition for star cluster formation, but clearly more is required since other clumps of similar large mass in the complex are not associated with any cluster. The additional condition is likely to be the mass of high density gas as shown in L1630, where a high angular resolution multitransition study of four massive dense cores reveals that cores associated with embedded clusters have a larger fraction of their mass at densities exceeding $\sim 10^5 \, \text{cm}^{-3}$ [9].

In summary, the large–scale properties of the GMCs are related to their small scale properties and to their star formation efficiency, so that understanding star formation must be placed on an equal footing with understanding the formation of massive filaments and dense cores. The hierarchy of scales seems to extend far below the dense core scale of a few 0.1 pc, once thought to be the homogeneous entities in which gravitational instability drives collapse and fragmentation. It is not clear which physical processes govern such an organization, but it is likely that it is an interplay of many.

In the next section, we give the characteristics of the hierarchy of dense molecular clouds, as observed in our Galaxy, with the perspective of providing the main constraints on the nature of the physical processes which govern the entire structure.

4. THE HIERARCHY OF MOLECULAR CLOUDS

Despite the existence of well–defined structures like the massive filaments described in the previous section, molecular clouds exhibit remarkable scale–free properties. Many clues to the physics and evolution of molecular clouds are likely to be hidden in these scalings laws but, as will be shown, they must be regarded with caution. There are many pros and cons to such a description, the main pro being that it is a rewarding step to be able to reduce the vast diversity of structures and morphologies to scaling laws extending over an impressive dynamical range. The main cons, though, are: *(i)* the scaling laws mix many physical effects of importance, as discussed below, *(ii)* the procedures followed to extract the cloud parameters are subject to criticism and the nature of the entities identified by these procedures has to be understood.

The cloud parameters are those defined for three–dimensional structures isolated in the four–dimensional space of the molecular line data sets $T_L(x, y, v_z)$, the line brightness temperature being a function of two spatial dimensions, the coordinates of the line of sight on the sky, and one spectral dimension, the projected velocity in the line of sight direction. In this 4D space, 3D structures are isolated following different methods [48, 49, 50, 25, 51].

The first example displays the H_2 mass versus the size (Fig 8). The gas mass and size are those given by the authors, unless some rescaling was necessary to take the different size definitions into account (for instance, the size at half-power of peak emission or the size at the 3σ level emission, see [52]). The masses are mostly derived from ^{12}CO, ^{13}CO or $C^{18}O$ data. The structures are identified in molecular surveys of the central parts of the Galaxy (stars, [53], open triangles [54]) and of the third quadrant (open hexagons, [55]), in the Rosette (crosses) and Maddalena (open squares) molecular clouds [49], in non-star-forming clouds (solid triangles [50], solid squares [51], tripods [57]), in ρ Ophiuchus (solid hexagons [25]) and in a high latitude cloud (starred triangles [39]). In addition, the average values obtained for a set of 10 representative dense cores correspond to the large solid square, [58]. We have added masses and sizes derived from submillimeter continuum observations of the dust for another set of low mass dense cores (solid squares of intermediate size, [59]) and those derived from CS(5-4) observations of massive dense cores in Orion (large open squares, [9]).

The second diagram displays the run of the non–thermal velocity dispersion of the structures versus their size (Fig 9). Note that all the velocity dispersions are much larger than the sound velocity c_s in H_2 at temperatures between 10K

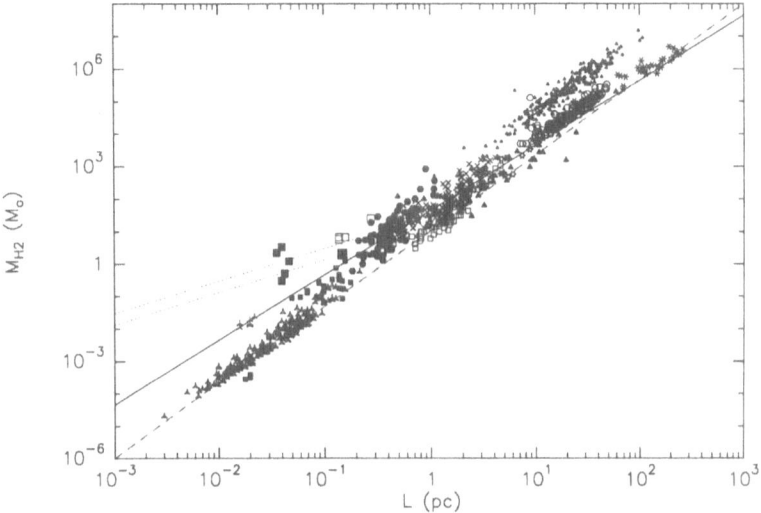

Fig. 8. — The scaling of H_2 gas mass versus size for an ensemble of galactic clouds (the meaning of the symbols is given in the text).

and 50K, the estimated kinetic temperature range of dense gas. The isothermal sound velocity is $c_s = (k_B T_k / \mu m_H)^{1/2}$ where $\mu = 2.33$ takes He into account in the mean mass per particle.

The departure from virial balance between self–gravity and non–thermal energy density is illustrated in Fig 10 which displays the quantity $270 \times |\, \Omega \,| / 2T = M/(L \Delta v_{NT}^2) = 600 \times (\tau_{dyn}/\tau_{ff})^2$ versus mass. This figure shows that in each complex, the most massive structures are the closest to virial balance between self–gravitating energy and internal kinetic energy (the full line in Fig 10), a fact already recognized in ρ Ophiuchus [60] and in Orion B, the Rosette and Cepheus OB3 [62]. The dependence $|\, \Omega \,| / 2T \propto M^{2/3}$ for all the substructures within a single cloud (dashed line in Fig 10) is a signature of the gradual passage from pressure–confined structures to self–gravitating structures as the mass increases, [62].

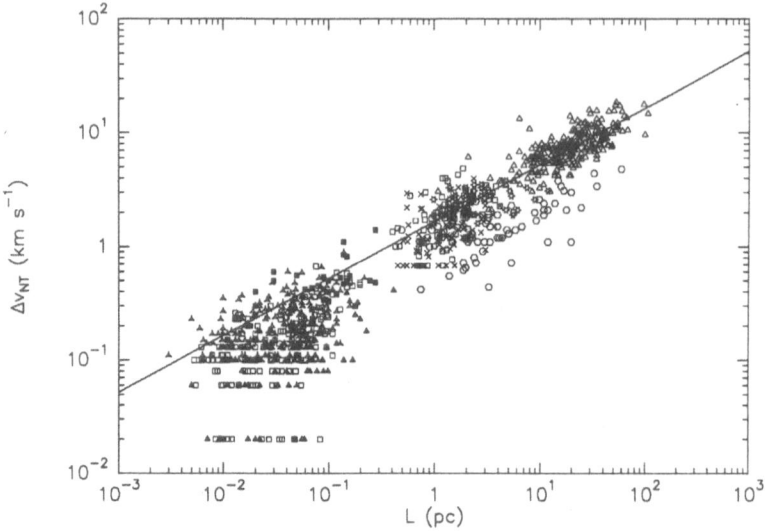

Fig. 9. — The scaling of the non–thermal velocity component with size scale. The symbols are the same as in Fig 8.

In Fig 8 and Fig 9, the full lines trace

$$\frac{M}{L^2} = 44.5 \left(\frac{P_0/k_B}{3 \times 10^4 \, \text{cm}^{-3} \, \text{K}} \right)^{0.5} \text{M}_\odot \text{pc}^{-2}$$

and

$$\frac{\Delta v}{L^{1/2}} = 1.64 \left(\frac{P_0/k_B}{3 \times 10^4 \, \text{cm}^{-3} \, \text{K}} \right)^{0.25} \text{km s}^{-1} \, \text{pc}^{-1/2}$$

which are the loci of isothermal self–gravitating polytropes in a state of critical stability when the specific internal energy density is varied, each polytrope being bounded by a uniform pervading pressure, P_0 [36]. Here the pervading pressure, $P_0/k_B = 3 \times 10^4 \, \text{cm}^{-3} \, \text{K}$, is due to the weight of the gas layer at mid–plane in the Galaxy [63, 64] also equal to the observed average non–thermal kinetic pressure at small scale in non star forming regions of the Solar Neighborhood [51].

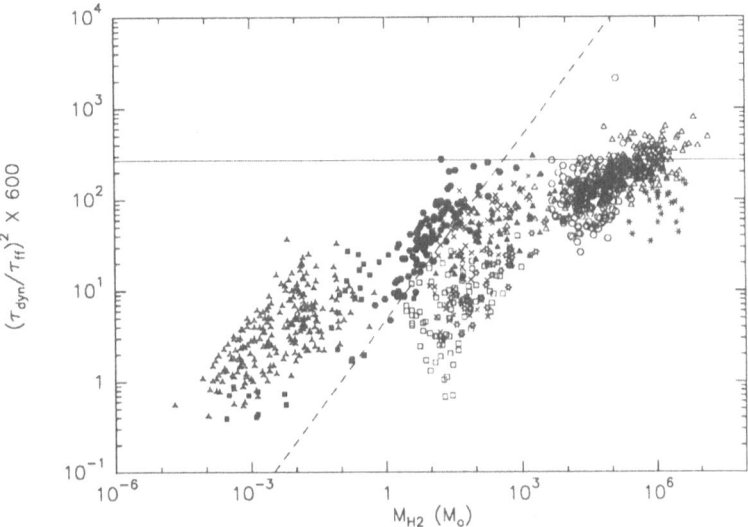

Fig. 10. — The scaling of $(\tau_{dyn}/\tau_{ff})^2 \times 600$ versus H$_2$ mass for the same ensemble of galactic clouds.

The two dotted lines in Fig 8 represent the locus of self–gravitating isothermal spheres, $\frac{3}{5}GM/(L/2) = 3c_s^2$, drawn for $T_k = 10$K and 50K. The factor 3 follows from $\overline{v^2} = 3\sigma_v^2$ for a Gaussian velocity distribution of one–dimensional dispersion $\sigma_v = c_s$. Note that in $M_\odot$ km s^{-1} pc units, $G = 1/232$. The dashed line in Fig 8 has slope 2.3, interpreted by [52] as a signature of the fractal structure of molecular clouds. It is also the slope of the mass–size relation found by [39] among structures in a high latitude complex which span two orders of magnitude in size.

The comparison of these theoretical predictions with the data suggests a few remarks: (i) the locus of self–gravitating structures in the Solar Neighborhood (see Fig 8) would rather be a factor of ~ 4 above the full line, suggesting that the equation of state of the molecular cloud is not isothermal but has a polytropic index close to $n = 3.2$ or $\gamma \sim 4/3$ according to [36]. In the present description, the internal energy is no longer thermal but non–thermal and we discuss in Section 5.1.2 the concept of equation of state of a turbulent medium,

(ii) a large number of structures have masses orders of magnitude smaller than the critically stable self–gravitating structures of the same size. Their internal pressure has therefore small gradients and they are in pressure balance with the external pressure, (iii) a number of structures lie above this curve, in particular the densest regions of dense cores in Orion or in nearby clouds, suggesting that these cores, as expected, are not bound by the ambient pressure of the Solar Neighborhood but by a larger pressure exerted by their self–gravitating environment, (iv) the data corresponding to the molecular ring complexes [54] lie well above the line corresponding to the Solar Neighborhood, suggesting that the ambient pervading pressure in the molecular ring and galactic center regions is larger than that in the Solar Neighborhood.

There is little observational evidence for self–gravitating structures at scales much below that of the dense cores $\sim 0.1\,pc$. If the mass estimates for the smallest scales are correct and the existence of non self–gravitating structures is confirmed, it will prove that a process other than fragmentation driven by gravitational instability is at work in shaping the medium. This illustrates the vital importance of a correct analysis of all the biases introduced in the mass determinations discussed in Section 2.

5. THE NON–THERMAL ENERGY DENSITIES OF MOLECU- LAR COMPLEXES

The previous section has shown that the internal energy density of molecular clouds is essentially non–thermal. It may be due to supersonic turbulent motions, or to the coupling of the gas to magneto–hydrodynamical waves. An interesting aspect of these two processes is that they both provide a way to self–initiated star formation, as opposed to the majority of the scenarios which require some triggering process generated by a previous generation of stars, like shocks associated with HII regions or supernovae remnants. Turbulence decays due to viscous dissipation and magnetic flux vanishes due to ambipolar diffusion. We discuss here recent advances in these two fields.

5.1. The Turbulent Velocity Field

In Section 4, the scaling laws between mass, size and internal velocity dispersion of molecular clouds have been compared to the theoretical predictions made in the study of gravitational stability of polytropic masses of gas. In doing so, the thermal energy has been implicitly replaced by a non–thermal energy, which is indeed the specific kinetic energy contained in the supersonic motions of the gas. These motions are described as turbulent because the Reynolds numbers which express the ratio of the transport due to advection to that due to viscosity are extremely large

$$Re = Lv/\nu \sim 10^8 - 10^{12}.$$

Profound differences exist between random thermal motions and turbulent motions: first, the power spectrum of turbulence is not flat and second, coherent structures exist in a turbulent velocity field. In what follows, we discuss the impact of these properties of turbulence on the physics and dynamics of molecular clouds. An important point to note is that the searches conducted so far to determine a velocity correlation length in the velocity field of molecular clouds have all been negative, one of the most recent being [65]. This does not mean that the flows in molecular clouds are not turbulent, but presumably that the outer scale is yet larger than the largest scales sampled (*i.e.* ~ 20 pc) and the inner scale smaller than the smallest scales sampled, ~ 0.01 pc.

5.1.1. The Jeans Criterion in a Turbulent Medium

In the classical Jeans analysis [1], the criterion for the exponential growth of gravitational unstable modes of wavenumber k and pulsation ω is given by the dispersion relation

$$\omega^2 = c_s^2 k^2 - 4\pi G \rho_0$$

which states that the unstable modes, those for which $\omega^2 < 0$, have wave numbers smaller than

$$k_J = \frac{(4\pi G \rho_0)^{1/2}}{c_s}$$

Note that the initial state (infinite isothermal medium of uniform density ρ_0) is not a solution of the equations of motion and Poisson equation. The gas internal energy is thermal, and the most unstable scales are those larger than a threshold, called the Jeans length, which depends only on the initial density and temperature.

An interesting consequence of the fact that turbulence seems to be the dominant support of clouds against self–gravity is that the classical Jeans criterion can be inverted, *i.e.* the largest scales may be stabilized against self–gravity because the power spectrum of turbulence is not flat. In a turbulent medium, the analog of the previous dispersion relation in an isothermal gas may be written, [66, 67]

$$\omega^2 = k^2 [c_s^2 + \frac{1}{3} u^2(k)] - 4\pi G \rho_0$$

where the internal energy density of scale k due to turbulent motions is

$$u^2(K) = \int_{k_m(K)}^{\infty} \epsilon(k) dk$$

[68] where k_m is slightly larger than K and $\epsilon(k) = Ak^{-\alpha}$ is the power spectrum of the turbulent velocity field. In incompressible turbulence, $\alpha = 5/3$. The dependence of the additional term on the wavenumber k shows that for a given density and temperature the range of unstable scales is reduced. For supersonic turbulence, the second term on the right hand side dominates the thermal pressure term and, for a steep enough energy power spectrum ($\alpha > 3$),

the criterion is reversed since in that case the most unstable scales become the smallest, a behaviour in better agreement with the observations. It also shows that there is a domain of densities in which no gravitational instability can grow, and for densities larger than this threshold, only a small range of scales is gravitationally unstable.

5.1.2. The Effective Adiabatic Exponent of Turbulence

The criterion of gravitational stability to a wave perturbation of a polytropic sphere of gas depends on the polytropic index n related to the adiabatic index $\gamma = 1+1/n = d\log P/d\log \rho$ which describes the loss of internal energy induced by a pressure perturbation. Isothermal spheres ($\gamma = 1, n = \infty$) are always unstable while polytropes with $-1 < n < 3$ are unconditionally stable [36]. Similarly, it is of critical importance to be able to describe how a turbulent mass of gas keeps or loses its internal turbulent energy following a pressure perturbation. This analysis has been performed analytically and numerically [69]. The major results are that a gradient of turbulent pressure builds up and stays correlated to the density gradient for all the duration of the run, and that an adiabatic exponent can be built by analogy with $d\log P/d\log \rho$ in which thermal pressure is replaced by the turbulent pressure. This exponent is $\gamma_{eff} \sim 2$ at earlier times after the perturbation is applied and then decreases with time down to $\gamma_{eff} \sim 1$ after about a few times the turn–over time of the large scales at which the energy is injected. For these two reasons, turbulence is able to stabilize molecular clouds against gravitational instability, over a timescale comparable to their lifetime.

5.1.3. The Space–time Intermittency of Turbulence Dissipation

In addition to its power spectrum, the velocity field in a turbulent fluid has the property to exhibit more spatial structure, than a random velocity field, in any of the quantities related to increments or derivatives of the velocity field. A comparison of the topology of the regions of large vorticity has been made for a random and a turbulent velocity field of the same power spectrum [70]. Fig 11 shows that spatial structures exist in a random velocity field and vorticity filaments are present. In a turbulent field, though, the intensity of these structures is much larger than in a random field and the structures are more coherent. These structures are thought to be responsible for the property of intermittency of turbulence, at the origin of the non–Gaussian probability distributions of the velocity increments, derivatives and vorticity [71, 72]. There is not a sharp cut–off between the non–Gaussian features due to the Kolmogorov power spectrum itself and those attributed to intermittency. In a turbulent field, the local enhancements of vorticity are just more prominent and more coherent in space and time than in a random field.

To verify or disprove the idea that the flows in molecular clouds are turbulent, the signatures of the regions of enhanced vorticity have been searched for in molecular line profiles. Line profiles have been synthesized at different epochs

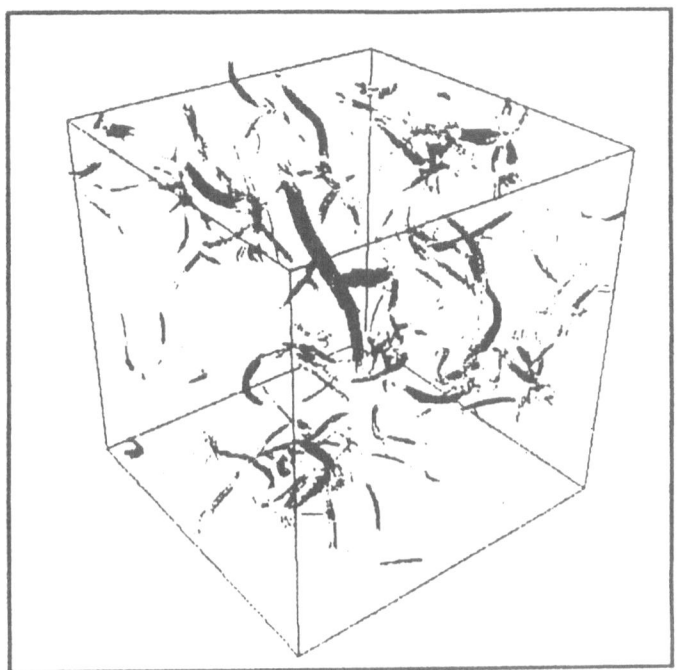

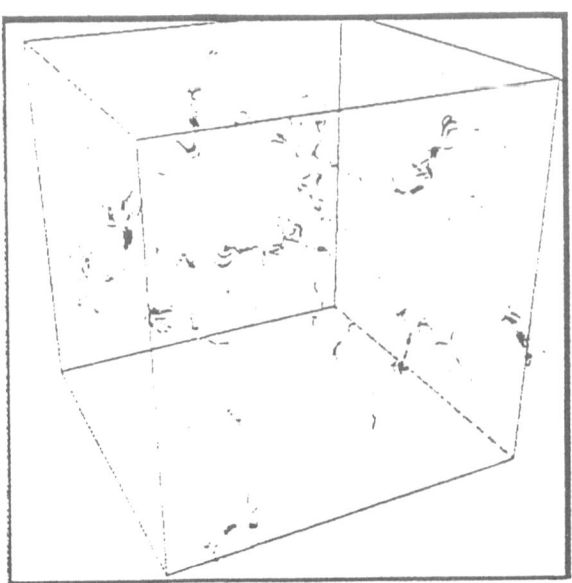

Fig. 11. — Regions where the vorticity exceeds a given threshold in (a) a random velocity field and (b) turbulent fluid. In both cases the power spectrum is the same Kolmogorov spectrum.

in the evolution of a decaying run of compressible turbulence [73] and in a random velocity field with a Kolmogorov power spectrum [74]. The results are very comparable except for the existence of very broad and weak line wings in the profiles present in the turbulent case. These weak line wings are associated with regions of large vorticity within the beam [75].

An interesting property of the spectra simulated in the turbulent data cube is the change of the line shapes with time. The three epochs selected in the simulations were 1) when strong shocks have just formed, 2) once shock interactions have generated vorticity, transferring most of the kinetic energy to non–compressible modes and 3) in the phase of self–similar decay of the turbulence. Only in the two latest phases are the profiles statistically similar to observed line profiles. This result suggests that the velocity field of molecular clouds is not dominated by shocks, either because shocks interact very rapidly once formed, transferring the energy contained in compressible modes to solenoidal modes, or because a magnetic field is present generating magnetic precursors which reduce the velocity discontinuities [76].

The regions of large vorticity in turbulent flows are closely associated with those of large dissipation rate of kinetic energy and the two subregions of space where these quantities are concentrated are highly structured and intermittent [71, 72]. The dissipation of turbulent kinetic energy via the viscous process of elastic collisions between particles eventually turns this energy into heat. The rate at which the gas is heated by this process is as large as the intensity of the intermittent event is large, $i.e.$ the local dissipation rate, $\epsilon_D = \eta/2 \sum_{i,j} (\partial_j v_i + \partial_i v_j)^2$, where $\eta = \rho\nu$ is the dynamical viscosity and ρ the mass density, may reach large values for events far in the non–Gaussian wings of the probability distribution functions (PDFs) of velocity increments, like those reported in [77, 78]. Indeed, experimental results obtained in flow experiments have been scaled to the interstellar medium, under the assumption that the statistical flow properties in turbulence only depend on the kinetic energy transfer rate. The local heating rate of the interstellar gas due to bursts of turbulent viscous dissipation has been found to exceed by orders of magnitude all the other heating rates in atomic clouds, those due to UV photons, to coupling to dust grains, or to cosmic rays ionisation. The lifetime of the intermittent structures is long enough to allow the gas to reach thermal balance and the ensuing temperature rise is large enough to trigger chemical reactions with activation barriers and endothermic reactions [80]. Time–dependent calculations in atomic gas show that, as expected, the abundances of molecules like OH, H_2O, CH^+, HCO^+, and therefore CO, in these regions can be orders of magnitude larger than in the surrounding colder gas [81]. The recent discovery of unexpected amounts of such molecules in interstellar gas of low extinction might be seen as a signature of intermittent dissipation of turbulence [82]. The intermittency of turbulence tends to create singularities in interstellar gas. They are heated by bursts of viscous dissipation and bear specific chemical signatures potentially subject to subsequent rapid radiative cooling, inducing condensation.

5.2. The Coupling of the Bulk of Gas to the Magnetic Field

A very thorough review has been recently published on all the processes linked to the coupling of the gas in molecular clouds to the magnetic field [83]. In particular, the important process of ambipolar diffusion is discussed in detail. This process is the drift of the neutral particles across the field lines as soon as the collision rate of the ionized components of the medium (the ions and most importantly small charged dust grains) with the neutrals is not large enough to maintain the coupling between the two populations, and therefore that of the magnetic field to the neutrals.

When the drift is induced by self–gravity acting on the bulk of the gas and therefore the neutrals, the important result, most relevant to star formation, is the constancy of the ratio between the ambipolar diffusion time and the free–fall time

$$\frac{\tau_{AD}}{\tau_{ff}} = (\frac{6}{\pi^3 G m_H})^{1/2} \langle \sigma v \rangle K_i$$

where $K_i = x_i n_H^{1/2} \sim 10^{-5} \text{cm}^{-3/2}$ relates the ionization degree of the gas x_i to its density when the primary ionization process is due to cosmic–rays, and $\langle \sigma v \rangle \simeq 1.5 \times 10^{-9} \text{cm}^3 \text{ s}^{-1}$ is the ion–neutral collision rate so that $\nu_{ni} = n_i \langle \sigma v \rangle$ is the collision rate of a neutral in a sea of ions. In hydromagnetic waves, decoupling between the ions and the neutrals causes the damping of the waves, above a cut–off frequency $\omega = 2\nu_{ni}$ (see [84]).

6. SCENARIOS FOR GENERATING CYCLES OF STAR FORMATION ACTIVITY

A large number of scenarios have been developed to couple the large and small–scale dynamics, and therefore large–scale evolution of the interstellar medium to the star formation process. Those discussed below are not representative of the diversity of approaches followed. They have been selected because they all lead to bursts of star formation activity although in quite different physical contexts. In each case, the large–scale structure is in the range 500 pc to 1kpc.

In the first scenario, the nonlinear response of a cloud fluid to an external perturbation is computed. The evolution of the cloud population is described by a kinetic equation in the position-internal velocity dispersion-mass phase space [85]. The perturbation is a nonlinear increase of the internal velocity dispersion of the cloud population caused by a galaxy encounter [86]. The system of clouds considered has a constant mass, the cloud mass transformed into stars being assumed negligible. Clouds collisions lead either to coalescence at low relative velocities, or to shredding at large relative velocities, and the feedback of star formation to the cloud population is an increase of its internal velocity dispersion. Clouds more massive than a threshold collapse and form stars, converting a fixed fraction of the cloud mass into stars and returning the rest into the cloud population. The key parameter in this scenario is the time delay t_d between massive cloud formation to cloud disruption due to star formation, compared to a characteristic cloud collisional time t_c. When $t_d \sim 0$, the equilibrium cloud configuration is stable to any increase of the velocity dispersion. As t_d/t_c is increased, the system exhibits limit cycles and a transition to chaotic behavior, with bursts of star formation. The reason for this behaviour is simple: the time lag before cloud disruption allows the formation of more and more massive clouds which in turn form a larger and larger mass of stars, eventually disrupting the clouds.

A completely different approach has been followed in [87] where the evolution of the mass exchanges between three components of the interstellar medium, the cold HI gas, the dusty molecular clouds and the young star population and their associated HII regions, is described. The system is open but keeps a constant mass, the rate of fresh gas inflow being equal to the mass loss rate in the form of low mass stars, which incidentally constrains the star formation rate to increase with the infall rate. There is an unlimited supply of fresh gas. The star formation rate here is a power of the molecular gas density (or mass since volume is constant), and spontaneous and triggered star formation mechanisms are introduced. The system cannot reach equilibrium and keeps accumulating matter in dead stellar remnants. Radiative cooling by atoms and molecules is introduced because it amplifies gas condensation. There are of course many parameters in this description and the time evolution of the mass fraction in each reservoir shows limit cycles within a broad range of parameters, while in some other cases these fractions remain stable.

The last approach relies on two–dimensional numerical simulations of self–gravitating turbulent compressible flows, representing the interstellar medium at the kpc scale [88]. Heating and cooling of the gas are described as a function of gas density, temperature and star formation activity. Star formation turns on a heating center each time the local density exceeds a threshold value, for a period of about the lifetime of OB associations. Heating due to star formation provides the energy input required to feed supersonic turbulence, which is in turn responsible for cloud formation through ram pressure. The system exhibits a self–sustaining cycle with bursts of star formation. A large number of clouds are out of virial balance between self–gravity and kinetic plus thermal internal energy, the longest lived clouds being the closest to virial balance.

7. CONCLUSIONS AND PERSPECTIVES

A consensus does not exist among the possible interpretations one can build on the impressive set of observations already accumulated on molecular complexes and star forming regions. We think though that a few robust statements emerge. First, all scales, from the protostellar scales up to those of GMCs and possibly further up, seem to be involved in the process of star formation. It is illustrated by the scaling of the mass of the substructures observed in the GMCs (mass per unit length of the massive filaments, mass of the dense cores) with the mass of the GMCs and that of its most massive star. Second, even though the nature of the process which couples the scales is not understood, the role of turbulence and magnetic field, in addition to that of self–gravity, seems undeniable. Third, the scaling laws observed between the masses, sizes and internal velocity dispersions of clouds allow to distinguish two families of clouds, those approximately in virial balance between non–thermal kinetic energy and self–gravity, and the others of about uniform (non–thermal) pressure. Unless the gas masses at small scales are systematically underestimated, this result suggests that self–gravity is not the only process at work in shaping the medium. Last, star formation seems to be controlled by a threshold process, the critical parameter being the mass fraction of gas denser than $\sim 10^5$-10^6 cm^{-3} in dense cores.

Acknowledgments

I would like to thank John Bally for providing me with his ^{13}CO(J=1–0) maps of OriA and OriB. This manuscript has benefited from the helpful comments of Steve Balbus, François Boulanger and Jean-Loup Puget.

References

[1] Jeans J.H., *Phil. Trans. A* **199** (1902) 1.
[2] Hoyle, F. *ApJ* **118** (1953) 513.
[3] Draine B., Bertoldi F., *ApJ* **333** (1996) 33.
[4] Habing H. S., *Ap. J.* **5** (1968) 95.
[5] Burton, Hollenbach, Tielens., *ApJ* **18** (1990) 162.
[6] Rouan D., Lemaire J.-L., Field D., Lai O., Pineau des Forêts G., Falgarone E., Deltorn J.-M., *MNRAS* **284** (1997) 395.
[7] van der Werf P.P., Stutzki J., Sternberg A., Krabbe A., *A&A* **313** (1996) 633.
[8] Bergin E.A., Snell R.L., Goldsmith P.F., *ApJ* **460** (1996) 343.
[9] Lada E.A., Evans N.J. II, Falgarone E. 1997 ApJ in press.
[10] Bergin E.A., Goldsmith P.F., Snell R.L., Ungerechts H., *ApJ* **431** (1994) 674.
[11] Hunter S.D., Digel S.W., de Geus E.J., Kanbach G., *ApJ* **436** (1994) 216.
[12] Digel S.W., Hunter S.D., Mukherjee R., *ApJ* **441** (1995) 270.
[13] Wall W.F., Reach W.T., Hauser M.G. et al., *ApJ* **456** (1996) 566.
[14] Martin H.M., Sanders D.B., Hills R., *MNRAS* **208** (1984) 35.
[15] Albrecht M.A., Kegel W.H., *A&A* **176** (1987) 317.
[16] Kegel W.H., Piehler G., Albrecht M.A., *A&A* **270** (1993) 407.
[17] Abergel A., Boulanger F., Mizuno A., Fukui Y., *ApJL* **423** (1994) L59.
[18] Draine B.T, Lee H.M., *ApJ* **285** (1984) 89.
[19] Andriesse C.D. *Vistas in Astronomy* **21** (1979) 107

[20] Mathis J.S., Rumpl W., Nordsieck K.H., *ApJ* **217** (1977) 425.

[21] Mezger P.G., Wink J.E., Zylka R., *A&A* **228** (1990) 95.

[22] Désert F.-X., Boulanger F., Puget J.-L., *A&A* **237** (1990) 215.

[23] Puget J.-L., Léger A., *Ann. Review Astr. Astrophys.* **27** (1989) 161.

[24] Ungerechts H., Thaddeus P., *ApJS* **63** (1987) 645.

[25] Loren R.B., *ApJ* **338** (1989) 902.

[26] Shu F.H. Radiation (University Science Books, Mill Valley, 1991) p. 280.

[27] Heiles C., Goodman A.A., McKee C.F., Zweibel E.G., Protostars and Planets III (The University of Arizona Press, Tucson &London, 1993) p. 279.

[28] Crutcher R.M., Troland T.H., Lazareff B., Kazes I., *ApJ* **456** (1996) 217

[29] Roberts D.A., Crutcher R.M., Troland T.H., *ApJ* **442** (1995) 208

[30] Maddalena R.J., Morris M., Moscovitz J., Thaddeus P., *ApJ* **303** (1986) 375.

[31] de Geus E.J., Bronfman L., Thaddeus P. *A&A* **231** (1990) 150.

[32] Boulanger F., Bronfman L., Dame T.M., Thaddeus P., 1997, ApJ, in press.

[33] Vogel S.N., Boulanger F., Ball R., *ApJL* **321** (1987) L145.

[34] Nordh L. et al., *A&A* **315** (1996) L185.

[35] Bally J., Langer W.D., Stark A.A., Wilson R.W. *ApJL* **312** (1987) 45.

[36] Chièze J.-P., *A&A* **171** (1987) 225.

[37] Langer W.D., Velusamy T., Kuiper T.B.H., Levin S., Olsen E., *ApJ* **453** (1995) 293.

[38] Falgarone E., Panis, J.-F., Heithausen A., Pérault M., Stutzki J., Puget J.-L., Bensch F., A&A, 1997, in press.

[39] Heithausen A., Bensch F., Stutzki J., Falgarone E., Panis J.-F., 1997, A&A, submitted.

[40] Heithausen A., Corneliussen U., Grossmann V., *A&A* **301** (1995) 941.

[41] Henriksen R., André P., Bontemps S., 1997 A&A, in press

[42] Loren, R.B., Wootten A., Wilking B.A., *ApJ* **365** (1990) 265.

[43] Abergel A. et al., *A&A* **315** (1996) L329.

[44] Falgarone E., Puget J.-L., *A&A* **142** (1985) 157.

[45] Phelps R.L., Lada E.A., *ApJ* **477** (1997) 176.

[46] Lada E.A., Strom K.M., Myers P.C., Protostars and Planets III (The University of Arizona Press, Tucson &London, 1993) p. 245.

[47] Zinnecker H., McCaughrean M.J., Wilking B.A., Protostars and Planets III (The University of Arizona Press, Tucson &London, 1993) p. 429.

[48] Stutzki J., Güsten R., *A&A* **356** (1990) 513.

[49] Williams J.P., de Geus E.J., Blitz L. *ApJ* **428** (1994) 693.

[50] Falgarone E., Pérault M., Physical processes in interstellar clouds (Reidel, Dordrecht, 1987) p. 59.

[51] Falgarone E., Puget J.-L., Pérault M. *A&A* **257** (1992) 715

[52] Elmegreen B.G., Falgarone E., *ApJ* **471** (1996) 816.

[53] Dame T.M., Elmegreen B.G., Cohen R.S., Thaddeus P., *ApJ* **305** (1986) 892.

[54] Solomon P.M., Rivolo A.R., Barrett J., Yahil A., *ApJ* **319** (1987) 730.

[55] May J., Alvarez, H. Bronfman L., 1997 A&A in press.

[56] Herbertz R., Ungerechts H., Winnewisser G., *A&A* **249** (1991) 483.

[57] Lemme C., Walmsley C.M., Wilson T.L., Muders D. *A&A* **302** (1995) 509.

[58] Wang Y., Evans N.J. II, Zhou S., Clemens D.P., *ApJ* **454** (1995) 217.

[59] Ward–Thompson D., Scott P.F., Hills R.E., André P., *MNRAS* **268** (1994) 276.

[60] Loren R.B., *ApJ* **338** (1989) 925.

[61] Henriksen R.N., *ApJ* **377** (1991) 500.

[62] Bertoldi F., McKee C., *ApJ* **395** (1992) 140.

[63] Elmegreen B.G., *ApJ* **338** (1989) 178.

[64] Boulares A., Cox D., *ApJ* **365** (1990) 544.

[65] Miesch M.S., Bally J., *ApJ* **429** (1994) 645.

[66] Bonazzola S., Falgarone E., Heyvaerts J., Pérault M., Puget J.-L., *A&A* **172** (1987) 293.

[67] Bonazzola S., Pérault M., Puget J.-L., Heyvaerts J., Falgarone E., Panis J.-F., *J. Fluid Mech.* **245** (1990) 1.

[68] Chandrasekhar S., *Proc. Roy Soc. A.,* **210** (1951) 18.

[69] Panis J.-F., Pérault M., 1997, Phys. Fluids A, submitted.

[70] She Z.S., Jackson E., Orszag S.A., *Nature* **344** (1990) 226.

[71] Vincent A., Meneguzzi M., *J. Fluid Mech.* **225** (1991) 1.

[72] Porter D.H., Pouquet A., Woodward P.R., *Phys. Fluids* **6** (1994) 2133.

[73] Falgarone E., Lis D.C., Phillips T.G., Pouquet A., Porter D.H., Woodward P.R., *ApJ* **436** (1994) 728.

[74] Dubinski J., Narayan R., Phillips T.G., *ApJ* **448** (1995) 226.

[75] Lis D.C., Pety J., Phillips T.G., Falgarone E., *ApJ* **463** (1996) 623.

[76] Pineau des Forêts G., Roueff E., Flower D., *MNRAS* **244** (1990) 688.

[77] Anselmet, F., Gagne Y., Hopfinger E.J., Antonia R.A., *J. Fluid Mech.* **140** (1984) 63.

[78] Tabeling P., Zocchi G., Belin F., Maurer J., Willaime H., *Phys. Rev. E* **53** (1996) 1613

[79] Douady S., Couder Y., Brachet M.E., *Phys. Rev. Letters* **67** (1991) 983.

[80] Falgarone E., Puget J.-L., *A&A* **293** (1995) 840

[81] Falgarone E., Pineau des Forêts G., Roueff E., *A&A* **300** (1995) 870.

[82] Lucas R., Liszt H., *A&A* **307** (1996) 237

[83] McKee C.F., Zweibel E.G., Goodman A.A., Heiles C., Protostars and Planets III (The University of Arizona Press, Tucson &London, 1993) p. 327.

[84] Kulsrud R.M., Pearce W.P., *ApJ* **156** (1969) 445.

[85] Scalo J.M., Struck–Marcell C., *ApJ* **276** (1984) 60.

[86] Scalo J.M., Struck–Marcell C., *ApJ* **301** (1986) 77.

[87] Bodifée G., de Loore C., *A&A* **142** (1985) 297.

[88] Vàzquez-Semadeni E., Passot T., Pouquet A., *ApJ* **473** (1996) 881.

Elements of Hydrodynamics Applied to the Interstellar Medium

J.–P. Chièze

DSM/DAPNIA/Service d'Astrophysique, CEA Saclay
F-91191 Gif-sur-Yvette Cedex 01, France

---

1. EQUATIONS OF FLUID MOTION

We present here some basis of the dynamics of astrophysical flows. Since hydrodynamical equations are certainly familiar to many readers, we try to present here some aspects of the derivation of the basic equations of hydrodynamics, focusing on their physical meaning. In particular, we present a unified expression of the Euler's equation of motion in which the viscous stress, magnetic field and gravitation enter formally in the same form. Regarding the numerical treatment of shock waves, we recast the widespread method of the pseudo-viscosity in its physical context. The structure of the stress applied to any Newtonian fluid element provides a general formulation of the classical method of the pseudo-viscosity. We summarize some basic plasma processes which play a major role in the evolution of the interstellar medium. The important process of gas cooling in an inhomogeneous medium, leading to the formation of distinct phases with different specific entropies is finally discussed.

1.1. Mass Conservation

The equation of continuity expresses the law of conservation of mass in any fluid element. The variation of the mass contained in a given volume V is governed by the flow through its boundary surface S

$$\frac{\partial}{\partial t} \int_V \rho \, dV = - \int_S \rho \mathbf{u} \cdot d\mathbf{S} \tag{1}$$

By applying Green's formula, one obtains

$$\frac{\partial \rho}{\partial t} = -\nabla_k \left(\rho u^k \right). \tag{2}$$

1.2. Momentum Conservation

According to the Newton law, the variation rate of the momentum of a parcel of fluid is equal to the total external force applied to it. Consider a volume V of fluid enclosed by a surface S. Among the forces acting on this portion of fluid, one distinguishes the bulk or body forces and surface forces. Body forces act directly on each particle of a fluid element, such as gravitational or magnetic forces. We shall note the body forces acting on a parcel of fluid of volume dV as $\mathbf{f}dV$ where $\mathbf{f}$ is the body force per unit volume. The fluid exterior to S exerts, on each surface element dS, a force, or stress, $\mathbf{T}dS$. If the force (per unit surface area) points towards the interior of S, it is a pressure, otherwise it is a tension. Since that force depends on the orientation of the surface element, it is described by a tensor $\mathbf{T}$. If $\mathbf{n}$ is the normal to the surface element, the components of $\mathbf{T}$ are $T^i = T^{ji} n_j$. With no loss of generality, we shall now derive the equation of momentum conservation, retaining, for the moment, only the surface forces. This restriction is motivated by the fact that the introduction, at that stage, of the body forces spoils the symmetry of the resulting equation. However, we show later that this symmetry can be indeed restored even when magnetic fields and gravitation are taken into account.

Consider some finite volume V bound by a closed surface S. The principles of conservation of the linear momentum and of the angular momentum applied to the fluid enclosed by S require that the resultant made of the applied external forces and the inertial force is null. If γ is the acceleration of a fluid element of volume dV, the inertial force is just $-\rho \gamma dV$. Accordingly, one has successively :

$$-\int_V \rho \gamma^i dV + \int_S T^{ij} dS_j = 0 \tag{3}$$

$$-\int_V \rho \left(x^i \gamma^j - \gamma^i x^j \right) dV + \int_S \left(x^i T^{kj} - T^{ki} x^j \right) dS_k = 0 \tag{4}$$

In virtue of the Green formula, the first equation reduces to :

$$\int_V \left(\partial_k T^{ik} - \rho \gamma^i \right) dV = 0 \tag{5}$$

We thus obtain the Euler's equation of motion :

$$\rho \gamma^i - \partial_k T^{ik} = 0 \tag{6}$$

The covariant analogue of this expression is :

$$\rho \gamma^i - \nabla_k T^{ik} = 0 \tag{7}$$

which reduces, in Cartesian coordinates, to Eq. 6. This is thus the covariant formulation of the equation of the motion. It is now useful to write the equation of momentum conservation in a form which identifies the flux of momentum. We must derive first the expression of the acceleration γ in a general coordinate system.

1.2.1. Eulerian, Total and Absolute Time Derivatives

When the time variation of any physical quantity f of a flow is viewed from some fixed position in a fixed reference frame, it is described by the time derivative $\partial f/\partial t$. Alternatively, one may evaluate the time variation of a flow variable, following locally (in space and time) the flow itself. The corresponding time derivative, noted Df/Dt is thus the time variation of the quantity f attached to a specified fluid element. These two time derivatives are related by the expression :

$$\frac{Df}{Dt} = \frac{\partial f}{\partial t} + v^i \frac{\partial f}{\partial x^i} \qquad (8)$$

a formulation only valid in *Cartesian* coordinates. The relevant expression in general curvilinear coordinates is the covariant form discussed below.

1.2.2. Covariant Form : Lagrangian and Absolute Time Derivatives

In a general curvilinear system of coordinates, the local basis $\{e_i\}$ varies with the position. The covariant expression of Df/Dt is now

$$\frac{Df}{Dt} = \frac{\partial f}{\partial t} + v^i \nabla_i f \qquad (9)$$

where $\nabla_i f$ is the covariant space derivative of f. Consider for example a vector $\mathbf{v}$ with contravariant coordinates v^i in some curvilinear system of coordinate, in which the natural basis is $\{e_i\}$, so that

$$\mathbf{v} = v^i e_i \qquad (10)$$

The differential of this vector :

$$d\mathbf{v} = dv^i e_i + v^i de_i \qquad (11)$$

combines the variations of the vector components v^i and the variations of the local basis itself. For a small displacement $d\mathbf{x}$ the vectors of the displaced basis can be expressed as :

$$de_i = \Gamma_k{}^i{}_j dx^k e_i \qquad (12)$$

so that :

$$d\mathbf{v} = dv^i e_i + v^j \Gamma_k{}^i{}_j dx^k e_i = \left\{ \partial_k v^i + v^j \Gamma_k{}^i{}_j \right\} dx^k e_i \equiv \nabla v^i e_i \qquad (13)$$

The quantities $\partial_k v^i + v^j \Gamma_k{}^i{}_j \equiv \nabla_k v^i$ are the components of the covariant space derivative of $\mathbf{v}$.

1.2.3. Acceleration of a Fluid Element

The component of the acceleration of a fluid element are :

$$\gamma^i \equiv \frac{\nabla v^i}{dt} = \frac{dv^i}{dt} + v^j \Gamma_k{}^i{}_j \frac{dx^k}{dt} = \frac{dv^i}{dt} + \Gamma_k{}^i{}_j v^k v^j \qquad (14)$$

The expression dv^i/dt is just the *total* time derivative of the velocity component v^i (and not the *absolute* time derivative), given by :

$$\frac{dv^i}{dt} = \frac{\partial v^i}{\partial t} + v^k \partial_k v^i \qquad (15)$$

so that the components of the acceleration are

$$\gamma^i = \frac{\partial v^i}{\partial t} + v^k \partial_k v^i + \Gamma_k{}^i{}_j v^k v^j = \frac{\partial v^i}{\partial t} + v^k \nabla_k v^i \qquad (16)$$

which is precisely the expression for Dv^i/Dt. In other words, the covariant expression $Df/Dt = \partial f/\partial t + v^i \nabla_i f$, for a scalar, a vector or a tensor, is precisely the absolute time derivative of f.

1.2.4. The Hydrodynamical Momentum Density Flux Tensor

Writing the *identity*

$$\rho \gamma^i = \frac{\partial}{\partial t}\left(\rho v^i\right) + \nabla_k \left(\rho v^k v^i\right) - v^i \left[\frac{\partial \rho}{\partial t} + \nabla_k \left(\rho v^k\right)\right] \qquad (17)$$

in order to single out the continuity equation, Euler's equation can be written in the conservative form

$$\frac{\partial(\rho v^i)}{\partial t} = -\nabla_j \left(\mathcal{T}^{ij} + \rho v^i v^j\right) \qquad (18)$$

which expresses momentum conservation in a natural way. The tensor

$$\mathcal{P}^{ij} = \mathcal{T}^{ij} + \rho v^i v^j \qquad (19)$$

represents the momentum density flux associated with the hydrodynamical flow. The first term represents the microscopic momentum flux, including the (static) pressure, tension and viscous forces, and the second term, the macroscopic momentum flux due to advection.

1.2.5. General Properties of Surface Forces

We can now prove that the tensor describing the surface forces is symmetric. The second equality can be written as

$$\int_V \left\{x^i \left(f^j - \rho \gamma^j\right) - \left(f^i - \rho \gamma^i\right) x^j\right\} dV + \int_V \partial_k \left(x^i \mathcal{T}^{kj} - \mathcal{T}^{ki} x^j\right) dV = 0 \quad (20)$$

or again, according to the Euler's equation

$$\int_V \left(\mathcal{T}^{kj}\delta^i_{\ k} - \delta^j_{\ k}\mathcal{T}^{ki} \right) dV = 0 \tag{21}$$

where δ^α_β is the Kronecker tensor. This is the proof that $\mathbf{T}$ is symmetric: $\mathcal{T}^{ij} = \mathcal{T}^{ji}$. This means that no intrinsic spin can be induced at the microscopic level by the surface forces. Consider an horizontal flow, stratified in the z direction and the surface forces applied on a cube of infinitesimal volume. The stresses acting on the horizontal surfaces parallel to the flow are antiparallel. Two equal and opposite forces are also applied to the faces perpendicular to the flow. Thus, the total stress acting on this fluid element exerts no force and no torque. This is a general property of Newtonian fluids.

1.3. The Stress Tensor and Viscosity

In presence of velocity gradients, momentum is irreversibly transferred by collisions among particles from high velocity regions to those of lower velocity, over scales of the order of the particle mean free path. Regarding a fluid element, two distinct groups of time scales can be derived : the first relates to the microscopic properties of the fluid and the second to the local properties of the macroscopic flow. In the first group, we find in particular the mean collision time τ_{coll} between particles, and the relaxation time scale τ_{rel} of the various internal degrees of freedom of particles, if any. In the second group, two distinct time scales characterize the rate at which the *shape* of the fluid element changes, and the rate at which its *volume* varies. Thermodynamical equilibrium requires that the internal time scales (those of the first group) are very short compared to the dynamical time scales (second group). The ratios $\tau_{(1)}/\tau_{(2)}$ of time scales of different groups is a measure of the departure from thermodynamical equilibrium, which results in the appearance of a viscous stress of magnitude $\sigma \approx p\,\tau_{(1)}/\tau_{(2)}$, where p is the usual (static) thermodynamical pressure as given by the equation of state.

Departures from thermodynamical equilibrium resulting from the finite relaxation time of internal degrees of freedom give rise to the so-called bulk viscosity, leading to a *scalar* correction to the thermodynamical pressure proportional to $\nabla \cdot \mathbf{u}$, the variation rate of the volume of a fluid element. The usual shear stress, *tensorial* in nature, is proportional to the rate at which the shape of a fluid element is altered. This means that rigid rotation or isotropic motions can not generate any viscous stress.

By definition, the stress component $\mathcal{T}_{ij}$ is the force acting in the direction x_i on a unit surface normal to x_j. Since internal viscous forces result from the relative motions of the fluid elements, the stress is a function of the space derivatives of the velocity field :

$$\mathcal{T}_{ij} = -p\delta_{ij} + \mu\left(\partial_j u_i + \partial_i u_j\right) + \lambda\partial_l u_l \delta_{ij}. \tag{22}$$

The coefficients μ, the dynamical viscosity, and λ, the second coefficient of viscosity are scalars for an isotropic fluid. Actually, the stress can be put in a form which separates the effects of the shear from the (possible) effects of dilatation and compressions

$$T_{ij} = -p\,\delta_{ij} + \zeta\,\partial_l u_l \delta_{ij} + \mu\left(\partial_j u_i + \partial_i u_j - \frac{2}{3}\partial_l u_l \delta_{ij}\right) \equiv -p\,\delta_{ij} + \sigma_{ij} \quad (23)$$

Here, the combination of the viscosity coefficients :

$$\zeta = \lambda + \frac{2}{3}\mu \quad (24)$$

is the *coefficient of bulk viscosity*, and p is the pressure in the fluid at rest, as given by the equation of state of the fluid in local thermodynamical equilibrium. The isotropic part of the stress acting under dynamical conditions, $-(p - \zeta\,\nabla\cdot\mathbf{u})$, may, in general, differ from the thermodynamical, hydrostatic pressure. However, the coefficient of bulk viscosity, ζ, is actually zero for monatomic gas, and the conjecture that it should be true for any fluid is known as the Stokes conjecture. Fluids for which $\zeta = 0$, or $\lambda = -2/3\ \mu$, are called Maxwellian fluids. But in fact, for sudden expansions or compressions, the relaxation time scales between the translational and the internal degrees of freedom of molecules (rotation, vibration, excitation, ionization) can be comparable or slower than the dynamical time scale of the flow, so that, in this non-equilibrium situation, the gas pressure is no longer given by the thermodynamical equilibrium expression of the equation of state. The bulk viscosity coefficient is thus a measure of the departure from thermodynamical equilibrium induced by relaxation processes (Zel'dovich & Raizer 1967, Landau & Lifshitz 1982, Mihalas & Mihalas 1984). It is generally discarded in most astrophysical applications.

We give here the expression for the shear stress and the equation of motion in the case of a spherically symmetric flow. General expressions in curvilinear systems can be found, for example, in Mihalas and Mihalas (1984). Recast into spherical coordinates, the expressions for the shear stress tensor are :

$$\sigma_{rr} = 2\mu\left(\frac{\partial u}{\partial r} - \frac{1}{3}\nabla\cdot\mathbf{u}\right) \quad (25)$$

$$\sigma_{\theta\theta} = \sigma_{\phi\phi} = -\frac{1}{2}\sigma_{rr} \quad (26)$$

It is clear that there is no viscous force when the flow is locally homologous, since in that case one has precisely $\nabla u = 3u/r$. This property should be kept in mind for the derivation of the so-called artificial viscosity discussed later, a procedure which allows for a simple numerical treatment of shock waves. The equation of the motion (without external force) has now the form

$$\rho\frac{du}{dt} = -\frac{\partial p}{\partial r} + \frac{\partial\sigma_{rr}}{\partial r} + \frac{3}{r}\sigma_{rr} \quad (27)$$

The radial component of the stress can be either positive (the stress acts as a tension) or negative (it acts as a pressure).

1.3.1. Pseudo-viscosity

In normal fluids, the viscosity coefficient is independent of the particle density n, and, for a given n, proportional to the particle mean free path. This means that the viscosity coefficient μ is inversely proportional to the collisional cross section, σ, of particles. Since the natural width of a shock front is a few particle mean free paths, it cannot be sampled by usual numerical techniques, except for very low density gas. The physical basis of the pseudo-viscosity method consists in *decreasing* the particle cross section, in order to adapt the rescaled mean free path to the resolution of the numerical grid.

In one dimensional flows, the dissipation rate can be written in the form

$$\dot{q}_{visc} = -p_{visc}\frac{\partial u}{\partial x} \sim p\,\frac{t_{coll}}{t_{dyn}}\frac{1}{t_{dyn}}, \tag{28}$$

where the local dynamical time scale is defined as $t_{dyn} = 1/(\partial u/\partial x)$. Accordingly, the energy dissipated by viscous forces per unit volume in a shock front moving at the velocity $\mathcal{D}$, scales as

$$\dot{q}_{visc}t_{dyn} \sim p\frac{t_{coll}}{t_{dyn}} \sim p_{visc} \sim \rho\mathcal{D}^2, \tag{29}$$

which implies, if L is the width of the shock

$$p_{visc} = -\mu\frac{\partial u}{\partial x} \sim \frac{1}{3}nm\bar{v}\lambda\frac{\mathcal{D}}{L} \sim nm\mathcal{D}^2, \tag{30}$$

or, to within a factor about unity

$$\frac{\bar{v}}{\mathcal{D}}\frac{\lambda}{L} \approx 1. \tag{31}$$

Since the mean particle velocity behind a shock, $\bar{v}$, is of the order of the shock velocity itself, the preceding relation establishes that the width L of a shock transition is of the order of the particle mean free path λ.

Another important point is that the downstream conditions (postshock density, velocity, temperature I) are independent of the viscosity coefficient, for the following reason. Sufficiently far from the shock transition, downstream and upstream conditions are related by the conservation laws of mass, momentum and energy, namely the Rankine-Hugoniot relations, which are independent of the nature of the microscopic processes by which the bulk kinetic energy of the upstream flow is converted into heat.

Accordingly, an (artificial) increase of the particle mean free path $\lambda \to \Lambda$ does not change the magnitude of the "viscous" pressure, since one has just

$$p_{visc} \to p_{visc}\frac{\lambda}{\bar{v}}\frac{\mathcal{D}}{\Lambda} \sim p_{visc}. \tag{32}$$

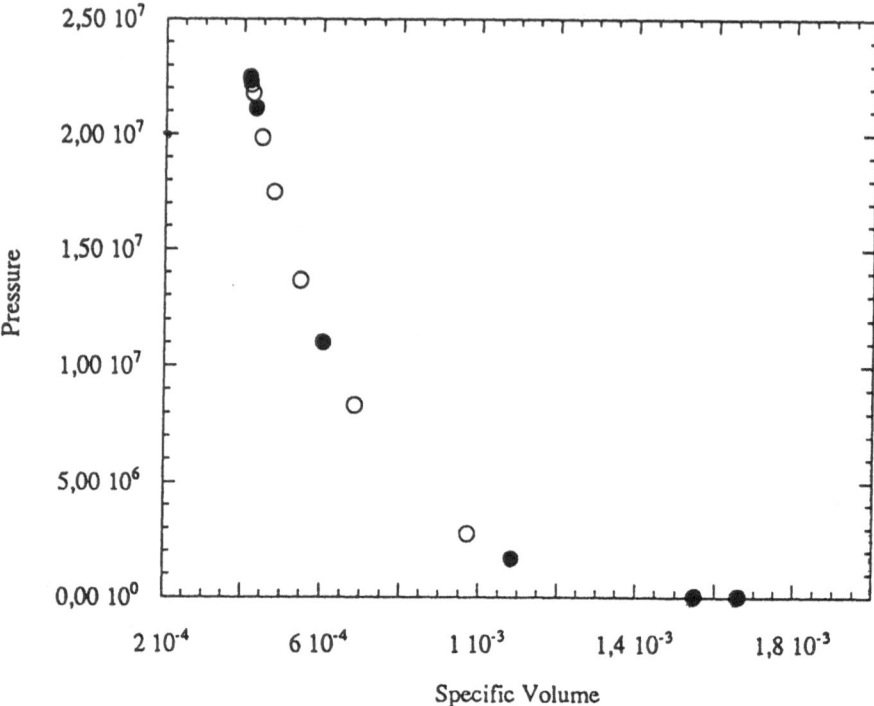

Fig. 1. — Identity of the (p,V) diagrams in a shock transition, calculated with natural and artificial viscosity.

In other words, an arbitrary increase of the mean free path (which actually means a decrease of the physical elastic collision cross section) is compensated by the decrease of the calculated velocity gradient, so that the resulting, calculated, "artificial" viscous stress remains equal to the actual one. Moreover, this method guarantees the correctness of the shock adiabat, which relates the pressure and the specific volume through the shock transition. This is the reason for the success of the widely used method of the pseudo-viscosity. This is shown in Fig. 1. Here, two calculations are compared. In the first one, the mesh resolution is a fraction of the particle mean free path (for which $\sigma = 10^{-15} cm^2$). The full set of Navier-Stokes equations is then solved. In the second one, the spatial resolution is downgraded by a factor 10^4, and consequently, the elastic cross section has been reduced by the same factor. One should keep in mind, however, that the Navier-Stokes equations give only an approximate, fluid description of the shock transition, generally sufficient for most astrophysical purposes. A more correct description involves a particle description of the fluid.

A second caveat concerns the fact that the shock transition is actually stretched both in space and in time : the time required for a fluid element to cross a shock transition is proportional to the viscosity coefficient. If it is increased artificially, the fluid spends proportionally more time to run through the shock adiabat. This may alter the output of other rapid inelastic collisional process, such as chemical reactions, occuring in the shock dissipative layer. One may correct for this by a straightforward generalization of the pseudo-viscosity method, which consists in reducing the values of *all* the relevant cross sections, elastic or inelastic, by the same factor, in order to obtain the correct reaction yields.

Finally, one should notice that it is important to adopt a formulation of the artificial viscous stress which respects the physical dependence on the flow variables illustrated by Eq. 23 or Eq. 25. For example, it would be misleading to adopt, in spherical flows, an expression of the viscous stress proportional to *div* **u**.

1.4. Body Forces

In the presence of body forces **f**, the equation of motion (Eq. 7) becomes :

$$\rho\gamma^i - \nabla_j \mathcal{T}^{ij} = f^i \tag{33}$$

We shall examine here the most important forces for astrophysical applications : the magnetic field and gravitation. We show that it is possible to express the variation of the fluid momentum due to these two important body forces as the divergence of a quantity analogous to a momentum density flux, in order to restore the symmetry of Eq. 6, lost in Eq. 33.

1.4.1. Magnetic Field

We suppose here that the magnetic field is sufficiently weak or the ionization degree sufficiently high, so that a single density and a single velocity can adequately describe the ionized fluid. In other words, the length of magnetic precursors (see Section XX), compared to the other characteristic length scales of the flow, can be ignored.

Neglecting the displacement current, Maxwell's equations involving the magnetic field are

$$\nabla \cdot \mathbf{B} = 0 \tag{34}$$

$$\nabla \times \mathbf{B} = 4\pi\mathbf{j} \tag{35}$$

$$\nabla \times \mathbf{E} = -\mu\frac{\partial \mathbf{B}}{\partial t} \tag{36}$$

the current density is :

$$\mathbf{j} = \sigma\left(\mathbf{E} + \mu\mathbf{v} \times \mathbf{B}\right) \tag{37}$$

and the Lorentz force per unit volume writes

$$\mathcal{F}_{\mathcal{L}} = \mu \mathbf{j} \times \mathbf{B} = \frac{\mu}{4\pi} (\nabla \times \mathbf{B}) \times \mathbf{B} \tag{38}$$

Using the formula :

$$\nabla (\mathbf{A} \cdot \mathbf{B}) = \mathbf{A} \times \nabla \times \mathbf{B} + \mathbf{B} \times \nabla \times \mathbf{A} + \mathbf{B} \cdot \nabla \mathbf{A} + \mathbf{A} \cdot \nabla \mathbf{B} \tag{39}$$

one obtain the expression of the Lorentz force, as the sum of a pressure and a tension :

$$\mathcal{F}_{\mathcal{L}} = -\frac{\mu}{8\pi} \nabla B^2 + \frac{\mu}{4\pi} \mathbf{B} \cdot \nabla \mathbf{B} \tag{40}$$

with Cartesian components :

$$\mathcal{F}_{\mathcal{L}}^i = -\partial_j \left(\frac{\mu}{8\pi} B^2 \delta^{ij} - \frac{\mu}{4\pi} B^i B^j \right) \equiv -\partial_j \mathcal{B}^{ij} \tag{41}$$

We deduce from this conservation equation the expression for the momentum density flux tensor associated with the magnetic field :

$$\mathcal{B}^{ij} = -\frac{\mu}{8\pi} B^2 \delta^{ij} - \frac{\mu}{4\pi} B^i B^j \tag{42}$$

Note the close connection of this expression with the expression of the hydro-dynamical momentum density flux tensor of the flow.

1.5. Gravitation

The gravitational force density $\mathcal{F}_G = \rho \mathbf{g}$, can be recast in a form equivalent to Eq 41, derived for the Lorentz force density. The quantities $\mu^{1/2} B$ and $G^{-1/2} \mathbf{g}$, where G is the gravitational constant, play formally an equivalent role. If Φ is the gravitational potential, the components of the gravitational acceleration are given by $g^i = -\partial_i \Phi$. According to the Poisson equation, the local density of the fluid can be expressed through the gravitational field in the form :

$$\rho = -\frac{1}{4\pi G} \partial_k g^k \tag{43}$$

Thus, the gravitational force density can be expressed as

$$\mathcal{F}_G^i = \rho g^i = -\frac{1}{4\pi G} g^i \partial_k g^k = -\frac{1}{4\pi G} \left\{ \partial_k \left(g^i g^k \right) - g^k \partial_k g^i \right\} \tag{44}$$

But the gravitational field is irrotational, so that the equalities $\partial_k g^i = \partial_i g^k$ lead to the expression

$$f_G^i = -\frac{1}{4\pi G} \left\{ \partial_k \left(g^i g^k \right) - g^k \partial_i g^k \right\} = -\frac{1}{4\pi G} \left\{ \partial_k \left(g^i g^k \right) - \frac{1}{2} \partial_k g^2 \delta^{ik} \right\} \tag{45}$$

from which one can define the momentum density flux tensor associated with gravitation :

$$\mathcal{G}^{ij} = \frac{1}{8\pi G}g^2\delta^{ik} - \frac{1}{4\pi G}g^i g^k \qquad (46)$$

It is interesting to note the formal equivalence between the hydrodynamical, magnetic and gravitational density momentum fluxes.

According to these expressions, the general covariant Euler equation of motion can be written in the following general form, in which there is no more distinction between surface forces and body forces :

$$\rho\gamma^i - \nabla_j\left(\mathcal{T}^{ij} + \mathcal{B}^{ij} + \mathcal{G}^{ij}\right) = 0 \qquad (47)$$

which expresses that a fluid element is in equilibrium against the inertial force and the external forces applied on it. In Cartesian coordinates, one has equivalently :

$$\frac{\partial}{\partial t}\left(\rho u^i\right) = -\partial_j\left(-\mathcal{P}^{ij} - \mathcal{B}^{ij} - \mathcal{G}^{ij}\right) \equiv -\partial_j\mathcal{M}^{ij} \qquad (48)$$

where the generalized momentum flux density tensor $\mathcal{M}$ is the sum $\mathcal{M}^{ij} = \mathcal{P}^{ij} + \mathcal{B}^{ij} + \mathcal{G}^{ij}$, where the stress $\mathcal{P}$ is defined by Eq. 19.

1.6. The Energy Equation and Dissipation

Considering the fluid flowing through a fixed volume V, one can readily obtain from the continuity and momentum equation, the following expression

$$\frac{\partial}{\partial t}\int_V \frac{1}{2}\rho u^2\, dV + \int_V \mathcal{M}_{ik}\frac{\partial u_i}{x_k}\, dV = -\int_S \left\{\frac{1}{2}\rho u^2 u_k - u_i\mathcal{M}_{ik}\right\}dS_k \qquad (49)$$

which is valid in Cartesian coordinates. This is an equality between volume terms and surface terms. The right hand side expression represents the rate at which energy flows into the volume (advection) plus the rate at which the total stress does work on the surface of the system. If we consider an isolated system, no energy enters the system and the r.h.s. vanishes. One has the quadrature

$$\frac{\partial}{\partial t}\int_V \frac{1}{2}\rho u^2\, dV + \int_V \mathcal{B}_{ik}\frac{\partial u_i}{x_k}\, dV + \int_V \mathcal{G}_{ik}\frac{\partial u_i}{x_k}\, dV + \int_V \mathcal{P}_{ik}\frac{\partial u_i}{x_k}\, dV = 0 \quad (50)$$

in which the first term represents the variation of the kinetic energy of the fluid, the second and the third, respectively the variations of the magnetic energy and gravitational energy. Accordingly, the last term,

$$\int_V \mathcal{P}_{ik}\frac{\partial u_i}{x_k}\, dV = -\int_V p\,\nabla\cdot\mathbf{u}\, dV + \int_V \sigma^{ij}\frac{\partial u_i}{\partial x_j}dV \qquad (51)$$

represents the variation of the internal energy of the system. It is made up of two terms. The first one represents the work done by compression, that is the increase of the internal energy which would result from an adiabatic, reversible process. The second term represents therefore the increase of the internal energy due to work of the viscous stress, which is irreversibly converted into heat. The dissipation rate per unit volume due to viscosity, is thus : $\dot{q} = \nabla_i u_j \sigma^{ij}$.

1.7. Entropy Evolution

The entropy of a fluid element can vary due to the heat exchange with neighbouring fluid elements due to conduction, radiation transport or convection, or due to irreversible processes like the dissipation of kinetic energy by viscous forces, particularly in shock waves, or the presence of energy sources, like chemical reactions.

Finally, the equation describing the entropy evolution of the fluid can be written as follows:

$$\frac{\partial \rho s}{\partial t} = -\nabla_k \left(\rho s u^k\right) + \frac{1}{T} \left\{ \dot{Q} + \nabla_i u_j t'^{ij} \right\} \tag{52}$$

1.8. Fragmented Systems of Uniform Energy Density

Consider a system of scale L and mass M_L, concentrated in N_L fragments of scale l and mass m_l. The total surface area Σ_L of the system scales as $\Sigma_L \propto N_L l^2$. When a quasistationary state is achieved, Eq 52 implies the equality

$$\int_{V_L} \mathcal{M}_{ij}\partial_j u^i dV = \int_{\Sigma_L} u_i \mathcal{M}_{ij} dS_j \tag{53}$$

and the corresponding scaling for the energy flowing through is

$$\langle \mathcal{M} \rangle_L \frac{\langle u \rangle_L}{L} L^3 \propto \langle u \rangle_L \langle \mathcal{M} \rangle_L N_L l^2 \tag{54}$$

Since the total mass is just $M_L = N_L m_l$, one obtains the relationship which must be obeyed between adjacent scales

$$\frac{M_L}{m_l} \propto \left(\frac{L}{l}\right)^2 \tag{55}$$

It is instructive to write Eq. 53 in a slightly different form. If $\mathcal{E} \propto \langle \mathcal{M} \rangle L^3$ is the total energy of the system (the total stress has the dimensions of an energy density), the volume and the surface terms are respectively $\mathcal{E} \times u/L$ and $\mathcal{E} \times n l^2 u$, in which appear the dynamical time scale of the structure $t_{dyn} = u/L$, and the clump collision time $t_{coll} = n l^2 u$. Equilibrium is achieved when

$$\frac{\mathcal{E}}{t_{coll}} = \frac{\mathcal{E}}{t_{dyn}} \tag{56}$$

Henriksen and Turner (1987) first noticed the fact that the scaling laws observed in fragmented molecular clouds could be interpreted as a resonance between the mean collision time between clumps and their dynamical time scale. Molecular clouds, which are the most massive objects of the interstellar medium, are highly fragmented structures. The mass M and the linear size L of the various substructures obey a relation of the form $M \propto L^n$, with $n \sim 2$ (see Falgarone, this volume, and references therein). According to the previous results, this can be roughly interpreted as the mass distribution supporting a uniform energy flux, flowing from the largest scales down to the smallest ones.

2.   GAS MICROPHYSICS

2.1.   Coupling Coefficients between Plasma Components

A magnetic field, even weak, can lead to strong decoupling of the hydrodynamical evolution of the fluids of electrons, ions and neutrals. Elastic (and inelastic) collisions between charged and neutral particles drive momentum and energy exchanges between the plasma components. The dynamical friction between charged particles and neutrals heats up the fluid, and especially minor species (Mullan 1971, Draine, Roberge & Dalgarno 1983, Flower, Pineau des Forêts & Hartquist 1985)

The calculation of coupling coefficients through collisions between electrons, ions and neutrals is well established when the three fluids have no relative motions. When large drift velocities are present, the calculation is slightly more complicated (Draine 1986).

Momentum and energy exchanges are affected by the drift velocity, and the relative kinetic energy of the two groups of particles is also dissipated into heat. The corresponding rate coefficients must be integrated over the double velocity distribution of the pair of interacting particles, which cannot be reduced to a simple Maxwellian distribution.

We distinguish two types of interacting particles (i) and (j), of masses m_i and m_j and reduced mass μ_{ij}. We define the reference frame as the center of mass of the unit volume of the combined fluid. In this frame, the individual velocities of the particles are $\mathbf{w}_i$ and $\mathbf{w}_j$. Their collective or hydrodynamical velocities are defined as $\mathbf{W}_{i \ or \ j} = < \mathbf{w}_{i \ or \ j} >$ and the streaming velocity between the two groups of particles is noted $\mathbf{D}_{ij} = \mathbf{W}_i - \mathbf{W}_j$.

The total cross section, σ, for the interaction of a pair (i, j) of particles depends on the incident energy in the center of mass of the colliding pair of particles (which does not coincide with the reference frame). Let $f_{ij}(\mathbf{w_i}, \mathbf{w_j})$ be the distribution function of the velocities of the pairs of particles. If $\mathbf{w}_{ij} = \mathbf{w}_i - \mathbf{w}_j$ is the relative velocity of the interacting particles, the expression for the collision rate is

$$\mathcal{C} = \int \int f_{ij}(\mathbf{w}_i, \mathbf{w}_j) \ \sigma(w_{ij})w_{ij} \ d^3\mathbf{w}_i d^3\mathbf{w}_j \qquad (57)$$

where we have implicitly assumed that the two types of interacting particles are discernible.

2.2.   Decay Rate of the Fluid Momenta

Let $\mathbf{q}$ be the momentum transfered in an elastic collision from a particle (j) to a particle (i), and $\sigma'(\mathbf{q}, w_{ij})$ the corresponding differential cross section, normalized to the total cross section σ

$$\int \sigma'(\mathbf{q}, w_{ij}) \, d^3\mathbf{q} = \sigma(w_{ij}) \tag{58}$$

Then, the decay rate of the respective momenta of the slipping fluids can be written as

$$\dot{\mathbf{Q}}_i = -\int\int f_{ij}(\mathbf{w}_i, \mathbf{w}_j) \int \mathbf{q} \, \sigma'(\mathbf{q}, w_{ij}) w_{ij} \, d^3\mathbf{q} \, d^3\mathbf{w}_i d^3\mathbf{w}_j \tag{59}$$

By symmetry, the inner integration over the momentum transfer (i.e. over the impact parameters) gives a vector anti-parallel to the available momentum $\mu_{ij} \mathbf{w}_{ij}$ in the center of mass system of the colliding particles. This leads to the definition of the momentum transfer cross section $S(w_{ij})$ as

$$\int \mathbf{q} \, \sigma'(\mathbf{q}, w_{ij}) \, d^3\mathbf{q} \equiv -\mu_{ij} \mathbf{w}_{ij} S(w_{ij}) \tag{60}$$

With this definition of S, the loss rate of the momentum density of the fluid (i) is just

$$\dot{\mathbf{Q}}_i = -\mu_{ij} \int\int f_{ij}(\mathbf{w}_i, \mathbf{w}_j) \, \mathbf{w}_{ij} S(w_{ij}) w_{ij} \, d^3\mathbf{w}_i d^3\mathbf{w}_j \tag{61}$$

This expression is naturally antisymmetric in the exchange of the particles.

2.3.   Rate of Energy Exchange between the Two Fluids

For an elastic collision, the energy variation of the interacting particles *in the reference frame* can be expressed as the scalar product of the momentum transfered from species (j) to species (i) and the velocity $\mathbf{z}_{ij}$ of the center of mass of the colliding particles

$$\Delta\epsilon_i = \mathbf{q} \cdot \mathbf{z}_{ij} \tag{62}$$

so that the variation per unit time of the total energy $\mathcal{E}_i$ of the fluids is

$$\dot{\mathcal{E}}_i = \int\int f_{ij}(\mathbf{w}_i, \mathbf{w}_j) \, \mathbf{z}_{ij} \cdot \left[\int \mathbf{q}\sigma'(\mathbf{q}, w_{ij}) w_{ij} d^3\mathbf{q}\right] d^3\mathbf{w}_i d^3\mathbf{w}_j \tag{63}$$

or, according to Eq. 60

$$\dot{\mathcal{E}}_i = -\mu_{ij} \int\int f_{ij}(\mathbf{w}_i, \mathbf{w}_j) \, [\mathbf{z}_{ij} \cdot \mathbf{w}_{ij} \, S(w_{ij}) w_{ij}] \, d^3\mathbf{w}_i d^3\mathbf{w}_j \tag{64}$$

This expression is antisymmetric in the exchange of the interacting particles, since the total energy of the gas is conserved. The variation of the kinetic energy of the fluids is now given by

$$\dot{\mathcal{K}}_{iorj} = \dot{\mathbf{Q}}_{iorj} \cdot \mathbf{W}_{iorj} \tag{65}$$

and the variation of their internal energies (or thermal contents) reads

$$\dot{\mathcal{U}}_{iorj} = \dot{\mathcal{E}}_{iorj} - \dot{\mathcal{K}}_{iorj} \tag{66}$$

2.4. Rate Coefficients

Equations 65 and 66 completely define the exchanges between two slipping groups of particles, each one with an arbitrary velocity distribution. These expressions can be recast in terms of three distinct velocity moments of the collision rate per pair of particles,

$$\mathcal{C}_{ij} = n_i n_j \, < \sigma(w_{ij}) w_{ij} > \tag{67}$$

$$\dot{\mathbf{Q}}_i = - \mu_{ij} \, n_i n_j \, < S(w_{ij}) w_{ij} \mathbf{w}_{ij} > \tag{68}$$

$$\dot{\mathbf{Q}}_j = + \mu_{ij} \, n_i n_j \, < S(w_{ij}) w_{ij} \mathbf{w}_{ij} > \tag{69}$$

$$\dot{\mathcal{E}}_i = - \mu_{ij} \, n_i n_j \, < S(w_{ij}) w_{ij} \mathbf{z}_{ij} \cdot \mathbf{w}_{ij} > \tag{70}$$

$$\dot{\mathcal{E}}_j = + \mu_{ij} \, n_i n_j \, < S(w_{ij}) w_{ij} \mathbf{z}_{ij} \cdot \mathbf{w}_{ij} > \tag{71}$$

which can be calculated if one knows the velocity (or energy) dependence of the collision cross section and the actual form of the velocity distribution.

These general expressions are expressed in terms of three momenta, over the double velocity distribution functions, of the collision cross section $\sigma(w)$ and the *momentum transfer cross section*, $S(w)$. When these cross sections are known (from quantum mechanical calculations), these momenta can be calculated assuming two distinct Maxwellian velocity distribution function for the interacting species. One has in order

$$\mathcal{C}_{ij} = n_i n_j \, \Lambda_{ij}^{(0)} \tag{72}$$

$$\dot{\mathbf{Q}}_i = -\mu_{ij} \mathbf{D}_{ij} \, n_i n_j \Lambda_{ij}^{(1)} \tag{73}$$

$$\dot{\mathbf{Q}}_j = +\mu_{ij} \mathbf{D}_{ij} \, n_i n_j \Lambda_{ij}^{(1)} \tag{74}$$

The rate of variation of the *internal* energies of the fluids is the sum of two terms, which represent respectively the irreversible dissipation of the relative kinetic energy of the two fluids, due to the work of the friction (Langevin) forces, and the irreversible flow of energy from the "hot" to the "cold" fluid

$$\dot{\mathcal{U}}_i = 2 \, \frac{m_j T_i}{m_i T_j + m_j T_i} \, \frac{1}{2} \mu_{ij} D_{ij}^2 \, n_i n_j \Lambda_{ij}^{(1)} - 2 \, \frac{\mu_{ij}}{m_i + m_j} \, k \, (T_i - T_j) \, n_i n_j \Lambda_{ij}^{(2)} \tag{75}$$

$$\dot{\mathcal{U}}_j = 2 \frac{m_i T_j}{m_j T_i + m_i T_j} \frac{1}{2} \mu_{ij} D_{ij}^2 \, n_i n_j \Lambda_{ij}^{(1)} - 2 \frac{\mu_{ij}}{m_j + m_i} k \, (T_j - T_i) \, n_i n_j \Lambda_{ij}^{(2)} \quad (76)$$

The exchange of internal energy is Galilean invariant, while the exchange of total energy is not.

Analytical expressions can be obtained when the cross sections σ and S are either constant or inversely proportional to the relative energy. When the cross sections are constant one has the following result, expressed in terms of the function $\Psi(x) = \sqrt{\pi}/2 \, \text{Erf}(x)$, of the reduced c.m. kinetic energy $\kappa = 1/2 \mu_{ij} \beta_{ij} D^2$ and of the mean relative thermal velocity $\bar{v} = (8/\pi \beta_{ij} \mu_{ij})^{1/2}$

$$\Lambda^{(0)} = \sigma \bar{v} \left\{ \frac{1}{2} \kappa^{-\frac{1}{2}} \Psi(\kappa^{\frac{1}{2}}) + \kappa^{\frac{1}{2}} \Psi(\kappa^{\frac{1}{2}}) + \frac{1}{2} e^{-\kappa} \right\}$$

$$\Lambda^{(1)} = S \bar{v} \left\{ -\frac{1}{4} \kappa^{-\frac{3}{2}} \Psi(\kappa^{\frac{1}{2}}) + \kappa^{-\frac{1}{2}} \Psi(\kappa^{\frac{1}{2}}) + \kappa^{\frac{1}{2}} \Psi(\kappa^{\frac{1}{2}}) + \frac{1}{4} \kappa^{-1} e^{-\kappa} + \frac{1}{2} e^{-\kappa} \right\}$$

$$\Lambda^{(2)} = S \bar{v} \left\{ \frac{3}{4} \kappa^{-\frac{1}{2}} \Psi(\kappa^{\frac{1}{2}}) + 3 \kappa^{\frac{1}{2}} \Psi(\kappa^{\frac{1}{2}}) + \kappa^{\frac{3}{2}} \Psi(\kappa^{\frac{1}{2}}) + \frac{1}{2} \kappa e^{-\kappa} + \frac{5}{4} e^{-\kappa} \right\}$$

Finally, when the cross sections are inversely proportional to the velocity $(H, H^+$ at low velocity), one has simply, with $\sigma(w)w = \lambda_0$

$$\lambda = \lambda_0, \qquad \Lambda^{(1)} = \lambda_0, \qquad \Lambda^{(2)} = \lambda_0 \left(\frac{3}{2} + \kappa \right)$$

In the case of the Coulomb long range potential, the effects of close collisions are overwhelmed by the cumulative effect of the numerous distant encounters (Chandrasekhar 1942, Spitzer 1962). There is generally no significant drift between ions and electrons, and thus the single relevant rate is the thermal exchange rate which can be calculated using the ion-electron cross section given by

$$\sigma_{ie} = \frac{\pi e^4 Z_i^2}{k^2 T_{ei}^2} \, ln \Lambda \tag{77}$$

$$\Lambda = \frac{3}{2 Z_i e^3} \left(\frac{k^3 T_{ij}^3}{\pi n_i} \right)^{\frac{1}{2}} \tag{78}$$

$$\bar{v}_{ie} = \left(\frac{8}{\pi} \frac{k T_{ie}}{\mu_{ie}} \right)^{\frac{1}{2}} \tag{79}$$

$$T_{ie} = \frac{m_e T_i + m_i T_e}{m_e + m_i} \tag{80}$$

With these expressions, the thermal exchange rate is just :

$$\dot{\mathcal{U}}_i = -2 \frac{\mu_{ie}}{m_i + m_e} k \, (T_i - T_e) \, n_i n_e \, \sigma_{ie} \bar{v}_{ie} = -\dot{\mathcal{U}}_e \tag{81}$$

2.5. Post-shock Relaxation in a Low-density Medium

Behind a strong shock of velocity D, in a perfect gas with adiabatic exponent $\gamma = 5/3$, the post-shock temperature, T_2 of species of molecular weight μ is given by $T_2 = 3\mu/16kD^2$. Accordingly, just behind a shock, ions, neutral particles and electrons have widely different temperatures. Light electrons are much colder than heavy ions or neutral particles. However, the collisional exchange of energy gradually restore a common post-shock temperature, on a length scale of the order of $1/4Dt_{eq}$, where t_{eq} is the energy equipartition time scale (per particle). Note that the equipartition length varies as $D^4 n^{-1}$. As an example, Fig. 2 represents the departure from thermodynamical equilibrium between ions and electrons in a simulated cluster of galaxies. The X-ray emitting intracluster gas is heated up through an accretion shock, which is fed by outer gas infalling on the cluster. Due to the very low density of the gas, the postshock temperature relaxation time may be greater than the Hubble time.

Another illustration is provided by the structure of a C-type shock in a magnetized medium, Fig. 3. This type of flow is characterized by a large drift velocity, forced by the magnetic field, between ions and neutrals. Charged particles are set in motion by the magnetic stress, while neutrals lag behind them. Dynamical coupling is gradually achieved through collisions between ions/electrons and neutrals. In the region where the flows are decoupled, the work done by the friction force between the slipping species essentially heats up the ions, which are generally very under abundant relative to neutrals. Unshocked neutrals are gradually heated up in the magnetic precursor by collisions with the hot ions, but they efficiently radiate away their energy by collisional excitation.

2.6. Heat Conduction

When temperature gradients are present, heat is carried by electrons and, to a lesser extent, by ions and neutrals, from the hotter to the colder regions, in order to restore thermal equilibrium. The coefficient of thermal conductivity, established from kinetic theory, is equal to

$$\kappa = \frac{3}{2}nk\frac{\lambda}{3}\bar{v} \tag{82}$$

where n is the number density of particles carrying the heat flow, λ their mean free path, and $\bar{v}$ their mean velocity (Zel'dovich & Raizer, 1967, Landau & Lifshitz, 1982). The mean free path λ is independent of the particle mass, but does depend on their charge and their temperature. Since the mean velocity of electrons and ions are in the ratio $\bar{v}_e/\bar{v}_i = (m_i/m_e)^{1/2}$, the electronic thermal conductivity of electrons dominates that of ions by the factor $\kappa_e = (m_i/m_e)^{1/2}\kappa_i$. Behind a shock, ions have a mean thermal velocity of the order of the shock speed, so that they can't appreciably cross the shock front, while electrons, with higher thermal velocities, diffuse efficiently from

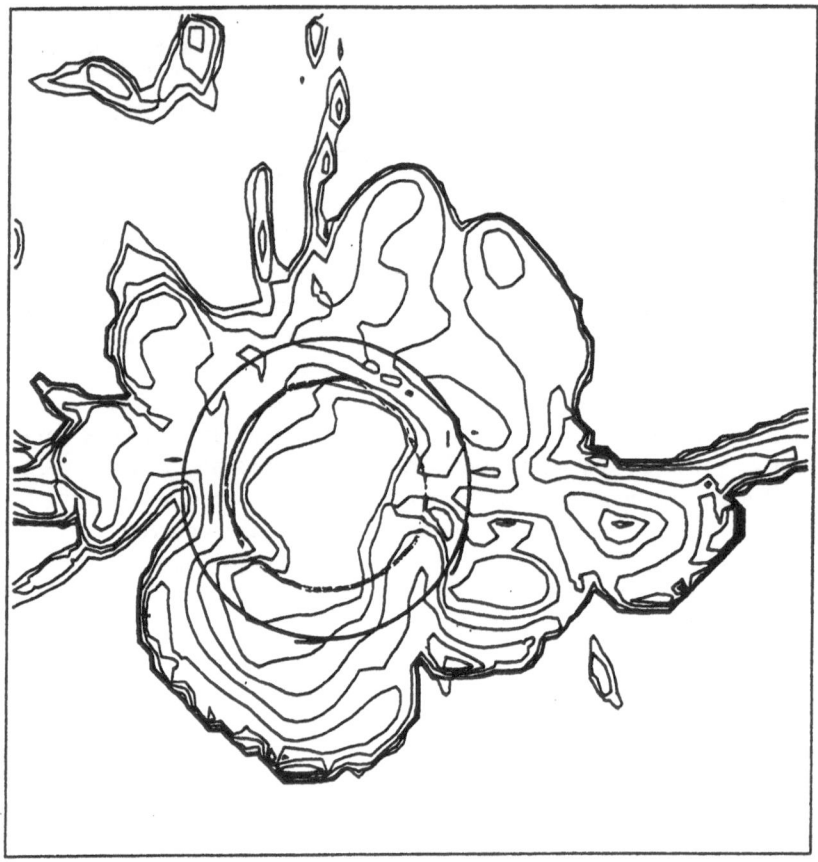

Fig. 2. — Departure from thermal equilibrium between ions and electrons in a simulated cluster of galaxies. The ratio $(T_i - T_e)/(T_i + T_e)$ is plotted with isocontours of 5, 10, 20, 40, 60 and 80 %, in a cut through the center of the cluster. The largest temperature decoupling occurs in the outer regions of the cluster, where the particle density is low.

the shocked region to the unshocked region. This leads to the formation of an electronic preheating precursor, which is able to ionize the gas far away ahead of the shock front. Ions and neutrals are gradually heated up in the hot electronic precursor by collisional heat transfer, and are then further shock heated by viscous dissipation in the viscous dissipative shock layer.

The situation where the electronic mean free path is locally smaller than the electronic temperature scale height corresponds to the diffusive regime of

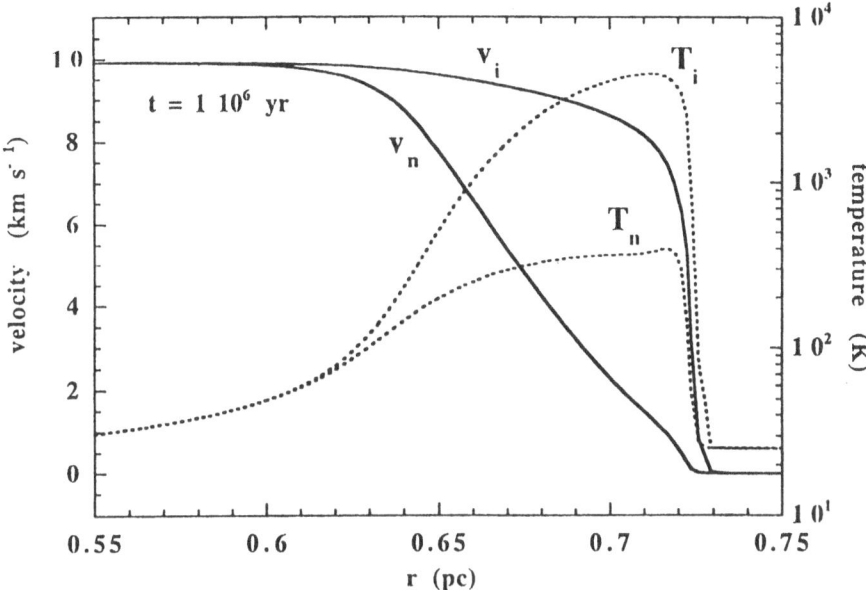

Fig. 3. — The structure of a C-shock. In presence of a magnetic field (here $B = 12\ \mu G$), the motions of charged and neutral particles are decoupled over scales much larger than the particle mean free path. The velocity of ions and electrons is v_i, and the velocity of neutrals is v_n. The energy dissipated by the friction between ions and electrons on one hand, and neutrals on the other (which gradually recouples the two fluids) essentially heats ions and electrons in the precursor, driven by the Lorentz force. In this example, the flow is stationary in a frame moving at a velocity of 10 $km\ s^{-1}$. T_i and T_n are respectively the temperatures of ions and neutrals. Electrons have virtually the same temperature as the ions. The rapid drop of the temperatures to about the unperturbed values, is due to efficient collisional cooling.

heat conduction. In this case, the coefficient of electronic conduction can be expressed as

$$\kappa_e \approx \frac{(kT_e)^{\frac{5}{2}}}{m_e^{\frac{1}{2}} Z e^4 \log \Lambda} \tag{83}$$

where m_e is the electron mass and Z the charge of ions. According to the Fourier law, only relevant to that regime, the heat flux carried by electrons is proportional to the temperature gradient

$$q_e = -\kappa_e\ \nabla T_e \tag{84}$$

Conversely, in the situation where the electron mean free path is comparable to the temperature scale height, or larger, the heat flux saturates, as no more heat than contained in half a Maxwellian can be transported by electrons of mean velocity $\bar{v}_e$. In this case, the heat flux can be expressed in the form

$$\mathbf{q}_e = -0.21 \left(\frac{kT_e}{m_e}\right)^{\frac{1}{2}} \mathcal{U}_e \frac{\nabla T_e}{|\nabla T_e|} \qquad (85)$$

where $\mathcal{U}_e$ is the internal energy density of electrons (Cowie & McKee 1977).

Fig. 4 shows an early time snapshot of the structure of a shock wave in a low density gas of cosmological composition. This calculation takes into account electronic conduction, energy exchanges between electrons, ions and neutral particles, the time dependent collisional ionization and recombination processes, and finally the relevant collisional line cooling terms, cooling by recombination and bremsstrahlung. In this high resolution calculation, the numerical mesh size is of the order of the shortest particle collisional mean free path, so that the viscous shock front, described by the Navier-Stokes equations, is actually resolved. The sharp front of the electronic precursor is due to the $T^{5/2}$ behaviour of the coefficient of electronic conduction.

The length Δ of the shock preheating layer can be evaluated at steady state by equating the outward conduction energy flux and the inward advected energy flux. Assuming, as it is generally the case, that the heat flux is unsaturated in most of the precursor, the resulting approximate formula is

$$\Delta_e = \frac{4}{15} \frac{\kappa_e(T_e)}{n_e kD} \approx 1.2 \ 10^4 \frac{T_e^{\frac{5}{2}}}{n_e \mathcal{D}_5} \ cm \qquad (86)$$

where $\mathcal{D}_5$ is the shock velocity expressed in kilometers per second, and T_e the electron temperature when they enter the shock.

3. INHOMOGENEOUS GAS COOLING

3.1. A Multiphase Model of the Gas

The gas distributed in a galaxy is an inhomogeneous system made up of cold, warm and hot regions, distributed in various proportions. Star formation from diffuse matter requires quite a large reduction of the gas entropy. Globally, bremsstrahlung and collisional cooling maintain a permanent cooling flow, streaming from the high entropy phase down to the low entropy, condensed phase, distributed in many cloud complexes. These clouds harbour star formation, the endpoint of the entropy cascade. Replenishment of the high entropy phase results from the injection, on large scales, of high entropy material by stellar winds and supernovae explosions. Note that for the same energy density, radiation has a much lower entropy than matter. In the Galaxy, the tenuous hot phase has an entropy (expressed per baryon and in unit of the Boltzmann constant) $s_1 \approx 80 \ k_B$, which drops to $s_2 \approx 57 \ k_B$ in HI clouds, $s_3 \approx 53 \ k_B$ in molecular clouds, $s_4 \approx 45 \ k_B$ in dense cores and finally to $s_5 \approx 20 \ k_B$ in stars themselves.

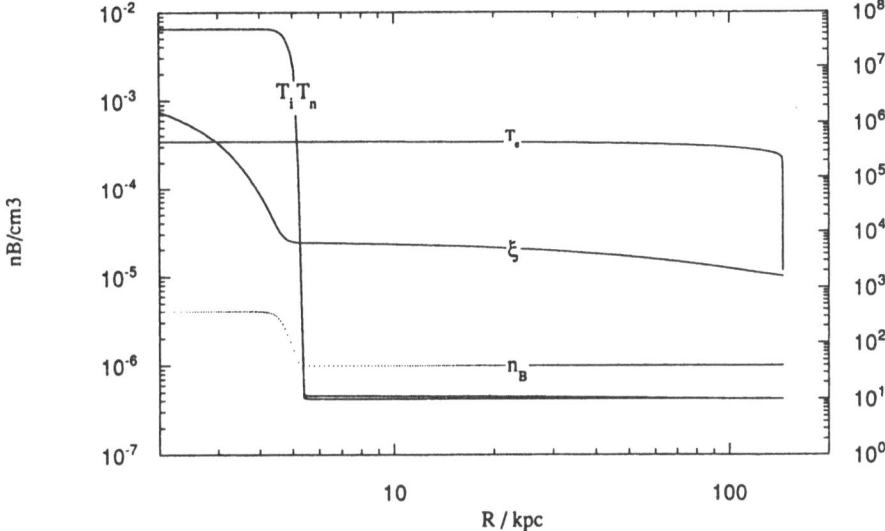

Fig. 4. — Detailed structure of a strong shock wave in a dilute medium, 10^7 years after the formation of the shock. The constant shock velocity is $10^3\,km\,s^{-1}$. The baryonic number density is n_B, and ξ the time dependent ionization fraction. The temperature of interacting electrons, ions and neutral particles are respectively denoted by T_e, T_i and T_n. The electron mean free path in the precursor is as large as $3 \times 10^2\,pc$, which corresponds to the largest mesh size adopted in this 1D Lagrangian Navier-Stokes calculation.

3.2. Mass-Entropy Spectrum

We discuss here a more precise description of the thermodynamical properties of a mass of gas M, sufficiently large to incorporate various gas phases such as clouds, and a warm or hot intercloud medium. A simple description of the local gas properties by a single pair of independent thermodynamical variables is inadequate, because any fluid elements of macroscopic scale is made up of a wide spectrum of thermodynamical states (Lioure & Chièze 1990, Reisenegger, Miralda-Escudé & Waxman 1996).

Consider the cumulative mass distribution $\mathcal{M}_s(s,t)$ in the entropy space, which represents the total mass of gas with specific entropy less than s. We will essentially make use of the differential mass distributions, defined as:

$$m(s,t) = \frac{\partial}{\partial s}\mathcal{M}(s,t) \qquad (87)$$

We adopt the usual hypothesis of pressure equilibrium among the various phases of the interstellar medium, which however breaks down for the densest regions bound by gravitation, as star forming regions. The various components of the system are at any time in pressure equilibrium, with a common pressure $p(t)$.

The time evolution of the pressure is easily obtained from the hypothesis of pressure equilibrium through the various components of the system. The total internal energy E of the system depends on the mass spectrum through the expression

$$E(t) = \int_s m_s(s,t) \, \epsilon \, (s,p) \, ds \tag{88}$$

where $\epsilon \, (s,p)$ is the specific energy of a small fluid element. The time variation of the internal energy of the system results both from cooling of its components and the mechanical work performed on it. We note $\dot{q} = \dot{q}_\epsilon \, (\epsilon, p) = \dot{q}_s \, (s, p)$, the local balance between the cooling and possible heating rates per unit mass, expressed in terms of the gas pressure and any independent pair of thermodynamical variables. The net energy loss of the entire system due to cooling reads:

$$\dot{Q}(t) = \int_s \dot{q}_s \, (s, p) \, m(s, t) \, ds \tag{89}$$

The variation rate of the total volume of the system is determined by its hydrodynamical evolution, and is

$$\omega(t) \; = \; \frac{1}{V} \frac{d\mathcal{V}}{dt} \tag{90}$$

Accordingly, the variation of the pressure of the system is determined by the equation:

$$\frac{1}{p}\frac{dp}{dt} \; = \; \frac{1}{E(t)} \int_s \dot{q}_s \, (s, p) \, m(s, t) \, ds - \frac{5}{3}\omega(t) \tag{91}$$

We establish now the equations governing the time evolution of the mass-entropy spectrum of the system, resulting only from thermal processes. The time variation of the specific entropy of any given small fluid element is a function of the variables $\{s, p(t)\}$, and is noted $\dot{s}$. The whole system of mass M can be viewed as a fluid flowing in the entropy space, at a rate $\dot{s}$ which depends only on the efficiency of the cooling and heating processes. The mass flux at any given entropy is:

$$\phi(s, p, t) \; = \; m(s, t) \, \dot{s} \tag{92}$$

Mass conservation can be expressed in the form of a continuity equation for the differential mass-entropy distribution, which reads:

$$\frac{\partial m(s, t)}{\partial t} = -\frac{\partial}{\partial s} \{m(s, t) \, \dot{s}\} \tag{93}$$

In this expression, the specific entropy s and the time t are considered as two independent variables.

Besides the entropy sink resulting from cooling, star formation provides a natural entropy source for the interstellar gas. Stars, end point of the entropy cascade, form in the coldest and densest gas phase, identified in the Galaxy as molecular clouds cores. No more than about 10% of the material condensed in a molecular cloud is converted into stars, which eventually destroy the cloud itself, by ionization, stellar winds or supernovae explosions. This feedback of high entropy material sustains a cycle between a hot, ionized, phase (HIM) to a dense cold and neutral phase (CNM). The mass locked by low mass star formation represents a weak mass leakage from that cycle, leading eventually to interstellar gas exhaustion. A quasi-stationary state can be reached, implying self-regulation of the star formation rate, in which the time scale of the variations of the global properties of the hot phase is governed by the star formation rate, or equivalently the gas condensation rate or the mass flux ϕ through the cycle. Defining $\psi(s,t)$ as the mass returned by stars at the specific entropy s, per unit time and per unit interval of s, the complete expression of the continuity equation for the mass-entropy spectrum is:

$$\frac{\partial m(s,t)}{\partial t} = -\frac{\partial}{\partial s}\{m(s,t)\,\dot{s}\} + \dot{\psi}(s,p,t) \tag{94}$$

Equations 93 and 94 determine completely the evolution of the thermodynamical properties of the system.

We consider now a current phase of the interstellar medium, defined as the total mass of gas with specific entropy lying in the definite, time independent range $[s_i, s_j]$. Its total mass is equal to

$$M_{[i,j]} = \int_{s_i}^{s_j} m(s,t)\, ds \tag{95}$$

and its time variation is obtained as:

$$\dot{M}_{[i,j]} = \phi(s_i,p,t) - \phi(s_j,p,t) + \int_{s_i}^{s_j} \dot{\psi}(s,p,t)\, ds \tag{96}$$

The variation of the total entropy of that phase can be calculated as:

$$\dot{S}_{[i,j]} = \int_{s_i}^{s_j} \frac{\partial m(s,t)}{\partial t}\, s\, ds \tag{97}$$

which can be expressed as:

$$\dot{S}_{[i,j]} = \phi(s_i,p,t)\, s_i - \phi(s_j,p,t)\, s_j + \int_{s_i}^{s_j} \phi(s,p,t)\, ds + \int_{s_i}^{s_j} \dot{\psi}(s,p,t)\, s\, ds \tag{98}$$

The first two terms of the r.h.s. respectively represent the entropy flowing out and flowing in the considered gas phase; the third term represents the entropy loss due to cooling, and the last term is the entropy brought by stellar ejecta.

References

[1] Balbus S.A., McKee, C.F. *ApJ* **252** (1982) 529
[2] Chandrasekhar, S. (1942) in "Principles of Stellar Dynamics", Univ. of Chicago Press
[3] Cowie L.L., McKee C.F. *ApJ* **445** (1977) 578
[4] Draine B.T. *ApJ* **241** (1980) 1021
[5] Draine B.T. *MNRAS* **220** (1986) 133
[6] Draine B.T., Roberge W.G., Dalgarno A. *ApJ* **264** (1983) 485
[7] Draine B.T., McKee C.F. *ARAA* **31** (1993) 373
[8] Flower D.R., Pineau des Forêts G., Hartquist T.W. *MNRAS* **216** (1985) 775
[9] Lioure A., Chièze J.P. *A&A* **235** (1990) 379
[10] Giuliani J.L. *ApJ* **277** (1984) 605
[11] Hénon M. (1973) in "Dynamical Structure and Evolution of Stellar Systems", eds. L. Martinet and M. Mayor, Geneva Observatory
[12] Henriksen X., Turner X. *ApJ* **445** (1977) 578
[13] Hunter G., Kuriyan M. *Proc. Roy. Soc. London* **A353** (1977) 575
[14] Landau L., Lifshitz X., 1982, "Statistical Physics" 3rd edition, Pergamon Press, Oxford
[15] Mihalas D. & Mihalas B.W. (1984) "Foundations of Radiation Hydrodynamics", New York : Oxford University Press
[16] Mullan D.J. *MNRAS* **153** (1971) 145
[17] Phelps, A.V. (1979) in "Electron-Molecule Scattering", ed. S.C. Brown, New York:Wiley, p81
[18] Reisenegger A., Miralda-Escudé, J. & Waxman, E. *ApJ* **457** (1996) L11
[19] Spitzer L. (1962) in "Physics of Fully Ionized Gases", ed. Wiley (N.Y.), 2nd rev. edition New York, Interscience Publishers
[20] Spitzer L. (1968) in "Diffuse Matter in Space", New York, Interscience Publishers
[21] Spitzer L. (1987) in "Dynamical Evolution of Globular Clusters", Princeton University Press
[22] Spitzer L., Hart, M.H. *ApJ* **164** (1971) 399
[23] Zel'dovich Ya .B., Raizer Yu. P. (1966) in "Physics of Shock Waves and High Temperature Hydrodynamics Phenomena", eds. W.D. Hayes & R.F. Probstein, New York, Academic Press

LECTURE 4

Isolated and Clustered Star Formation: Observations and Theoretical Models

F. Palla

Osservatorio Astrofisico di Arcetri
Largo E. Fermi, 5, 50125 Firenze, Italy

1. INTRODUCTION

The formation of stars occurs within the densest regions of molecular clouds. The detection of infrared point sources embedded in these dense cores shows unequivocally this fundamental property. However, star formation is not simply the result of clouds breaking apart into these dense substructures. The onset of collapse is rather a highly localized occurrence within large complexes, and the character of collapse dictates the structure of the nascent protostar. It is also true that individual collapses can occur over extensive regions of a complex which give birth to large collection of stars in multiple systems, groups, associations and clusters, in order of increasing hierarchy. The observational evidence is that most clouds are not in a state of collapse: support against gravity comes from a balance of forces that persists over long periods of time. The main support is provided partially from thermal pressure, but also from turbulent motions and the interstellar magnetic field, especially on the largest scales. Typically, the star formation efficiency, defined as the ratio of the mass in stars and the total mass (gas+stars), is of order of few percent in giant molecular clouds. In localized dense clumps, the conversion of gas into stars can be very efficient, reaching values greater than 20-30%, and gravitationally bound clusters of stars can form. The dual character of star formation that gives rise to either isolated single and binary stars or to groups and clusters has come out clearly from a large body of observations of the most spectacular nearby star forming regions.

Other lectures in this volume deal in great detail with some of the aspects outlined before. For example, the physical state of molecular clouds, their dynamical balance, the role of turbulence and magnetic fields, and the properties of dense cores are described in the lectures by E. Falgarone. Also, the basic equations and numerical methods necessary to follow the complex hydro-dynamical processes that occur during gravitational collapse and protostellar evolution are reviewed by J.P. Chièze. It is the purpose of these lectures to provide a general description of young stellar objects, both from a theoretical and observational perspective and over a wide range of stellar masses. I begin by briefly describing the modes and distribution of star formation in molecular clouds. Section 3 presents the main observational properties of young stellar objects and the characteristics of infall and outflow. The analysis of gravitational collapse of dense cores and of the protostellar accretion phase is given in Section 4, together with a discussion of the impact of the results of protostellar evolution on the following stage of pre–main-sequence contraction. The last Section focuses on the mode of star formation in clusters.

2. MODES AND DISTRIBUTION OF STAR FORMATION

The knowledge of the distribution of star forming regions within giant molecular clouds (GMCs) provides a clue to understanding the star formation mechanism. Schematically, two extreme situations have been envisaged as being the dominant modes of star formation: a *spontaneous* mode, in which the formation process occurs without external influence, and an externally *stimulated* mode. Which of the two is prevalent in molecular clouds, however, is still not well understood. Let us consider each of these schemes in some detail.

2.1. Stimulated Star Formation

The variety of physical agents that can induce star formation under external influences has been thoroughly presented by Elmegreen [19]. The basic idea is that external forces, represented by the action of stellar winds, ionization fronts, and supernova explosions, result in a compression of pre-exisiting clouds, altering their stable configuration and making them suitable for gravitational collapse. The propagation of shock fronts raises the gas density and enhances the local rate of turbulent, magnetic, or rotational energy dissipation, leaving the gas in a state close to virialization. Clouds can then collapse, and, because of the higher initial density, they do so on a faster time scale. On the other hand, the presence of strong shocks also influences the gas bulk motions, enhancing the kinetic energy of the clouds. Their overall stability against collapse is thus favored, and in some cases, shocks and ionization fronts can even induce the evaporation of a large fraction of the original interstellar material. Numerical simulations of the interaction of ionized radiation and expanding HII regions with globules have shown the twofold effects on the interstellar condensations:

an increase both in the *mass loss* rate, by evaporation of the neutral matter from the surfaces, and in the *mass accretion* rate, by compression of the gas to higher densities ([31] [9]).

From an observational viewpoint, the outcome of the triggered mode is that star formation should take place predominantly near sites of previous generations of massive stars, and not throughout a GMC. The best example of such a situation is provided by the linear sequences of stellar subgroups in OB associations [10]. But stimulated star formation can also affect the large scale distribution of the interstellar gas, creating holes, shells and bubbles. On an even larger scale, strong dynamical effects can induce the formation of galactic fountains and shape the structure of a galaxy as a whole. More convincing examples of the *small scale* range of induced star formation are the bright rimmed globules ([72]). These globules have the same physical characteristics as the quiescent dense cores, but there is evidence that the embedded IRAS point sources could represent stars of intermediate mass. Thus, more massive stars are produced by this process than would otherwise if the clouds were in isolation.

2.2. Spontaneous Star Formation

According to this idea, the evolution of an individual star forming cloud is determined solely by the balance of the destabilizing effects of self-gravity and the restoring forces due to thermal pressure, plus turbulence, rotation and magnetic fields. Each of these energy reservoirs has to be dissipated *before* gravitational collapse can occur. Thus, the typical timescale for star formation is not governed by the gravitational free-fall time, that at the typical densities of molecular clouds is of order 10^{5-6} yr, but by the dissipation time, that can amount to $\sim 10^7$ yr. The physical processes that govern spontaneous star formation are conducive of a situation where a dense core gradually builds up inside the larger molecular cloud, presumably via the action of ambipolar diffusion [43]. This stage then corresponds to the first phase of the evolutionary scenario outlined above, and represents the initial conditions for the actual phase of gravitational collapse.

Observationally, the spontaneous mode predicts a more or less uniform spatial distribution of stars throughout a GMC complex. Examples of the location of young stars in several regions were first discussed by Larson [40], who noted that while in Taurus most emission line stars are scattered through the cloud (and similarly in the streamers of Ophiucus), the opposite is true in Orion: the majority of the stars is in large clusters, with high degree of central condensations. The two situations shall be referred to in the following as the *isolated* and *clustered* modes of star formation. The data used originally by Larson were representative of limited portions of the molecular clouds in Orion, and definite conclusions about the nature of star formation could not be drawn at that time. The results of recent unbiased, systematic surveys in individual molecular clouds in the Orion complex have revealed that the clustered mode is indeed prevalent (cf. Lada et al.[37]). Similar results have been found in several other complexes via near-infrared imaging which has revealed large populations of low-mass stars always associated with the luminous OB stars. If these initial results represent the actual conditions in an average GMC, then the clustered mode should account for the formation of the bulk of the stars in our Galaxy, independent of their mass. The problem of star formation is thus shifted to the question of how massive cores form in the first place, and produce rich clusters of stars.

In the rest of these lectures, I will concentrate on the properties of spontaneous star formation for which observations are most circumstantial and a consistent theoretical framework has been developed and tested. The next Section describes the results of the observational studies of young stellar objects (YSOs) both in low-mass and high-mass star forming regions.

3.   THE QUEST FOR LOW-MASS PROTOSTARS

3.1.   Evolutionary Properties of YSOs

Studies of the *shape* of the spectral energy distribution (SED) of YSOs have proved very useful to determine their nature and evolutionary state. Since the emitted spectrum of a YSO depends on the distribution and physical properties (density, temperature and composition) of the surrounding dust and gas, it is natural to expect a dependence of its shape on the evolutionary state of the exciting source. A protostellar object deeply embedded in the parent cloud should has an infrared signature markedly different from that of a mature PMS star, for which most of the circumstellar material has been accumulated onto the central object.

Various ways of characterizing YSOs have been devised in recent years. The most successful classification scheme has been developed by Lada and collaborators based on the slope of the SED longward of 2.2 μm ([35] [80]). For each SED, a spectral index $\alpha \equiv -d\log(\lambda F_\lambda)/d\log\lambda$, with F_λ the flux density at wavelength λ, is computed between λ=2.2 and 10−25 μm, and the resulting morphological classes can be broadly classified in three distinct groups. Each class in turn corresponds to a well defined evolutionary stage, according to current models of protostellar and early stellar evolution (e.g. [69]). Class I sources have $\alpha > 0$, indicating a rise in the SED all the way up to $\lambda \sim$100 μm. The IR excess is very conspicuous and the SED is much broader than that of a single temperature blackbody function. From the evolutionary viewpoint, Class I sources are thought to represent accreting protostars, surrounded by luminous disks, with radii of $\sim 100 - 1000$AU, and by infalling, extended envelopes with sizes of $\sim 10^4$AU.

Class II sources have $-2 < \alpha < 0$. Their SEDs fall towards longer wavelengths, but are still broad, due to a significant amount of circumstellar dust. A class II source results from the clearing of the circumstellar envelope, due to the action of a powerful stellar wind, and corresponds to a classical T Tauri or Herbig Ae/Be star surrounded by a spatially thin, optically thick circumstellar disk of radius ~ 100AU. Finally, class III sources have $\alpha < -2$, and the SED resembles that of a normal, reddened stellar photosphere. The absence of an IR excess indicates the disappearance of the circumstellar structures, disks and envelopes, and the approach to the conditions of a normal main sequence star.

In recent years, most of the observational and theoretical effort has been directed at testing the goodness of this scheme, and in particular class I sources have been studied in the hope of finding among them the elusive protostars. The reason for the lack of success in the identification of a *bona fide* protostar is now becoming more clear. It has been suggested that Class I sources are in fact more evolved objects than true protostars, having already assembled most of the final stellar mass. Support to this view comes from studies at millimeter and sub-millimeter wavelengths of deeply embedded sources. In particular, dust continuum mapping of molecular cloud cores has revealed the presence of numerous cold, centrally condensed protostellar clumps which often remain totally undetected even by IRAS ([50]; [2]). Their spectral energy distribution is blackbody-like SED at temperatures of 15-30 K, as illustrated in Figure 1 in the case of HH24 MMS in the Orion B molecular cloud. Typically, these cores have masses between 0.5 and 20 $M_{\odot}$, sizes between 0.01 and 0.1 pc, which imply extremely high number densities, 10^7-10^8 cm^{-3} ([17]). In some cases, they are associated with embedded YSOs and have been named *Class 0* sources, to distinguish them from the more evolved Class I objects ([2]). In a Class 0 source is the central hydrostatic object has already formed, but not yet accreted the majority of its final mass, most of the mass still residing in the circumstellar envelope/disk. Although unbiased surveys have not yet been carried out, the small number of Class 0 so far identified in regions such as ρ Ophiuchi indicates that they have a very short lifetime. André and collaborators estimate that their probable age is $\approx 10^4$ yr, i.e. an order of magnitude shorter than the typical Class I sources. Since star formation in a cloud lasts for several million years, one would expect to find a fraction of Class 0 objects in a young cluster of only 1 percent (cf. [5]), thus explaining the relative paucity.

3.2.   Kinematic Signatures of Collapse

Continuum observations in the far-infrared and submillimeter have identified the best protostellar candidates in nearby star forming regions. However, this information alone is not enough to directly show that the sources are indeed undergoing gravitational collapse. In order to demonstrate that, one has to find kinematic signatures of the infalling gas. This requires high angular and spectral resolution observations with millimeter interferometers which can reach spatial resolutions $\lesssim 0.02$ pc in regions such as Taurus, Ophiucus, and Chamaeleon. At such distance from a protostar of ~ 1 $M_{\odot}$, the free-fall velocity of the gas is about 0.5 km s^{-1}, greater than the sound speed of 0.2-0.3 km s^{-1}

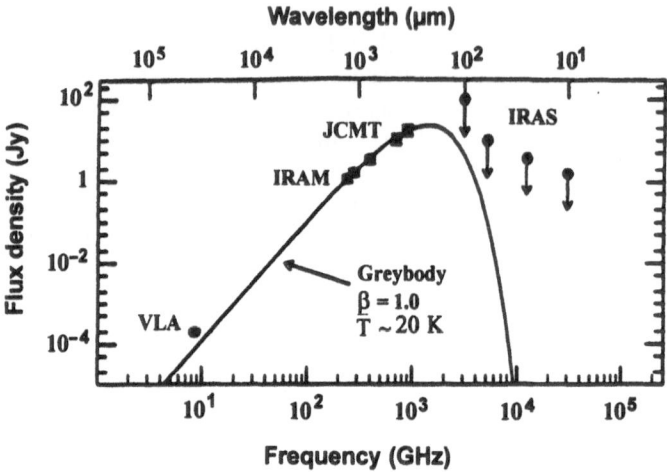

Fig. 1. — *Spectral Energy Distribution of the Class O source HH24 MMS in Orion B ([78]).*

(for $T_g \sim 10\text{-}20$ K), and comparable to the typical line width observed in cold cores. Thus, the profile of a spectral line is expected to contain direct evidence of the infall motions. Although straightforward in principle, the observational searches for such signatures have yielded controversial results for many years due to the non-uniqueness of the interpretation. The situation has greatly improved recently, and clear examples of infall have been reported in the case of several sources via observations of CS, H_2CO and HCO^+.

The first indication of collapse is the presence of an *asymmetry* in the line profile with stronger blue-shifted emission and red-shifted self-absorption. This asymmetry was predicted by radiative transfer models of collapsing clouds (e.g. [42]; see also the review by [83]) and can be qualitatively understood in the following way. If the gas temperature increases towards the center of a collapsing cloud, as expected from heating by the central protostar, the external colder layers of gas will appear in absorption against the bright core interior. Any infalling motion in the layers will distinctively shift the absorption feature towards the red producing an asymmetry in the line profile. Thus, the conditions necessary for the asymmetry to indicate infall are the increase of both the excitation temperature and velocity towards the center of the cloud.

The presence of self-absorption alone is not sufficient to prove infall, since several other effects can give rise to the same feature. Among others, fortuitous interlopers along the line of sight whose blue components happen to be stronger; anomalous gas motions, like cloud rotation and, more likely, an outflow source with a stronger blue lobe. In order to eliminate these uncertainties, one can observe a second line (either an isotopic line or a high excitation line) which is optically thin. In such a case, the line profile should be symmetric and peaked at the absorption dip of the optically thick line. Such a situation is observed in Figure 2 which shows the spectra toward B335, an isolated globule at a distance of 250 pc associated with a far-infrared source, in various transitions of CS and H_2CO. Note how the CS J=5-4 line which has the highest critical density does not show the deep self-absorption from the overlying static envelope. In addition, higher J lines are broader than lower J lines, an indication that the velocity increases toward the center. Such an increase is predicted in all theoretical models of collapse and the dashed line shown in Figure 2 is the result of a fit using such models. In addition to the kinematics, these fits allow to determine important physical quantities such as the infalling radius (~ 0.02 pc), the infall age ($\sim 10^5$yr), and the mass accretion rate (few$\times 10^{-6}$ $M_\odot$ yr^{-1}). Similar results have been already obtained in several other clouds, IRAS 16293-2422 ([76]), L1527 and CB54 ([84]), and NGC1333-IRAS2 ([79]). A systematic study of 26 Class 0 sources observed in two transitions of HCO$^+$ by [24] has led to the identification of six other sources with the correct blue spectral asymmetry in both lines. It is natural to predict that the number of real protostellar sources will increase substantially in the next few years.

3.3. Outflows and Jets

If gravitational infall is hard to find observationally, mass ejections from YSOs are readily detected in almost *all* sources [3]. Molecular outflows are characterized by the presence close to the star of high velocity gas seen in various molecular lines, including ^{12}CO, HCO$^+$, NH$_3$ and SiO. The CO spectra present the usual Gaussian shape plus high velocity emission (up to 100 km s^{-1}, i.e. much larger than the escape velocities), which is generally faint (from 0.1 to 5 K). The high velocity gas is optically thin with a ^{12}CO/^{13}CO intensity ratio greater, by a factor about 10, than in static clouds. Since radio observations only provide the radial velocity of the gas, it is difficult to determine its kinematics. But, given the high velocities involved, rotation and gravitational collapse can be excluded, since they would require implausibly massive central stars. In addition, direct evidence for outflowing material comes from the proper motion of the Herbig-Haro objects associated with them, from the presence of optical and molecular jets, and from the bipolar morphology of the blue-shifted and red-shifted gas, suggestive of motion in opposite direction from the central star.

The physical parameters that characterize the majority of molecular outflows are: R=0.1-0.2 pc; V_{CO} ~4-60 km s^{-1}; t_{dyn} ~0.6-40$\times 10^4$ yr; molecular mass M_{CO}=0.1-100 $M_\odot$ (e.g. [20]). The large masses and velocities of the outflows indicate that they are very energetic phenomena. Also, since $M_{CO} \gg M_*$, most of the flow is ambient material entrained by the flow.

The momentum and kinetic energy of the outflows vary in the range $P_{CO} = M_{CO}V_{CO} \sim 1 - 100$ $M_\odot$ km s^{-1} and $\mathcal{T}_{CO} = 1/2M_{CO}V_{CO}^2 \sim 5 \times 10^{44} - 10^{47}$

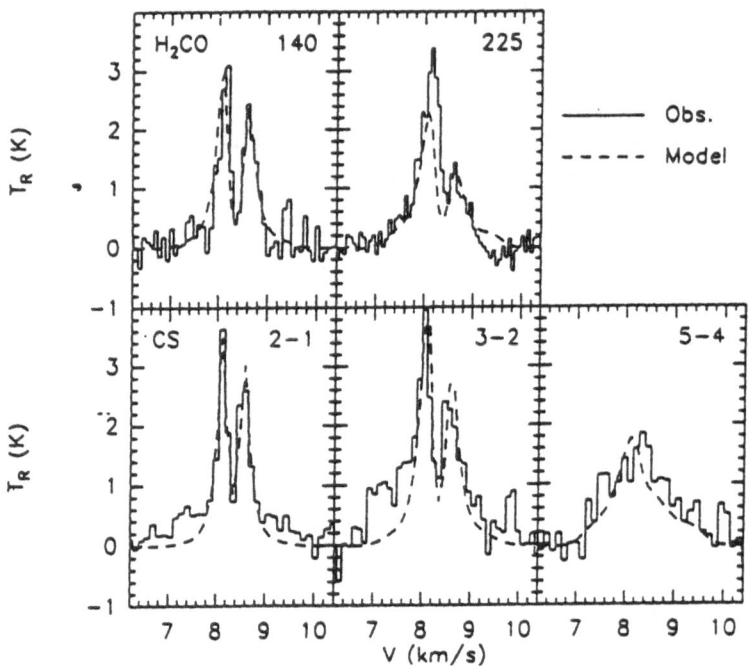

Fig. 2. — *Evidence for infall in B335 as shown by the asymmetric line profiles with the blue peak brighter than the red peak both in H_2CO and CS transitions. Note how the self-absorption disappears for highest J transition of CS ([83]).*

erg. The latter is comparable to the total kinetic energy of the clumps within molecular clouds, associated with the nonthermal motions of the gas, $T_{cl} = M_{cl}\sigma^2$ with $M_{cl} \sim 10 - 10^3 M_\odot$ and $\sigma \sim 1$km s^{-1}. Thus, outflows can provide turbulent energy to support molecular clouds ([55]; [48]).

A good correlation exists between the force required to drive the flow and the bolometric luminosity of the central source that holds over a large luminosity interval ($L_{bol} \sim 1$ to $10^5 L_\odot$). Typically,

$$log \left(\frac{F_{CO}}{M_\odot \, km \, s^{-1} \, yr^{-1}} \right) = -5.0 + 0.8 \, log \left(\frac{L_{bol}}{L_\odot} \right) . \qquad (1)$$

Since the slope of the correlation is near unity, there is a linear relationship between F_{CO} and L_{bol}. This in turn indicates that the outflow efficiency, defined as the mechanical to the radiative momentum flux ratio, $F_{CO}/(L_{bol}/c)$, is of order 100 (e.g. [14]). These large values exclude any mass-loss mechanism similar to those that occur in main sequence stars (radiation pressure, coronal ejections, etc.). Recently, Bontemps et al. [12] have shown that this ratio is much higher, $\sim$1000 on average, for Class 0 stars, a result that does not depend

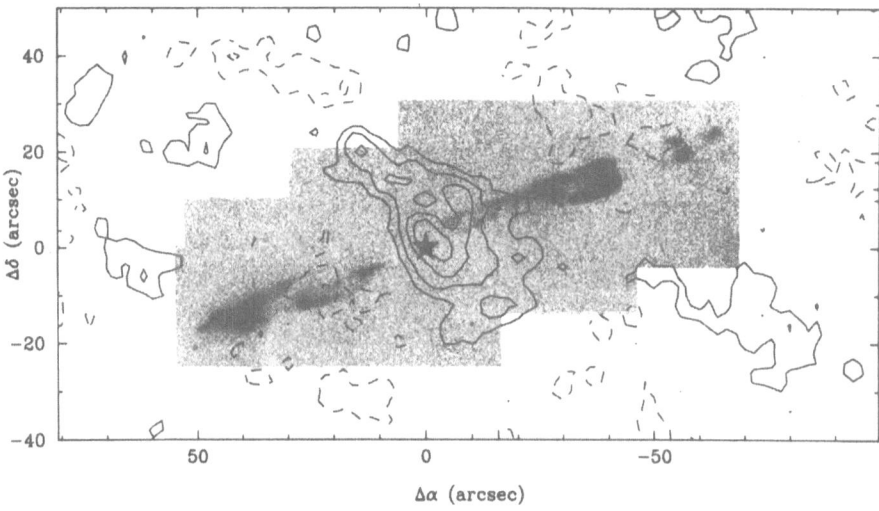

Fig. 3. — *The spectacular jet HH 212 seen in the H_2 $v = 1 - 0S(1)$ line at 2.12 μm. The solid contours show the emission observed in $NH_3(1,1)$. The star marks the position of the jet source HH 212-mm ([3]).*

on inclination effects (large viewing angles would underestimate L_{bol}). Such a decrease in the outflow efficiency is suggestive of an evolutionary effect whereby the outflow activity declines during the main accretion phase going from Class 0 to Class I sources.

Kinematic and dynamical models show that the physical mechanism for the origin and maintenance of outflows is the presence of a collimated jet-like component, possibly linked to the optical jet observed in most YSOs. The momentum associated with the optical jet itself is not enough to drive the outflow, but the neutral component with velocities up to $\sim$100 km s^{-1} can actually accomplish the task (cf. [44]; [45]). These high velocity jet components have been detected both in atomic (HI) and molecular species (H_2, HCO$^+$, SiO), and are characterized by the presence of warm gas (T$\sim$25-50 K) and small, bullet-like clumps (M$\sim$10^{-4}-10^{-3} M$_\odot$). These jets are highly supersonic with Mach numbers >20 and very well collimated, at least several hundred AU from the source. Often they consist of a series of quasi-periodically separated knots (internal *working surfaces*, e.g. [63]) and their length can vary from a few times 10^2 AU to several tenths of a parsec (even bigger structures of about 1.5 pc have been found in the giant HH34 complex, see [4]). Particularly interesting are the jets observed in H_2, the most dramatic example being HH212, a symmetric two-sided jet excited by a low-luminosity protostellar source in Orion

([47]). The H_2 emission does not trace the bulk jet fluid itself, but the radiative losses from post-shocked regions associated with jet propagation. The luminosity emitted in H_2 lines is often comparable to the mechanical luminosity in the CO outflow in the youngest sources, but it decreases to much smaller values ($\sim 1/20$) in more evolved YSOs, as expected if the jet propagates into less dense material.

The fact that outflows are always observed in YSOs indicates that infall and outflow are a natural outcome of gravitational collapse. Theoretical models which include the action of rotation and magnetic field can account for this fundamental character of star formation. Schematically, the central protostar provides the deep gravitational well which is the ultimate energy source for the outflow. The small angular momentum of the infalling gas is responsible for the formation of a circumstellar disk in Keplerian rotation about the center. Matter can be accreted through the disk via removal of angular momentum and a significant fraction of the gravitational energy, instead of being liberated as heat, can be converted into mechanical energy of the flow by magnetic torques exerted on the disk by the wind/jet (e.g. [32] [61]). Numerical MHD simulations of the onset and collimation of outflows from the surface of accretion disks show that the flow can either be self-collimated into stationary jet-like flows or produce episodic knots within the jet ([58]). Observational constraints on the models are provided by recent observations with the Hubble Space Telescope of a number of jets and Herbig-Haro objects that have allowed to resolve the jet in the direction perpendicular to the axis of propagation very close to the source ([64]). These observations show that the opening angles appear very large, $\sim 30°$-$60°$ near the source (few tens of AU), suggesting that the jet is poorly focussed at the origin before being collimated at larger distances (several hundred AU away).

4.   THE FORMATION OF HIGH-MASS STARS

The formation of massive stars also takes place in dense cores within molecular cloud complexes. A powerful two color selection criterion has been developed by Wood & Churchwell[82] to identify massive YSOs still embedded inside molecular clouds. These sources are characterized by the largest flux densities at 100 μm, implying that the emitting dust is quite cool ($T_d \sim 30$ K); their spectra peak at $\sim 100\mu$m and the shape does not show an appreciable variation from source to source. The population of embedded massive stars, and their associated ultracompact (UC) HII regions, have distinctive far-infrared colors which have been used to estimate their total number and distribution in the Galaxy, the timescale of the embedded phase (typically ~ 10-20% of the main-sequence lifetime), and the current rate of massive star formation ($\sim 3 \times 10^{-3}$ O stars yr^{-1}).

The study at high angular resolution of individual UC HII regions reveals complex morphologies since massive stars never appear alone. Different strategies have been followed in order to identify massive protostars in these fields. Some groups have observed the molecular environment in the close vicinity of UC HII regions ([56][57][15] with the aim of finding a protostar in the same cluster as the early type star. Other groups have instead selected their targets among water masers *without* free-free continuum emission ([16]: the association with H_2O masers guarantees the presence of high-mass stars, while the lack of free-free emission should identify the youngest stellar objects. Both approaches have been very successful in identifying *hot cores*, namely small (~ 0.1 pc), dense ($\sim 10^7$ cm^{-3}), hot (> 100 K) molecular condensations which very likely host high mass stars still too young to develop an UC HII region. A beautiful example of the complex environment of a hot core is shown in Figure 4 for the case of G31.41+0.3. The peak of the high density gas as traced by the $NH_3(4,4)$ line emission is not spatially coincident with the location of the UC HII region. Most of the activity in the field is associated with the embedded sources in the hot core, which excites both OH and H_2O maser emission, and *not* with the massive star ionizing the circumstellar gas.

4.1. The Evolution of Ultracompact HII Regions

The presence of an UC HII region is the earliest manifestation of massive stars of spectral type earlier than B3. Observationally, UC HII regions are small ($r \lesssim 10^{17}$ cm), dense ($n \gtrsim 10^4$ cm^{-3}), high pressure ($nT \gtrsim 10^8$ cm^{-3}K) HII regions found in about 10% of the O stars in the Galaxy (cf. [82]). This frequency of occurrence is quite high and leads to the following paradox. The lifetime of an UC HII region is approximately given by

$$t_{UC} \sim 0.1\, t_{O6} \sim 3 \times 10^5 \text{ yr}, \qquad (2)$$

where $t_{O6} = 3 \times 10^6$ yr represents the main-sequence lifetime of an O star. The dynamical time is

$$t_{dyn} = \frac{r}{a_s} \sim 3 \times 10^3 \text{ yr}, \qquad (3)$$

where a_s is the sound speed, here taken to be ~ 10 km s^{-1}. Thus, the expansion of the UC HII region should take place very quickly and very few of them should be observed at any given time.

Several explanations have been proposed to explain the long duration of the ultracompact phase. However, before discussing models in some detail, let us first see the basic fact that the UC regions cannot be classical HII regions (cf. [18]). The main reason is that the observed radii are much smaller than the typical size of a developed HII region. To see why, consider the radius of the Strömgren sphere

$$r_S^{in} = \left(\frac{3\phi}{4\pi n_o^2 \beta_2} \right)^{1/3} \sim 0.02\, \phi_{48}^{1/3}\, n_5^{-2/3} \text{ pc} \qquad (4)$$

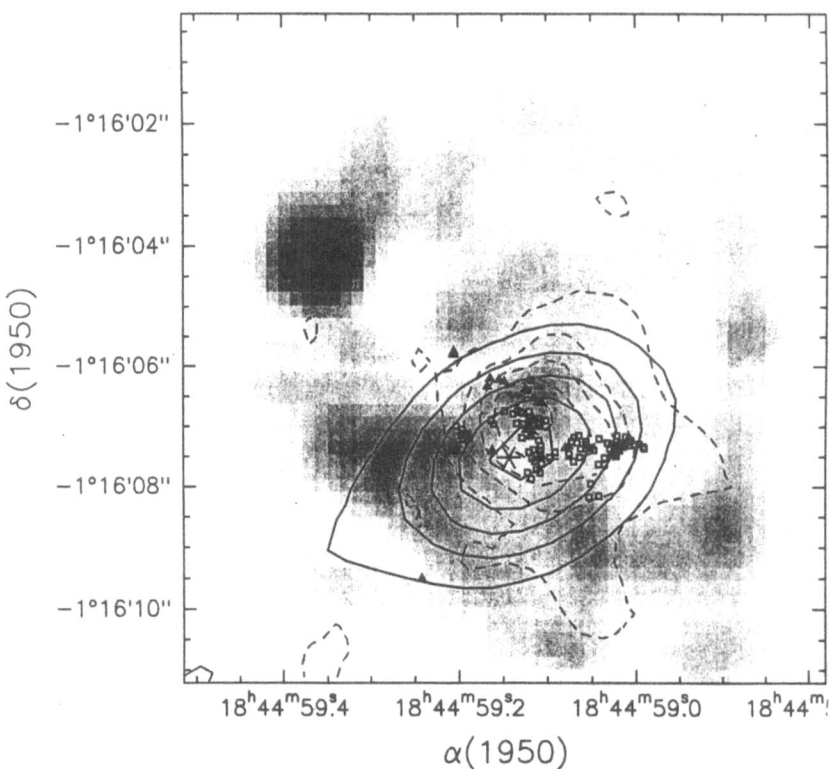

Fig. 4. — *The environment of the hot core associated with G31.41+0.3. The radio continuum emission at 1.3 cm is shown as a grey scale. The location of the UC HII region coincides with the peak in the upper left of the figure. The hot core is traced by the NH₃(4,4) (dashed contours) and CH₃CN (6,5) line emission. The asterisk at the peak of the center of the core marks the position of millimeter source. Triangles and squares show the distribution of the H₂O and OH maser spots, respectively ([16]).*

where ϕ is the stellar ionizing photon rate (s^{-1}), $\phi_{48} \equiv \phi/10^{48}$ s^{-1}, β_2 is the hydrogen recombination rate coefficient to levels $n \gtrsim 2$, n_o is the number density and $n_5 \equiv n_o/10^5$ cm^{-3}. The evolution of the Strömgren radius with time is given by

$$r_s = r_s^{\text{in}} \left(1 + \frac{7}{4}\frac{a_s t}{r_s^i}\right)^{4/7} \tag{5}$$

If we take as a lower limit to the lifetime of the ultracompact region $t = 10^5$ yr, then eq. (5) indicates that $r_s \sim 10^{18}$ cm~ 0.3 pc, a factor of at least 5 larger than the observed size of an UC HII region.

One possibility to solve the problem of the lifetimes is that the *ram pressure* from the infalling gas provides pressure to confine the gas. Although very interesting, and initially supported by the observations of OH lines which seemed to indicate collapse ([65]), this explanation however fails since the balance between ram and thermal pressure is unstable to small perturbations ([29]). A simple argument can show why this is true. The thermal pressure is given by $P_{HII} \sim n_e T_e$. Now, the ionizing flux is approximately $\phi \sim n_e^2 r^3$; since $\phi \sim$const, then $n_e \sim r_s^{-3/2}$. The electron temperature is $T_e \sim 10^4$K$\sim$ *const* and finally

$$P_{\text{th}} \propto n_e T_e \sim r^{-3/2} \qquad (6)$$

On the other hand, the dependence of the ram pressure of the infalling gas on the density can be estimated in the following way. The ram pressure is equal to $P_r = \rho V_{\text{inf}}^2$, where V_{inf} is the infall velocity. From the equation of continuity, the mass accretion rate is given by $\dot{m} = 4\pi r^2 n m_H V_{\text{inf}}$. Since $V_{\text{inf}} \propto r^{-1/2}$ and assuming $\dot{m} =$const, it follows that $n \propto r^{-3/2}$. Thus, the ram pressure is

$$P_r \propto n V_{\text{inf}}^2 \sim r^{-5/2} \qquad (7)$$

Now, we can see why a small perturbation will make the equilibrium unstable. In fact, a displacement to larger r will produce a larger decrease in P_r than in P_{th} and the HII region will expand freely. In the opposite limit, the ram pressure can overcome the thermal pressure and the infalling matter will crash the HII region. We will use this result below to derive a critical mass accretion rate for choking-off the radio emission from a massive protostar.

More realistic models have been proposed to solve the lifetime problem. Among them, the possibility that the UC HII region is formed in bow shocks caused by the stellar motion through dense molecular gas ([75]). In this model, the equilibrium is established between the pressure of the stellar wind coming from the O star and the ram pressure of the infalling gas. Unlike the previous case, such a balance is stable. This model successfully accounts for the cometary UC HII regions which represent about 20% of the Wood & Churchwell sample. However, a major limitation is that the required stellar velocities ($v_* \sim$5-10 km s^{-1}) appear uncomfortably high to explain *all* the observed ultracompact HII regions, considering that the typical velocity dispersions of few km s^{-1} observed in young stars.

Finally, an interesting model has been suggested by Hollenbach et al. [29] based on the idea of photoevaporation of accretion disks surrounding O stars. Here, the diffuse radiation field due to hydrogen recombination in the photoionized gas produces a steady evaporation at the surface of the disk, producing a *disk wind*. In this case, the lifetime of the UC HII region depends on the mass of the disk. Using typical parameters of disks, $r_d \gtrsim 10^{15}$-10^{16} cm and $M_d \sim 0.3 M_* \sim$3-10 M$_\odot$, the characteristic lifetime of the photoevaporating disk is

$$t_d = \frac{M_d}{M_*} \sim 7 \times 10^4 \, \phi_{49}^{-1/2} \left(\frac{M_*}{10 M_\odot}\right) \, yr \qquad (8)$$

Then, if $M_d \sim$2-10 $M_\odot$, $t_{UC} \gtrsim 10^5$ yr, in agreement with the observational constraints.

4.2. The Search for Massive Protostars

Interestingly, the first claim of the detection of infalling motions in a molecular core was found in the case of a massive star by Ho & Hascick [28] who observed an inverse P-Cygni profile (characterized by redshifted absorption and blueshifted emission) in HCO$^+$ towards the UC HII region W49. However, recent observations at high-angular resolution ($\sim 5''$) have shown that the velocity field is quite complex and most likely not due to gravitational infall ([81]). In general, the search is made difficult both for the fact that the UC HII regions are usually distant (typically more than 5 kpc), and for the presence of high-velocity molecular outflows. A statistical study of 122 high-mass star forming regions made by Shepherd & Churchwell [67] has shown that indeed outflows are a common property of newly formed massive stars, similar to the case for low-mass stars.

The ideal massive protostar is a source that has high infrared luminosity as measured by IRAS (in excess of $\sim$10^4 $L_\odot$), is associated with high density molecular gas (as traced by NH$_3$ observations), but does not have radio continuum emission detectable with the VLA interferometer. In other words, one should look for a precursor to an UC HII region that has not had time to develop an ionized Strömgren sphere, even though its mass is appropriate. Such examples have been found in recent surveys ([52], [15]) and raise the following interesting theoretical problem: can accretion stop the expansion of the HII region? And if so, what is the required mass accretion rate?

Following Walmsley [77], the minimum accretion rate required to choke-off the expanding HII region is given by

$$\dot{M}_{cr} = \left(4 \pi N_{Lyc} G M_* m_H^2 \beta_2\right)^{1/2} \quad M_\odot yr^{-1} \qquad (9)$$

where N_{Lyc} is the Lyman continuum photon luminosity, and β_2 has been already defined. Values of $\dot{M}_{cr}$ are given in Table I for three cases. Also listed is the accretion luminosity ($L_{acc} = G M_* \dot{M}/R_*$), computed for two values of the mass accretion rate, $\dot{M}_{acc} = 10^{-4}$ and 10^{-5} $M_\odot$ yr^{-1} respectively. One can immediately see that, contrary to expectation, a modest mass accretion can choke-off the HII region: values of few$\times 10^{-6}$ $M_\odot$ yr^{-1} are considered lower limits to the actual accretion rates in low-mass stars; thus, high-mass stars should have even higher accretion rates which can therefore easily reverse the expansion of the ionized region, making the sources undetectable in the radio. Note also in Table I that stars more massive than $\sim$ 15 $M_\odot$ have always $L_* \gg L_{acc}$, for reasonable values of $\dot{M}_{acc}$. Thus, massive YSOs may still be

Table I. — Critical mass accretion rates for massive stars

Sp. Type	M_* $(M_\odot)$	L_* $(L_\odot)$	$\log N_{Lyc}$ (s^{-1})	$\dot{M}_{cr}$ $(M_\odot \ yr^{-1})$	$L_{acc}(-5)$ $(L_\odot)$	$L_{acc}(-4)$ $(L_\odot)$
O5	60	8×10^5	49.7	1×10^{-4}	2×10^3	1.5×10^4
O8	23	1×10^5	48.6	1×10^{-5}	8×10^2	8×10^3
B0	18	5×10^4	47.6	4×10^{-6}	7×10^2	7×10^3
B2	10	5×10^3	43.6	6×10^{-8}	5×10^2	5×10^3

in the accretion phase even though the underlying source is already a main-sequence star.

5. ISOLATED STAR FORMATION

The formation of a protostar from the quiescent conditions typical of dense cores occurs through the gradual accumulation of the interstellar gas onto an accreting core. This process requires a large decrease in gravitational potential energy. A large fraction of this energy is radiated away in radial or disk accretion shocks that form as a result of the abrupt change in the velocity of the freely-falling gas. If no net energy is absorbed by the circumstellar material, the resulting luminosity is given to a good approximation by

$$L_{proto} = \frac{GM_* \dot{M}_{acc}}{R_*} , \tag{10}$$

where M_* and R_* are the instantaneous mass and radius of the protostellar core, and $\dot{M}_{acc} = dM_*/dt$ the mass accretion rate. Thus, estimates of the luminosity emitted during this phase rely on the knowledge of two fundamental quantities: *the mass accretion rate*, and the *mass-radius relation*. The former is determined by the dynamics of the gravitational collapse, while the relation between the mass and the radius are established by processes occuring in the protostellar interior. The radiation produced at the shock is absorbed, reradiated and thermalized in the optically thick dusty infalling envelope. Most of the observable radiation is emitted at mid- and far-infrared wavelengths. The exact shape of the emergent spectrum depends on the density and temperature distribution of the dust component, which are set by the dynamics of collapse. This section describes how current models of protostellar formation and evolution have constrained both the value of the mass accretion rate, and the structural properties of the central core. Some relevant observational tests will be presented at the end.

5.1. Protostellar Collapse: the Mass Accretion Rate

An estimate of $\dot{M}_{acc}$ requires the solution of the dynamical collapse problem. In the idealized case of the collapse of a marginally unstable cloud, such a solution has been found semi-analytically by Shu[68] . In this theory, the rate at which a protostar is built up is

$$\dot{M}_{acc} = \alpha \, \frac{a_{eff}^3}{G} \, . \tag{11}$$

In eq. (11), a_{eff} is the effective isothermal sound speed, G the gravitational constant, and α a constant of order unity, If the cloud is supported only by thermal pressure, the expression for the sound speed is simply $a_{eff} = \sqrt{(kT/m)}$, where T is the gas kinetic temperature. Otherwise, a_{eff} should also include the contribution of magnetic and turbulent pressures. In the conditions found in dense cores with very little nonthermal support, the gas kinetic temperature varies between 10 and 30 K, corresponding to $a_{eff} = 0.1 - 0.3$ km s^{-1}. This yields an average accretion rate of order $\dot{M}_{acc} \sim$ several$\times 10^{-6} - 10^{-5}$ $M_\odot$yr^{-1}. Then, for (sub)solar protostars, the duration of the embedded phase is expected to last typically $t_{acc} = M_*/\dot{M} \gtrsim 10^5$yr. The remarkable property of the form of $\dot{M}_{acc}$ in eq. (11) is that it depends only on one parameter, the sound speed, which is controlled by the conditions in the parent cloud, and not by the properties of the central protostar. Eq. (11) also implies that the higher the temperature of the cloud, the faster the accretion, a result that goes against physical intuition. This situation is in fact radically different from the more familiar case of the Bondi-type solution of the accretion problem, in which the mass accretion rate does depend on the mass of the star, and is inversely proportional to the velocity (and hence the temperature) of the gas. In Shu's solution, a faster accretion rate merely implies higher initial density for the onset of gravitational collapse. The predictions of the *inside-out* collapse model, and the estimate of $\dot{M}_{acc}$, have been extensively used as a diagnostic tool in the interpretation of molecular cloud observations. In the following, I will give a brief account of its derivation, in the light of the results of numerical simulations of the collapse phase.

5.1.1. Classical Collapse Models

The onset of the dynamical collapse of a self-gravitating cloud requires the specification of the initial conditions. In the first numerical models (e.g. [38]), these were assumed to correspond to the physical conditions in interstellar clouds at the end of the fragmentation phase, a process by which a massive cloud breaks up into smaller and smaller subunits, clumps and fragments. Considering thermal and gravitational effects alone, the requirement that the clumps be self-gravitating leads to the determination of a critical mass, the well-known Jeans criterion. For an isothermal, non-magnetic, non-rotating sphere composed primarily of molecular hydrogen, neutral helium and heavy metals in solar proportions, the critical wavelength of a perturbation is given by

$$\lambda_J = \left(\frac{\pi a_s^2}{G\rho}\right)^{1/2} = 0.2 \text{ pc} \left(\frac{T}{10 \text{ K}}\right)^{1/2} \left(\frac{n_{H_2}}{10^4 \text{ cm}^{-3}}\right)^{-1/2} \quad (12)$$

where a_s is the sound speed, T the gas temperature, ρ and n the mass density and the total number of particles per unit volume. Those perturbations with wavelength exceeding this *Jeans length* have exponentially growing amplitudes. Written in terms of the mass enclosed by λ_J, the Jeans criterion becomes

$$M_J = \frac{m_1 a_T^3}{\rho^{1/2} G^{3/2}} = 1.6 M_\odot \left(\frac{T}{10K}\right)^{3/2} \left(\frac{n_{H_2}}{10^4 \text{ cm}^{-3}}\right)^{-1/2} \quad (13)$$

The numerical factors of the equations (12) (13) show that typical dense cores are close to the edge of gravitational instability. In fact, those with internal stars have already crossed this threshold, and at least some of these ought to show signs of ongoing collapse. Conversely, any structures with measured masses substantially less than M_J or sizes less than λ_J could be stable, given sufficient external pressure, or they could be temporary configurations of positive total energy that will soon cool or disperse. Thus, Larson suggested that protostellar clouds should have initially an almost uniform density, and that the initial density would be close to the critical one.

Starting with these initial conditions, the numerical simulations of protostellar collapse have identified three main stages: an initial *isothermal* contraction, during which the gravitational potential energy is effectively radiated away by the dust and the collapse proceeds at constant temperature; an *adiabatic* phase, corresponding to the transition from low to high optical depth, as a result of the increased density; and, finally, the formation of an *hydrostatic core*, that grows in mass as the collapse proceeds. The basic physical processes that govern each of these phases have been extensively reviewed and need not be repeated here (see [11]). Suffice to mention only the most general ones, and underline the important fact that they do not depend on the restrictive assumptions under which the numerical calculations were carried out originally (mainly, spherical symmetry). More realistic, and complex, 2- and 3-dimensional collapse models with less stringent geometry constraints have been developed in the meantime (e.g. [13]), but the main results of the simple case can still be used as useful guidelines.

Gravitational collapse occurs in a highly *non-homologous* way. The central regions of the cloud collapse faster than the outer parts, pressure gradients develop fast, and the non-homology sets in already during the initial isothermal phase. Independent of the initial conditions, whether uniform or centrally condensed, the matter develops a density distribution of the type $\rho \propto r^{-2}$, where r is the radial distance from the cloud center. In a time slightly longer than the free-fall time, the structure of the cloud resembles that of an asymptotically singular isothermal sphere, modified at the center for the presence of a stable hydrostatic core, which contains only few percent of the total cloud mass. The core accretes matter from a freely-falling inner envelope, characterized by a density and velocity profile of the type $\rho \propto r^{-3/2}$ and $v \propto r^{-1/2}$. The outer parts of the clouds are in nearly static equilibrium.

On the basis of these results, Shu has shown that the singular isothermal sphere,

$$\rho = \frac{a_{\mathrm{eff}}^2}{2\pi G r^2} , \tag{14}$$

has a self-similar analytic solution that reproduces well the dynamical collapse model. Since a power-law of this type does not have a typical scale associated with it, the solution does not define a characteristic mass scale to be associated with the formation of an ordinary star. What the self-similar solution defines is the *rate* at which the central star is built up by accretion. It is exactly this rate that appears in eq. (11). The collapse is said to proceed from *inside-out*, since the inner parts collapse first, and the outer parts later on after being reached by an expansion wave that propagates outward at the sound speed, a_{eff}.

5.1.2. More Realistic Models

The assumptions that enter in the derivation of $\dot{M}_{\mathrm{acc}}$ represent an oversimplification of the realistic case where forces other than thermal compete in maintaining the stability of molecular clouds. It is clear that molecular clouds cannot be collapsing as a whole on a dynamical time scale, as the Jeans criterion would predict, or the resulting star formation rate in the Galaxy would be far too high. Thus, the concept of the Jeans mass appears of little use in the definition of the appropriate initial conditions, and eq. (13) should be replaced by a more realistic one. Among the various mechanisms of cloud support, magnetic fields are the most likely to play the dominant role. A magnetic field of strength B can support a molecular cloud of radius R, provided its mass is less than a magnetic critical mass

$$M_{\mathrm{cr}}^B = 0.12 \frac{\Phi}{\sqrt{G}} \simeq 10^3 \left(\frac{B}{30\mu G}\right)\left(\frac{R}{2pc}\right)^2 M_\odot , \tag{15}$$

where Φ is the magnetic flux through the cloud ([53]). Depending on the cloud mass, M_{cl}, two situations can be envisaged. Supercritical clouds ($M_{\mathrm{cl}} \gg M_{\mathrm{cr}}$) cannot be supported by magnetic fields alone even if they were perfectly frozen in the gas (no magnetic dissipation), and they would collapse on a magnetic diluted free-fall time scale. Subcritical clouds ($M_{\mathrm{cl}} \ll M_{\mathrm{cr}}$) are supported by the magnetic field against collapse (even if the external pressure increases) and evolve on the timescale that characterizes the diffusion of the magnetic field. Observational evidence that magnetic fields of the magnitude required by eq. (15) are present in dense cores and in molecular cloud complexes has been obtained by Zeeman splitting measurements ([26]).

At the scale of the dense cores, the value of M_{cr}^B is of the same order as M_J, namely a few solar masses. The main difference between the two cases is in the time scale of evolution. In dense cores, the gas is lightly ionized (the fractional ionization is about 10^{-7}), and the relevant mechanism of magnetic diffusion is due to *ambipolar diffusion* ([49]), a process in which the fluid of charged particles can slowly drift with respect to the fluids of neutral particles, the two fluid being coupled by collisions of atomic and molecular species. Studies of the quasi-static evolution of molecular cores have shown that the configuration tends to acquire the density profile of the singular isothermal sphere at the time the central regions become gravitationally unstable and undergo dynamical collapse ([43]). Indeed, observations towards dense cores in the Taurus-Auriga region have revealed a high degree of central condensations. Collapse calculations starting from these new initial conditions have been presented by Galli & Shu ([21],[22]), who considered in detail the effects of magnetic field and plasma drift. The main result is that the collapse still takes place in an inside-out fashion, but now propagating outward as a fast magnetosonic wave, traveling faster in the direction parallel to the initial magnetic field. Surprisingly, the value of $\dot{M}_{acc}$ obtained in the non-magnetic case is *not* significantly altered by the presence of a magnetic field. The reason is that the gas is slowed down, because of the Lorentz force opposing gravity, but the information travels outward faster, because of the increased characteristic speed of the system. The two effects cancel out, and the numerical value of $\dot{M}_{acc}$ remains unchanged.

However, the dynamics of the gas is significantly affected by the magnetic field. The strong pinching Lorentz force deflects the infalling gas toward the equatorial plane, with the formation of an inner disequilibrium structure. The disk instantaneous radius (in AU) is given by

$$r_B = 0.12 \left(\frac{G^2 B^4}{a_{\text{eff}}} \right)^{1/3} t^{7/3} \simeq 600 \left(\frac{B_0}{30\mu G} \right)^{4/3} \left(\frac{a_{\text{eff}}}{0.35 \text{kms}^{-1}} \right)^{-1/3} \left(\frac{t}{10^5 \text{yr}} \right)^{7/3}.$$
(16)

Disk-like structures over scales of several hundred AU have been mapped recently in molecular line emission around several YSOs ([6]). The scale of the magnetic pseudo-disk is much larger than that of the centrifugal disk, that forms as a result of the effects of the initial small, but not negligible rotation of the molecular cloud. The value of the centrifugal radius (in AU) is estimated to be ([73])

$$r_C = 0.06 \left(a_{\text{eff}} \Omega^2 \right) t^3 \simeq 7 \left(\frac{a_{\text{eff}}}{0.35 \text{kms}^{-1}} \right) \left(\frac{\Omega}{4 \times 10^{-14} \text{s}^{-1}} \right)^2 \left(\frac{t}{10^5 \text{yr}} \right)^3, \quad (17)$$

where Ω is the initial rotation rate of the core. Inside r_C, infalling matter encounters a centrifugal barrier in the equatorial plane, and accumulates in a disk. The various elements that characterize the collapse of magnetized, rotating molecular cores over a range of different scales are schematically shown in Figure 6.

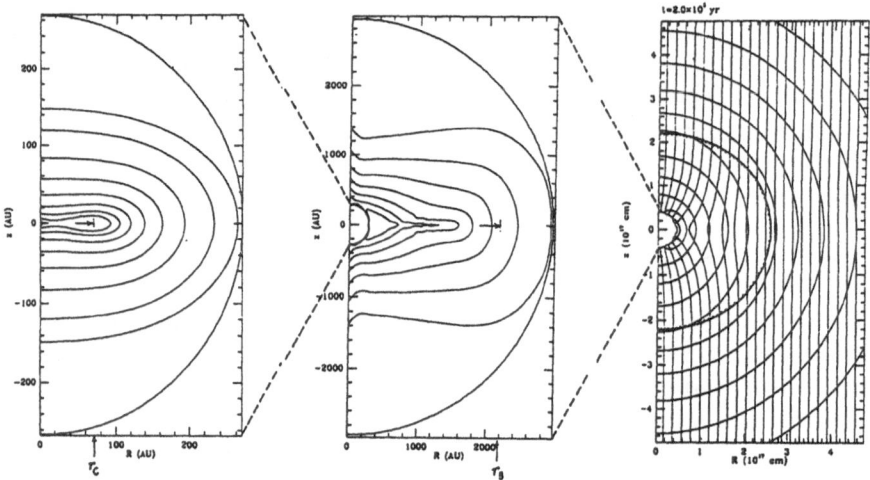

Fig. 5. — *Global view of the collapse of a magnetized, rotating dense core. The right panel shows the isodensity contours and magnetic field lines over a scale of about 0.1 pc. The dotted line represents the infall magnetosonic wave that propagates faster in the direction perpendicular to the field. In the middle panel, on scales of $\sim 10^3$ AU, the dynamics is dominated by magnetic and gravitational forces that shape the density distribution in the equatorial plane. The left panel shows the isodensity contours in the innermost $\sim 10^2$ AU. The positions of r_B and r_C (see text) are indicated in the middle and left panels ([21])*

5.2. Protostellar Evolution

Having established the magnitude of the mass accretion rate, we now discuss the properties of protostars, as they grow from the small hydrostatic core formed at the end of the adiabatic phase to a stellar sized object during the main accretion phase. The purpose here is to show how a *mass-radius* is established, that can then be used in eq. (10) to estimate the protostellar luminosity. In the inside-out collapse scenario, the structure and evolution of the protostar are solely determined by the specification of $\dot{M}_{acc}$. The evolution of the infalling envelope and the hydrostatic core are effectively decoupled, owing to the existence of a region of very low opacity (*the opacity gap*) in the outer parts of the envelope. The photons produced at the accretion shock can escape freely, instead of being pushed back into the core by the infalling matter. This results in a great simplification of the problem, that otherwise would require the joint solution of the full hydrodynamical equations for the dynamical collapse and detailed radiation transfer. Despite great efforts, fully self-consistent models do not yet exist (see Boss 1993). The quasi-hydrostatic evolution of the protostar can be treated as a problem in stellar structure, with the modifications introduced by a time varying mass and nonstandard surface boundary conditions provided by an accretion shock ([71]). These models have shown the key role of the nuclear burning of interstellar deuterium in determining the properties of the accreting core. In fact, the deuterium generated luminosity, L_D is given by

$$L_D = \dot{M}_{acc}\delta = 15L_\odot \left(\frac{\dot{M}_{acc}}{10^{-5} M_\odot \mathrm{yr}^{-1}} \right) , \qquad (18)$$

where δ is the nuclear energy per unit mass available in interstellar matter (assuming a [D/H] value of 2.5×10^{-5}). The extreme temperature sensitivity of the energy generation rate is such that deuterium acts as an effective thermostat, preventing the central temperature from rising above $T_c \sim 10^6$ K as the star gains mass. Since the amount of energy available in deuterium, δ, is comparable to the gravitational binding energy of the star, GM_*/R_*, the thermostatic effect results in a core radius that increases almost linearly with mass during the phase of active burning.

The protostellar evolution of low- and high-mass stars has been computed by several groups ([59] [7] [8] [70]). Despite the different techniques and input physics, the general trend of the mass-radius relation is the same in all calculations and an example is shown in Figure 6.

Note the large increase in radius at a mass of approximately 2 $M_\odot$. The transition occurs when the following condition on the relevant timescales, the accretion time and the Kelvin-Helmholtz time, is satisfied

$$t_{KH} \equiv \frac{GM_*^2}{R_* L_{rad}} \approx \frac{GM_*^2}{R_* L_D} \approx \frac{GM_*^2}{R_* \dot{M}\delta} \approx \frac{M_*}{\dot{M}} \equiv t_{acc}. \qquad (19)$$

The increase of the protostellar radius has very important effects. If nothing else happened in the interior, the rapid contraction would lead to the conditions appropriate for hydrogen burning in the center, and the star would then join the main-sequence while still accreting. As a consequence, there should be no stars of mass greater than $2 - 3\ M_\odot$ in the pre–main-sequence phase ([39]). This result is in clear contradiction with the observational evidence of the Herbig Ae/Be stars, whose location in the HR diagram is well above the ZAMS. The reason why this expected behavior does not occur is subtle, and is again related to the effects of deuterium burning. In simple terms, more massive protostars experience deuterium burning in a subsurface shell that, as more mass is gathered, slowly moves toward the surface. The release of nuclear energy in these low-density regions results in a dramatic swelling of the star. The expansion of the radius counteracts the gravitational pull, and delays the beginning of the phase of gravitational contraction. The fusion of ordinary

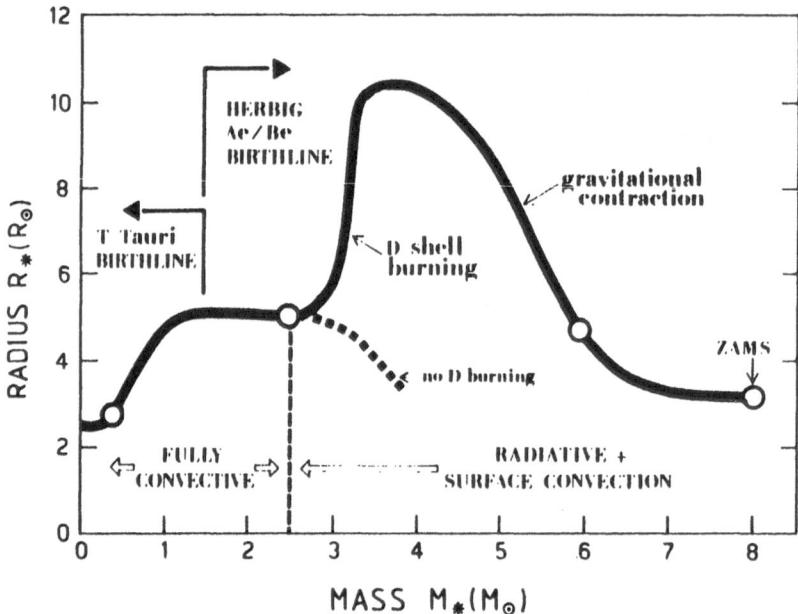

Fig. 6. — *The variation of the radius with mass in accreting protostars ([59]).*

hydrogen and the arrival on the Main Sequence are postponed to higher masses, $\simeq 8\,M_\odot$ in the specific case considered here, thus solving the inconsistency with the observations.

5.3. Early Stellar Evolution

Once the main phase of accretion is completed, the stellar core emerges as an optically visible star along the birthline. The physical process by which infall stops is still not known, although stellar winds and the associated bipolar flows are most likely to play a fundamental role. The circumstellar matter surrounding these young stars, partly distributed in a disk and the rest in an extended envelope, still emits copiously at infrared wavelengths: the emergent spectral energy distribution departs substantially from that of a normal stellar photosphere, showing "excess" emission extending from the near- to the far-infrared. The origin of this excess has been commonly attributed to the presence of accretion disks. Depending on the mass of the star, young visible stars are classified as T Tauri (low-mass, typically $\sim 1\,M_\odot$), or Herbig Ae/Be (intermediate-mass, $\lesssim 10 M_\odot$) stars.

The calculation of the buildup of a protostar to a mass M_* automatically yields the initial PMS structures for all stars of lower mass, provided only that the main accretion process ends relatively quickly (compared to the Kelvin-Helmholtz time scale). Since the contraction time is of order 10^7 yr for a solar-type star, while the typical clearing time due to a wind/outflow is 10^5 yr, this assumption is well satisfied.

The results of the calculations of PMS evolution incorporating the protostellar initial conditions are shown in the HR diagram of Figure 7a. for stars from 0.6 to 6.0 $M_\odot$. While low mass stars descend from the birthline (the starting locus from which stars begin the PMS phase) on convective paths, stars more massive than ~ 4 $M_\odot$ move to the left on radiative ones. Stars between 2 and 4 $M_\odot$ undergo thermal relaxation, looping back slightly behind the birthline and recrossing it before pursuing horizontal tracks. Compared to the classical set of PMS tracks (e.g. [30]), those shown in Figure 7a occupy a much more limited portion of the diagram. The reason is that the older tracks correspond to stellar radii too large to have attained during protostellar accretion. Thus, the starting luminosities are much lower than previously envisioned. In addition, the surface temperature of these stars begin *higher*. For example, a 5 $M_\odot$ star starts with T_{eff}=12000 K in contrast to the value of 4400 K obtained by [30]. Notice also that the birthline intersects the main sequence at a mass $M_* \sim 8$ $M_\odot$: this point represents the critical stellar configuration in which hydrogen burning has stopped gravitational contraction while the star is still growing in mass. *More massive stars, therefore, have no PMS phase at all, but appear directly on the main sequence once they are optically revealed.*

The most important result of the new calculations is that observed PMS stars should not be located to the right of the birthline. This prediction is well in accord with the distribution of T Tauri and Herbig Ae/Be stars in the HR diagram shown in Figure 7b. For both T Tauri and Herbig Ae/Be stars the agreement is excellent, despite the rather strong assumption of a constant mass accretion rate of 10^{-5} $M_\odot$ yr^{-1} used in the models. Also, the upper envelope of the stellar distribution is well matched by the theoretical birthline. This fact indicates that a limited dispersion in the properties of molecular cloud cores (which determine the magnitude of $\dot{M}_{acc}$) somehow produces an impressive range of stellar masses. The case for even more massive stars is complicated by the difficulty in determining the appropriate stellar parameters to place them in the HR diagram.

6.   COLLECTIVE STAR FORMATION

The last section deals with the collective properties of star formation, as revealed by the study of young stars and embedded star clusters. Two major results need to be emphasized at the outset: first, the majority of PMS stars form binary/multiple systems; second, most stars form in groups/clusters. The first statement finds its support from a series of studies that have been carried out in the last five years on the binary nature among T Tauri and Herbig

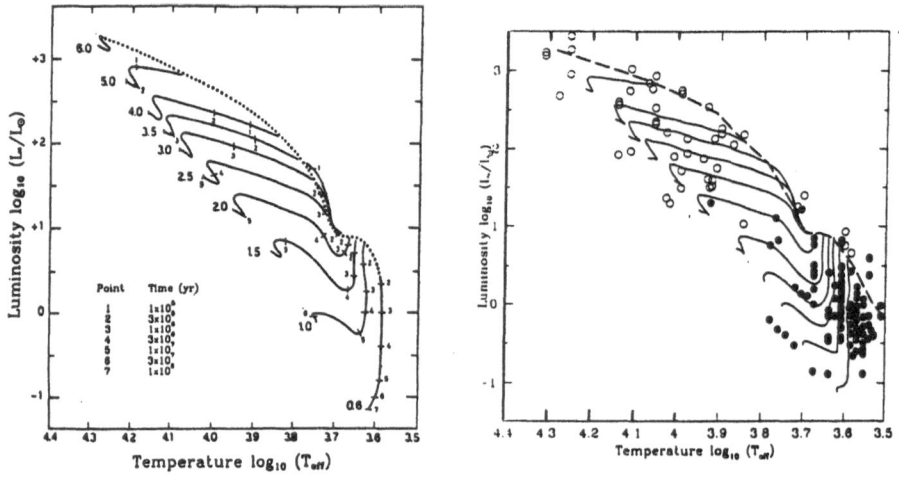

Fig. 7. — *Left Panel: Evolutionary tracks in the HR diagram for low- and intermediate-mass stars. Each track is labeled by the corresponding mass, in solar units. Tick marks indicate points in the evolution whose time is given in the table. For each track, the evolution starts at the birthline (dotted curve), and ends at the ZAMS (not shown) ([60]). Right Panel: Observed distribution of young stars in the HR diagram. Some evolutionary tracks are also shown.*

Ae/Be stars in nearby star forming regions. The topic has been covered in the excellent review by Mathieu [46]. It has been suggested that *all* stars form in binary systems. The lower frequency of main-sequence binaries is then the results of dynamical evolution during the phase of pre–main-sequence contraction. Despite the important implications of these studies on star formation, we will not discuss the properties of young binaries and mechanisms of binary formation further in this section. In the context of a School on *Starbursts*, it is more appropriate to concentrate on the mode of star formation where stars forms in groups and clusters!

Star clusters are the smallest systems in which the processes that generate a distribution of stellar masses, the Initial Mass Function (IMF), can be studied observationally. An enormous boost in this field has been provided by the

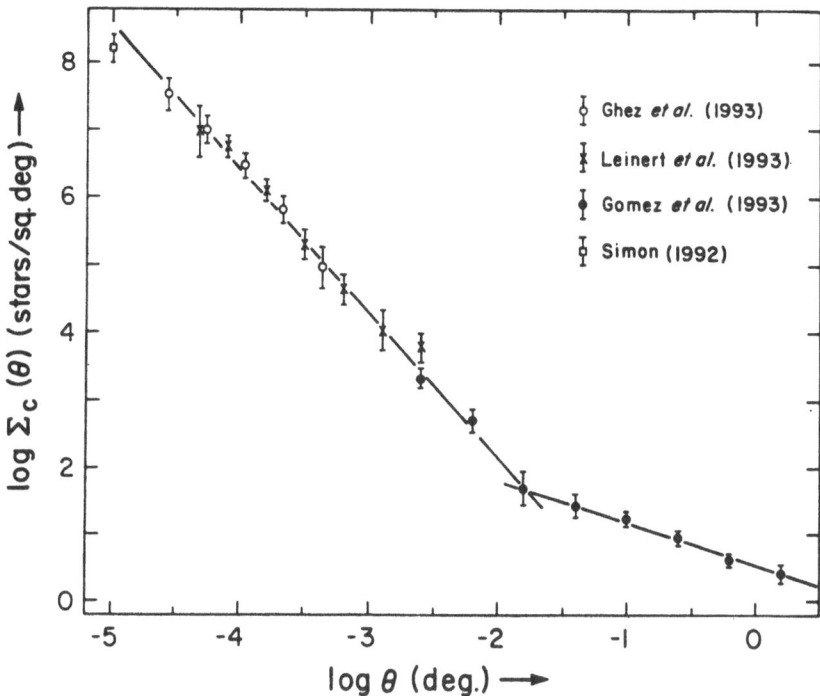

Fig. 8. — *Variation of the stellar surface density of companions with angular separa-tion. The two power-laws have exponents −2.2 for binaries (small separations) and −0.6 for groups (from [41]).*

availability of large format near-infrared array detectors of high sensitivity that allow to completely sample the young stellar population over an entire star forming cloud. The notion that stars tend to be formed in groups goes back to the pioneering study on the distribution of O and B stars by Roberts [66] who found that the great majority (if not all) of O stars are formed and are still present in clusters and associations, but only ∼10% of B stars belong to presently recognized associations/clusters. The latter point is also consistent with the fact that clusters and associations do expand into the general field a time scale less than ∼ 10^8 yr. We now know that the dynamical ages of OB associations are approximately ∼ 10^7 yr, greater than the lifetimes of individual O stars (∼ 10^6 yr). Since the lifetime of a B star is an order of magnitude longer than the lifetime of an association, it is clear that all present day B stars probably formed in OB associations.

Miller & Scalo [51] re-examined the problem and concluded that stellar as-sociations account for at least 70% of all stars more massive than 2–5 $M_\odot$,

but that the uncertainty is such that they can also account for *all* such stars
not produced by clusters - if the mass functions of associations is similar to
the field-star mass function. Note that galactic open clusters account only for
about 10% of all stars presently being formed.

6.1. The Evidence for Clustering

In this subsection, I will briefly discuss the observational evidence for clustering
in young stars of all masses. Beginning with the T Tauri stars, these are opti-
cally visible, low-mass ($M_* \lesssim 2\,M_\odot$) stars characterized by an age of $t_{TTS} \sim 10^6$
yr, low dispersion velocity (less than 2 km s^{-1}) and a strong spatial correlation with the dense molecular gas. Their scattered distribution in nearby star
forming regions has been always taken as evidence in support of the individual
mode of star formation. However, a recent study of the sample of T Tauri stars
in Taurus-Auriga by Gomez et al. [23] has shown that the spatial distribution
is in fact non-random: the two-point angular correlation function has a power-
law of the form $W(\theta) = A\theta^{-1.2}$, where θs is the angular separation, indicating
real clustering in small groups over a wide range of scales. It appears that
about half of the T Tauri stars can be assigned to groups, while the rest are
scattered. The mean projected separation is ~ 0.3 pc, quite similar to the size
of the dense cores (~ 0.1 pc). Groups have typically ~ 15 members within 0.5-1
pc, with a volume density of about 10-50 stars pc^{-3}. Gomez et al. conclude
that isolated star formation does not occur, and that even single stars form
with a relatively close companion (not necessarily gravitationally bound).

The richness of the stellar population associated with the intermediate-mass
Herbig Ae/Be stars has been studied by Hillenbrand ([27]) and Testi et al.
([74]). Although their samples are still small, the initial results are encouraging.
For example, there is a clear dependence of the richness of the cluster with the
spectral type of the Herbig star: the earlier the spectral type, the richer the
cluster. Also, the presence of clusters appear at a detectable level only for stars
earlier than B5-B7 where the average space density is about 100-300 star pc^{-3}.
For later types the density is similar to that found in T Tauri stars. If confirmed,
this result would indicate that there is a threshold mass for the appearance of
large clusters with obvious implications for star formation theories.

As we said at the beginning, it is long known that O stars form in clusters.
However, the optical studies have limited the knowledge of the stellar population only to the brightest members and basic issues, such as the shape of the
initial mass function at the low-mass end, have been left unanswered. The best
example of a nearby cluster associated with O stars is the Trapezium cluster in
Orion. Optical studies have indicated a stellar density in excess of 3000 pc^{-3} in
a region of 3×3 arcmin centered on $^1\theta$ Ori. The majority of the stars have ages
less than $\sim 10^6$ yr. By means of near-infrared imaging, over 500 members have
been identified [62]. In the central 1 arcmin (0.14 pc) diameter core, the stellar
number density exceeds 10^4 pc^{-3}, which implies a mean separation between
the clusters stars of ~ 0.04 pc, making it one of the densest known star forming

regions in the Galaxy. The infrared images go deep enough to almost completely sample the entire stellar population. The K-band luminosity function shows a strong peak at $12^m.0$, well above the detection limit of $15^m.5$. Whether this feature reflects a real turnover in the luminosity (and mass) function or not is still debated.

One of the most interesting results in the Trapezium cluster is that the OB stars are located right at the center, a property common to many other young clusters. It appears that the stars formed in the middle rather than arrived there by dynamical evolution. The one-dimensional rms velocity dispersion of the optical stars near the center is $\sim$2.5 km s^{-1}, enough for three crossings of the central parsec, assuming an age of 10^6 yr. The fact that there is no correlation of the velocity dispersions with mass and distance to the center supports the interpretation that mass segregation is largely due to initial conditions. These findings represent strong constraints on theories of massive star formation in clusters which cannot be considered simple extrapolations of the standard scenario described in the first sections.

6.2. Properties of Clustering

There are interesting implications of the observational evidence of clustering at different spatial scales and member richness. In the remainder, I will discuss three examples: clustering and binary formation; clustering and star formation efficiency; clustering and the IMF.

6.2.1. Clustering and Binary Formation

The first point deals with the important question: does clustering extend to all spatial scales? or, is there a characteristic threshold? In an interesting paper, Larson [41] has extended the analysis of the clustering of T Tauri stars by Gomez et al. to even smaller separations by using the rich database available on binary/multiple stars in Taurus-Auriga. Larson finds that binary stars exhibit self-similar clustering on the largest scales, but that there is a definite break from self-similarity at a scale of about 0.04 pc. This change separates the regime of binary and multiple systems on smaller scales from that of true clustering on larger scales. This trend is shown in Figure 8, which shows the surface density of companions on the sky as a function of the angular separation. At θ <0.04 pc, the power-law corresponds to a distribution of binary separations between 10 and 8000 AU. The break in the observed distribution may suggest that there is an intrinsic scale in the star formation process: Larson suggests that this scale can be identified with the Jeans length in a typical cloud core. However, as we have seen in section 4.1.1, λ_J $\sim$0.2 pc, a factor of at least five bigger than the break-up scale. The observed hierarchical clustering can be described as a fractal distribution with a dimension of about 1.4, possibly originating from pre-exisiting hierarchical structures in the progenitor molecular clouds (see also the discussion in Falgarone and Combes, this volume). Additional support to the findings in Taurus-Auriga has come from recent studies of the binary

population in Chamaeleon and Ophiucus that show a similar behavior of the
two-point correlation function. However, the situation is different in the Orion
cluster where the distribution of pairs is much shallower ($\theta^{-0.9}$) and dominated
by the filamentary structure of the cluster rather than by the self-similarity of
the cloud cores [25].

6.2.2. Clustering and Star Formation Efficiency

The second property of clustering bears directly on the star formation mechanisms in dense regions. It is clear that the standard model of isolated star
formation mediated by the action of ambipolar diffusion is hardly tenable in
these conditions. Therefore, other mechanisms must be invoked. It is well
known that, although most star formation occurs in molecular cores that form
large groups of stars, most main sequence stars are *not* found in clusters. Thus,
most of the embedded clusters that have been found in star forming regions
cannot survive for long as bound open clusters and the majority must be dis-
persed soon after their formation. The most likely physical reason for this
behavior is the low efficiency of star formation in clusters that do not remain
gravitationally bound after gas removal. Lada et al. [36] have shown that the
formation of a bound cluster requires at least a 30% efficiency of conversion of
gas into stars. This value is much higher than the typical 2-3% estimated in
molecular clouds, but it is also higher than that derived for the young cluster
in ρ Ophiuchi (about 20%). In the case of the Trapezium cluster, the situation
is more favorable since the total mass of stars is comparable to the mass of
gas in the region, so that the efficiency there ($\sim$50%) seems to have been high
enough to produce a bound cluster.

Models of the early evolution of young clusters still embedded in the parent
gas are needed to firm up the estimate of the minimum star formation efficiency
needed to form bound clusters. The important result of Lada et al. is based
on a simplified dynamical model in which the gas was assumed to be lost
to the system with an ad hoc time-law prescription. Recent improvements
[34] using N-body+ fully hydrodynamical codes allow to follow the radiative
and mechanical interaction between stars and gas in time and therefore the
mechanisms by which the binding energy is removed from the system. For
example, it is found that most of the mass is lost via the action of stellar
winds and outflows during the earliest phases of the evolution, before the most
massive stars evolve to supernovae. Also, there is a strong dependence on the
stellar mass distribution and on mass segregation. In fact, there is evidence
that massive stars are born at the center of the clusters and do not settle
there via dynamical friction and gravitational drag. The formation of the most
massive objects can thus occur only at the center of dense clusters, where the
deep potential well of the surrounding gas and interaction with close objects
may allow accretion to continue onto very luminous stars.

6.2.3. Clustering and the IMF

This point introduces the last question of what determines the mass of a star in a cluster. We have seen that in all YSOs infall and outflow occur simultaneously. Also, the amount of mass which is available from cloud cores is generally bigger than the final stellar mass, so that it is not because of lack of reservoir that stars stop accreting. It is quite natural then to identify in the outflows the main agents for reversing the infall and limiting the final stellar mass. This idea has been put forward initially by Norman & Silk [55] (see also the chapter by Silk in this volume) and more recently by other authors ([1] [54]). The main idea is that the mass outflow pushes out the surrounding core material, producing a bubble and a thin dense shell. The condition for stopping infall is achieved when the shell radius R_{sh} grows to the core radius R_c, or

$$R_{sh} \sim \left(\frac{\dot{M}_{out} V_{out}}{\rho_c} \right)^{1/4} t^{1/2} \approx R_c \sim \left(\frac{M_c}{\rho_c} \right)^{1/3} , \qquad (20)$$

where $\dot{M}_{out}$ and V_{out} are the mass outflow rate and outflow velocity, respectively. Using typical values for the outflow and core properties, Nakano et al. obtain a value of the typical mass of about 2 $M_\odot$ with a dependence on the mass infall and outflow rates $M_* \sim (\dot{M}_{out}/\dot{M}_{inf})^{7/6}$. The estimated star formation efficiency is found to be less than $\sim 5\%$, in agreement with the observations.

The main problem with the scenario in which the mass of a star is set by the outflow mass rate is that the condition in eq. (20) is achieved when the bubble has reached a value of ~ 0.2-0.4 pc, i.e. the core radius. Now in the conditions of dense clusters, the typical separation varies between 0.04 pc (Trapezium cluster) and 0.12 pc (dense cores in L1641). Both values are smaller than the shell and core radii. Thus, accretion is shut off not because of outflows, but because of lack of available material and by interaction with neighboring protostars. In other words, the small extent of the protostellar envelopes does not allow the outflow to propagate to large distances and therefore would limit the total amount of matter than can be accreted to much smaller values than the characteristic solar mass.

As a possible alternative, it is interesting to envisage a model in which the stellar mass is not set by the feedback effects of forming stars on their environment, but by the clumpy structure of molecular clouds. As discussed in the chapter by Falgarone, strong evidence exists for clumpy substructures down to the smallest scales. Also, the clump mass spectrum is a power-law $N(M_{cl}) \sim M_{cl}^{-1.6}$, for $M_{cl} \gtrsim 1$ $M_\odot$. Thus, it is natural to simply relate the clump mass spectrum to the stellar mass. Indeed, the stellar IMF is a power-law $N(M_*) \sim M_*^{-1.7}$ at subsolar masses and becomes much steeper at larger masses ($\sim M_*^{-2.3}$) ([33]). The similarity of the two power indexes suggests a linear relation between the mass of the clump and that of the star, with an efficiency of conversion of gas into stars of about 100%. At higher masses, the efficiency decreases rapidly, probably due to subfragmentation of more massive clumps, implying that it becomes more and more difficult to build up massive stars. In order to reconcile this scenario with the observed stellar IMF, the star formation efficiency should scale with the mass of the clump as $M_{cl}^{-0.7}$. This can be achieved if massive stars result from the coagulation of several protostellar cores in the same clump.

Acknowledgments

I wish to thank D. Galli, S. Stahler, P. Caselli, R. Cesaroni and Ph. André for numerous and useful discussions that have helped to clarify many concepts used in this work.

References

[1] Adams, F.C. & Fatuzzo, M. *ApJ* **464** (1996) 256
[2] André, P., Ward-Thompson, D. & Barsony, M. , *ApJ* **406** (1993) 122 *A&A* (1997) inpress.
[3] Bachiller, R. *ARAA* **34** (1996) 111
[4] Bally, J., Devine, D. & Alten, V. *ApJ* **473** (1996) 921
[5] Barsony, M. in Clouds Cores and Low Mass Star Formation (ASP Conf. Ser., D.P. Clemens and R. Barvainis eds., San Francisco, 1994) p. 197
[6] Beckwith, S.V.W. & Sargent, A. *Nature* **383** (1996) 139
[7] Beech, M. & Mitalas, R. *ApJS* **95** (1994) 517
[8] Bernasconi, P. & Maeder, A. *A&A* **307** (1996) 829
[9] Bertoldi, F. & McKee, C.F. *ApJ* **354** (1990) 529
[10] Blaauw, A.W. in The Physics of Star Formation and Early Stellar Evolution (C.J. Lada & N.D. Kylafis eds., Kluwer Academic Publ., Dordrecht, 1991) p. 125
[11] Bodenheimer, P. in Star Formation in Stellar Systems (G. Tenorio-Tagle et al. eds., Cambridge University Press, Cambridge, 1992) p. 1
[12] Bontemps, S., André, P., Terebey, S., Cabrit, S., *A&A* **311** (1996) 858
[13] Boss, A.P. *ApJ* **468** (1996) 231
[14] Cabrit, S. & Bertout, C., *A&A* **261** (1992) 274
[15] Cesaroni, R., Churchwell, E., Hofner, P., Walmsley, C.M. & Kurtz, S. *A&A* **288** (1994) 903
[16] Cesaroni, R., Felli, M., Testi, L., Walmsley, C.M. & Olmi, L. *A&A* (1997) in press
[17] Chini, R., Krügel, E., Haslam, C.G.T., Kreysa, E., Lemke, R., Reipurth, B., Sievers, A. & Ward-Thompson, D., *A&A* **272** (1993) L5
[18] Dyson, J.E. in Star Formation and Techniques in Infrared and mm-Wave Astronomy (T.P. Ray & S.V.W. Beckwith eds., Springer-Verlag, Berlin, 1994) p. 93
[19] Elmegreen, B.G. in Star Formation in Stellar Systems (G. Tenorio-Tagle et al. eds., Cambridge University Press, Cambridge, 1992) p. 383

[20] Fukui, Y., Iwata, T., Mizuno, A., Bally, J. & Lane, A.P., in Protostars and Planets III (E.H. Levy & J.I. Lunine eds., Univ. of Arizona Press, Tucson, 1993) p. 603

[21] Galli, D. & Shu, F.H. *ApJ* **417** (1993) 220

[22] Galli, D. & Shu, F.H. *ApJ* **417** (1993) 243

[23] Gomez, M., Hartmann, L., Kenyon, S.J. & Hewett, R. *AJ* **105** (1993) 1927

[24] Gregersen, E.K., Evans, N.J.,II, Zhou, S., & Choi, M., *ApJ* (1997) in press

[25] Hartmann, L. ,priv. communic.

[26] Heiles, C., Goodman, A., McKee, C.F. & Zweibel, E.G. in Protostars and Planets III (E.H. Levy & J.I. Lunine eds., Univ. of Arizona Press, Tucson, 1993) p. 279

[27] Hillenbrand, L., Meyer, M.R., Strom, S.E. & Skrutskie, M.F. *AJ* **109** (1995) 280

[28] Ho, P.T.P. & Hascick, A. *ApJ* **309** (1981) 514

[29] Hollenbach, D.J., Johnstone, D., Lizano, S. & Shu, F.H. *ApJ* **428** (1994) 654

[30] Iben, I. *ApJ* **141** (1965) 993

[31] Klein, R.I., Whitaker, R.W. & Sandford, II,M.T. in Protostars and Planets II (D.C. Black & M.S. Matthews eds., University of Arizona Press, Tucson, 1985) p. 340

[32] Königl, A., *ApJ* **342** (1989) 208

[33] Kroupa, P. *ApJ* **453** (1995) 350

[34] Kroupa, P. *MNRAS* **277** (1995) 152

[35] Lada, C.J. in Star Forming Regions (M. Peimbert & J. Jugaku eds., Reidel Publishing Co., Dordrecht, 1987) p. 1

[36] Lada, C.J., Margulis, M. & Dearborn, D. *ApJ* **285** (1984) 141

[37] Lada, E.A., Strom, K.M. & Myers, P.C. in Protostars and Planets III (E.H. Levy et al. eds., University of Arizona Press, Tucson, 1993) p. 245

[38] Larson, R.B., *MNRAS* **145** (1969) 271

[39] Larson, R.B., *MNRAS* **157** (1972) 121

[40] Larson, R.B., *MNRAS* **200** (1982) 159

[41] Larson, R.B., *MNRAS* **272** (1982) 213

[42] Leung, C.M. & Brown, R.L., *ApJ* **214** (1977) L73

[43] Lizano, S. & Shu, F.H. *ApJ* **342** (1989) 834

[44] Lizano, S., Heiles, C., Rodriguez, L. et al., *ApJ* **328** (1988) 76

[45] Masson, C. & Chernin, L.M., *ApJ* **414** (1993) 230

[46] Mathieu, R. *ARAA* **32** (1994) 465

[47] McCaughrean, M.J., Rayner, J.T. & Zinnecker, H., *ApJ* **436** (1994) L189

[48] McKee, C.F., *ApJ* **345** (1989) 782

[49] Mestel, L. & Spitzer, L. *MNRAS* **116** (1956) 505

[50] Mezger, P.G., *ApSS* **212** (1994) 197

[51] Miller, G.E. & Scalo, J.S. *PASP* **90** (1978) 506

[52] Molinari, S., Brand, J., Cesaroni, R. & Palla, F. *A&A* **308** (1996) 573

[53] Mouschovias, T.Ch. *ApJ* **206** (1976) 753
[54] Nakano, T., Hasegawa, T. & Norman, C. *ApJ* **450** (1995) 183
[55] Norman, C. & Silk, J., *ApJ* **238** (1980) 158
[56] Olmi, L., Cesaroni, R. & Walmsley, C.M. *A&A* **276** (1993) 489
[57] Olmi, L., Cesaroni, R., Neri, R. & Walmsley, C.M. *A&A* **315** (1996) 565
[58] Ouyed, R., Pudritz, R.E. & Stone, J.M., *Nature* **385** (1997) 409
[59] Palla, F. & Stahler, S.W. *ApJ* **375** (1991) 288
[60] Palla, F. & Stahler, S.W. *ApJ* **418** (1993) 414
[61] Pelletier, G. & Pudritz, R.E., *ApJ* **394** (1992) 117
[62] Prosser, C.F., Stauffer, J.R., Hartmann, L. et al. *ApJ* **421** (1994) 517
[63] Ray, T.P. & Mundt, R., in Astrophysical Jets (D. Burgarella et al. eds., Cambridge Univ. Press, Cambridge, 1993) p. 145
[64] Ray, T.P., Mundt, R., Dyson, J.E., Falle, S.A.E.G. & Raga, A.C., *ApJ* **468** (1996) L103
[65] Reid, M.J., Hascick, A., Burke, B.F., Moran, J.M., Johnston, K.J. & Swenson, G.W. *ApJ* **239** (1981) 89
[66] Roberts, M.S. *PASP* **70** (1958) 462
[67] Shepherd, D.S. & Churchwell, E. *ApJ* **472** (1996) 225
[68] Shu, F.H., *ApJ* **214** (1977) 488
[69] Shu, F.H., Najita, J., Galli, D., Ostriker, E., & Lizano, S., in Protostars and Planets III (E.H. Levy & J.I. Lunine eds., Univ. of Arizona Press, Tucson, 1993) p. 3
[70] Siess, L. & Forestini, M. *A&A* **308** (1996) 472
[71] Stahler, S.W., Shu, F.H. & Taam, R.A. *ApJ* **302** (1980) 590
[72] Sugitani, K. & Ogura, K. *ApJS* **92** (1994) 163
[73] Terebey, S., Shu, F.H. & Cassen P. *ApJ* **286** (1984) 529
[74] Testi, L., Palla, F., Prusti, T., Natta, A. & Maltagliati, S. *A&A* (1997) in press
[75] van Buren, D., Mac Low, M.-M., Wood, D.O.S. & Churchwell, E. *ApJ* **353** (1990) 570
[76] Walker, C.K., Lada, C.J., Young, E.T., Maloney, P.R., & Wilking, B.A., *ApJ* **309** (1986) L47
[77] Walmsley, C.M. *Rev. Mex. Astr. Astrophys.* **1** (1995) 137
[78] Ward-Thompson, D., Chini, R., Krügel, E., Andrè, Ph. & Bontemps, S. *MNRAS* **274** (1995) 1219
[79] Ward-Thompson, D., Buckley, H.D., Greaves, J.S., Holland, W.S. & André, P. *A&A* (1997) in press
[80] Wilking, B.A., Lada, C.J. & Young, E.T. *ApJ* **340** (1989) 823
[81] Wilner, D.J., Ho, P.T.P. & Zhang, Q. *ApJ* **462** (1996) 339
[82] Wood, D.O.S. & Churchwell, E. *ApJ* **340** (1989) 265
[83] Zhou, S. & Evans, N.J.,II in Clouds (Cores and Low Mass Star Formation, D.P. Clemens, and R. Barvainis eds., ASP Conf. Ser. San Francisco, 1994) p. 183
[84] Zhou, S., Evans, N.J.,II, Wang, Y., *ApJ* **466** (1996) 296

LECTURE 5

Stellar Physics and Starburst Evolution

G. Meynet

Geneva Observatory, CH-1290 Sauverny, Switzerland

1. INTRODUCTION

Starbursts regions offer a very interesting opportunity to study the links between stellar and galactic evolutions. Indeed starburst triggering is often related to dynamical processes. As mentioned in F. Combes' lectures, the most spectacular starbursts occur in tidally interacting galaxies and mergers. On the other hand, many observational features of starbursts, like their integrated luminosity, colors, spectral evolution, depend on stellar physics. In these lectures, we shall concentrate on this second aspect, trying to give some elements of response to the following question: how do photometric and chemical evolution of starburst regions depend on stellar physics ? The reader will find many interesting contributions on this subject in the books "From stars to Galaxies: The impact of Stellar Physics on Galaxy Evolution" (1996, ASP Conference Series, Volume 98, eds. Leitherer, Fritze-von Alvensleben & Huchra) and "The Interplay Between Massive Star Formation, the ISM and Galaxy Evolution" (1996, 11th IAP Astrophysics Meeting, eds. Kunth D., Guiderdoni B., Heydari-Malayeri M, Trinh Xuan Thuan, Editions Frontières). Moreover, he will find a database for galaxy evolution modeling in Leitherer et al. (1996). The CD-ROM accompanying this paper contains empirical stellar libraries, theoretical libraries of selected spectral indices, of stellar continuum fluxes, of WR energy distribution, libraries of star clusters and early type galaxies, evolutionary models and isochrones, evolutionary synthesis models and models of ionized gas.

The present lectures are organized as follows: in section 2, we shall recall some basic principles governing stellar evolution as well as the evolutionary scenarios followed by stars of different initial masses. We shall also describe the main physical ingredients incorporated into stellar models. Section 3 will be devoted to the presentation of observational tests of stellar models. This step is a necessary prerequisite in order to assess the performances as well as the limitations of the stellar models, further used as an ingredient of population syntheses models. The impact of some physical ingredients of the stellar models on the outputs of population syntheses models will be discussed in section 4. Some aspects of the chemical evolution of starburst regions will be presented in section 5. Finally, as a conclusion, we shall briefly propose some directions of research for future works.

2. STELLAR PHYSICS

Stars are sites where there are strong interactions between matter and radiation. Eddington was the first to point out the importance of the radiation pressure which, together with the gas pressure, supports the stellar material against the action of gravity. In a famous passage of his book on *The Internal Constitution of the Stars* (Eddington 1926, p. 16), he arrives at the conclusion that the observed range of stellar masses corresponds to the range of masses of gas spheres, in which, both sources of pressure significantly contribute to the preservation of the hydrostatic equilibrium.

Chandrasekhar (1984) proposed a more rational version of Eddington's argument which, in addition, provides a natural scale for the stellar masses. There is a general theorem by Chandrasekhar (1936) which shows that the actual pressure, P_c, at the center of a star of mass M in hydrostatic equilibrium is bounded by two values[1]: the actual central pressure is superior to the central pressure of a uniform density gas sphere having a density equal to the *mean* density of the star ($\bar{\rho}$), and inferior to the central pressure of a uniform density gas sphere having a density equal to the *central* density of the star (ρ_c). Thus one has

$$\frac{1}{2}G\left(\frac{4}{3}\pi\right)^{1/3}\bar{\rho}^{4/3}M^{2/3} \leq P_c \leq \frac{1}{2}G\left(\frac{4}{3}\pi\right)^{1/3}\rho_c^{4/3}M^{2/3}, \qquad (1)$$

where G is the constant of gravitation. Now supposing that the matter is in the state of a perfect gas, the total pressure can be written

$$P = \left[\left(\frac{k}{\mu m_p}\right)^4 \frac{3}{a}\frac{1-\beta}{\beta^4}\right]^{1/3}\rho^{4/3}, \qquad (2)$$

where k is the Boltzmann constant, μ the mean molecular weight, m_p the mass of a proton, a Stefan's radiation-constant and β the ratio of gas pressure to

---

[1] the mean density inside a radius r is assumed to decrease outwards.

the total pressure. Inserting this expression for P in the inequalities (1) and considering the right-hand side of the inequality, one obtains

$$M \geq \left(\frac{6}{\pi}\right)^{1/2} \left[\left(\frac{k}{\mu m_p}\right)^4 \frac{3}{a} \frac{1-\beta_c}{\beta_c^4}\right]^{1/2} \frac{1}{G^{3/2}}. \tag{3}$$

Now, replacing a by $\frac{8\pi^5 k^4}{15 h^3 c^3}$, where h is the Planck's constant and c the velocity of light and isolating the natural constants on one side of the inequality, one obtains

$$\mu^2 M \left(\frac{\beta_c^4}{1-\beta_c}\right)^{1/2} \geq 0.1873 \left(\frac{hc}{G}\right)^{3/2} \frac{1}{m_p^2}. \tag{4}$$

The right-hand side has the dimension of a mass and its value is equal to 5.48 $M_\odot$ $((hc/G)^{3/2} 1/m_p^2 = 29.2 M_\odot)$. Thus one sees that assuming the mechanical equilibrium is maintained by both the gas and the radiation pressure, one obtains a combination of physical constants providing a natural scale for the masses of the stars.

2.1. Recall of Some Important Processes in Stellar Evolution

The natural scale of stellar masses presented above involves the constant of gravitation and physical constants associated with other fundamental interactions of physics. This translates the fact that stars are a privileged place in the universe where microscopic physics interacts with macroscopic phenomena. If the long range force of gravity plays the main role in driving the birth of the stars, their life and sometimes their death, the other three interactions of physics govern the processes of production, transfer and loss of energy either under the form of electro-magnetic radiation or through neutrinos. In other words, if the luminosity of the stars is a consequence of their global hydrostatic equilibrium, their ability to maintain this equilibrium for many millions or billions years, depending on their initial mass, is a consequence of the microscopic structure of matter, more precisely of the atomic nuclei. Indeed, let us recall that in non degenerate stars, the pressure gradient, necessary to balance the gravitational force, implies a temperature gradient and therefore the existence of an energy flow from the hot inner parts of the star to the cool outer parts. Thus a sphere of (perfect) gas in mechanical equilibrium necessarily radiates energy. The star produces this energy alternatively by macroscopic contractions and microscopic ones (the thermonuclear reactions). While gravitational contractions can maintain mechanical equilibrium during a Kelvin-Helmholtz time scale which is, for the sun, of the order of a few tens of millions year, the nuclear energy source can sustain the stellar luminosity during much longer times ($\sim$3 orders of magnitude longer in case of the Sun).

The contraction, which follows the end of a nuclear burning phase, drives the central regions of the star towards zones of the temperature-density diagram where degeneracy effects are important. The fact that the electronic pressure,

in degenerate regime, does not depend on temperature has important consequences for stellar evolution. Very briefly, one can say that the two mechanisms
of energy production behave quite differently in non degenerate and degenerate
regimes (see the mathematical developments in Kippenhahn & Weigert 1990,
p. 243 and 266):

1) A homologous contracting sphere of perfect gas follows a straight line with
slope 1/3 in the central temperature - density plane ($\Delta \log T = 1/3 \Delta \log \rho$).
When the electrons become more and more degenerate, further contraction
cools the ions to supply energy to the electrons ($\Delta \log \rho > 0 \implies \Delta \log T < 0$).

2) While the nuclear burning is stable in non degenerate regimes, it is explosive
in degenerate conditions.

These are the main processes responsible for the different evolutionary scenarios
briefly resumed below.

2.2. The Evolutionary Scenarios

Schematically four ranges of stellar masses are considered (see Chiosi 1986, for
more details):

1) The first includes objects that undergo no nuclear burning stage (progenitors
of brown dwarfs).

2) The second includes stars that go only through the hydrogen burning phase
(progenitors of helium white dwarfs).

3) The third includes stars that go only through the hydrogen and helium
burning phases (progenitors of carbon-oxygen white dwarfs).

4) And finally the fourth consists of all stars evolving beyond the helium burning phase (progenitors of neutron stars, black holes or possibly leaving no remnant).

2.2.1. Progenitors of Brown Dwarfs

Objects having an initial mass below 0.08 $M_\odot$ enter the degenerate zone before
hydrogen ignition. Further contraction cools the ions and these objects never
enter the hydrogen burning phase (cf. D'Antona and Mazzitelli 1985; Dorman
et al. 1989).

2.2.2. Progenitors of Helium White Dwarfs

In the mass range between 0.08 and 0.5 $M_\odot$, hydrogen burning occurs but
the helium core becomes electron degenerate before becoming hot enough for
helium ignition. So, like the hydrogen in the previous stellar range, it never
burns. These stars end up as helium white dwarfs (Burrows et al. 1989; Dorman
et al. 1989). Their lifetime is much longer than the present age of the universe
and thus these stars are still burning hydrogen in their core. They constitute
the most numerous stellar component of the Galaxy.

2.2.3. Progenitors of Carbon-oxygen White Dwarfs

Between 0.5 and 8 $M_\odot$, stars go through the H and He-burning phases. Stars with an initial mass between ~ 0.5 and ~ 1.85 $M_\odot$ undergo the He-flash: after a stable H-burning phase, the star ascends the red giant branch. At the red giant tip, He ignites in a degenerate environment and a He-flash occurs. The He-flash removes the degeneracy in the core and permits the pursuit of the He-burning phase in non degenerate conditions. In the Hertzsprung-Russell (HR) diagram, after the He-flash, stars are removed from the tip of the red giant branch and relocated on the horizontal branch (at low metallicity) or on the so called red clump (at solar and higher metallicity) where core He-burning takes place. After the He-burning phase, stars evolve along the asymptotic giant branch until the outer envelope is removed by wind and thermal pulses. A planetary nebula forms around the white dwarf (see the review by Frost & Lattanzio 1995; for numerical simulations of these phases see Lattanzio 1986; Mazzitelli & D'Antona 1986; Bowen & Willson 1991; for lectures on white dwarfs see Kawaler 1997).

Stars with an initial mass between 1.85 and 8 $M_\odot$ evolve like those which undergo a He-flash. Only the end of their evolution could be different. Indeed these stars experience a prolonged thermally pulsing asymptotic giant branch phase, terminated either with the envelope ejection and the formation of a carbon-oxygen white dwarf (as in the case of "He-flash" stars), or if the carbon-oxygen core reaches the Chandrasekhar limit (~ 1.4 $M_\odot$), with carbon ignition in degenerate matter. In this latter case, the subsequent evolution is still subject to major uncertainties (see Chiosi 1986). Probably, the star will be completely disrupted by a supernova explosion (type I 1/2 supernova, Iben & Renzini 1983). Maeder & Meynet (1989) questioned the existence of such supernovae. Indeed, the best estimates for the maximum initial mass leading to the formation of a white dwarf, based on the presence of white dwarfs in clusters, give values around 8 $M_\odot$ (cf the review by Weidemann 1990). Such a high value leaves no room for carbon detonation supernovae in case the inferior mass limit for massive stars is equal to 8 $M_\odot$ (see below).

2.2.4. Progenitors of Neutron Stars and Black Holes

Hereafter we shall call massive stars, the stars that go through all nuclear burning phases (H, He, C, Ne, O and Si-burning) until the formation of an iron core. The minimum initial mass for a star to proceed through the entire sequence of nuclear burning phases is set at a value of 8 $M_\odot$ by Maeder & Meynet (1989) who took into account the effects of mass loss and core overshooting and to 10 $M_\odot$ by Becker and Iben (1979) and Nomoto (1984) who did not consider the two effects just mentioned. These stars are the progenitors of supernovae of type II, and are also likely to be the progenitors of at least a fraction of supernovae of type Ib, if these latter can be accounted for by the explosion of a Wolf-Rayet star (see the review by Weiler & Sramek 1988). Their final fate is a neutron star or a black hole depending on the mass of their iron core: if the core mass is below Oppenheimer-Volkoff's mass limit (around 2 $M_\odot$), the final product will be a neutron star, otherwise the collapse is not stopped by the pressure exerted by the degenerate neutrons and a black hole is formed (for lectures on neutron stars and black holes see Srinivasan 1997; Novikov 1997).

Let us add here that the most massive stars ($M > 100\ M_\odot$), having a mass of the helium core greater than 35 $M_\odot$, encounter pair instability during the oxygen burning phase. This instability occurs because, at temperatures around 2×10^9 K, a large part of the energy from gravitational contraction goes into the creation of pairs of electrons and positrons (Fowler & Hoyle 1964; Barkat et al. 1967). This process subtracts energy that otherwise would provide pressure support and therefore may trigger a dynamical collapse. The final fate of these stars is described in Woosley (1986). Let us note that this instability will never occur if, as an effect of strong mass loss by stellar winds, the mass of the He core is decreased below a value equal to $\sim 35\ M_\odot$. This may well be the case at solar metallicity, where all stars more massive than 40 $M_\odot$ end their stellar life with final masses between 5 and 10 $M_\odot$ (cf. Schaller et al. 1992). In that case, the pair instability would be encountered only by massive stars at low metallicity.

2.2.5. Evolution of Interacting Binaries

In a close binary, the evolutionary scenarios can be radically different from those just described above for single star evolution. In a close binary system, mass can be stripped from either star in a Roche-lobe overflow event and either transferred to the other star or lost from the system or both. We shall not review this vast subject here. The reader will find a comprehensive introduction to the field of interacting binaries in Pringle and Wade (1985), see also van den Heuvel (1994).

2.3. A Few Comments on Some Important Physical Ingredients of Stellar Models

The various inputs of stellar models can be classified in two main categories: the ingredients which comes from laboratory experiments and theoretical considerations and quantities which comes from astronomical observations. In the first class, one finds the opacities, the equation of state, the nuclear reaction rates, the neutrino emission rates. In the second category there are the initial mass of stars, their initial composition, the mass loss rates. The convection parameters are in some respect in between these two classes of parameters. Ideally one would like to have an ab initio theory of convection enabling us to predict the turbulence from whatever physical conditions. Since this theory does not exist, most stellar modelers use a phenomenological approach, introducing some free parameters calibrated on observed quantities sensitive to them.

2.3.1. The Opacities

The new tables of opacities obtained by the OPAL group (see e.g. Iglesias & Rogers 1993) represent a major advance in our knowledge of this important physical ingredient of the stellar models. With respect to the older Los Alamos Opacity Library (Huebner et al 1977), these new tables present two significant differences, which depend primarily on the adopted metal abundance. For solar abundances, a first enhancement of the opacity, which amounts to a factor of 3, occurs at temperatures of a few hundred thousand degrees. A second enhancement, of much smaller amplitude (about 20% only), occurs at temperatures of about one million degrees. Another team, known as the Opacity Project group (OP), obtained results generally in close agreement, leaving no doubt about the magnitude of these changes (Seaton et al. 1994). One of the great successes of these new opacities is that they permitted astrophysicists to understand what has been a major challenge of stellar pulsation theory, namely the cause of β-Cephei variability (see e.g. Dziembowski & Pamyatnykh, 1993). The problem is now solved thanks not to new astrophysical ideas but to progress in opacity calculations. It is interesting to note that such an enhancement of the opacities had been suggested some years ago in order to explain the driving mechanism active in these variable stars (Simon 1982). Will there be in the future new major improvements of the opacity tables ? It is not possible of course to discard such a possibility but we can note that the good agreement between the OPAL and OP groups, who used different techniques to compute the opacities, does not suggest, at the present time, such a perspective.

2.3.2. Nuclear Reaction Rates

Except for one reaction, all the rates of the nuclear reactions relevant to stars rely either completely or partially on theoretical calculations (see e.g. Rolfs & Rodney 1988). So far, there is only one measurement which reaches the Gamov-peak energy region. The ^{3}He$+^3$He reaction was measured down to $E = 24$ keV by Krauss et al. (1987). Presently, measurements of this reaction rate at still lower energy are being attempted in order to know if a narrow resonance is present (Trautvetter 1993). Let us note that such a resonance might be responsible for a part, if not all, of the famous solar neutrinos deficit. Can we have confidence on the other nuclear rates for which no direct experimental data are available in the Gamov peak region ? In general, at least for the H-burning phase, we can say yes and one reason for this optimistic statement relies on the observations of the surface abundances of initially massive stars, which have uncovered their stellar core by ejection of their envelope through stellar winds (Wolf-Rayet stars, see section 4.1). Some of these stars have surface abundances representative of material processed through the CNO cycle. The good agreement between the observed and predicted values of CNO equilibrium indicates the general correctness of our understanding of the CNO cycle and of the relevant nuclear data (Maeder 1983). For the reactions taking place in the He-burning phase, a still important uncertainty entails the rate of the reaction

$^{12}C(\alpha, \gamma)^{16}O$. Important results of stellar models depend on the adopted rate for this reaction which may fluctuate for calculations different authors by a factor 3. Indeed this reaction not only governs the proportion of carbon and oxygen in the core at the end of the He-burning phase but influences also the type of stellar remnant (black hole or neutron star) left over by massive stars (Woosley 1986). An improvement of our knowledge of this key nuclear reaction rate would represent a major step in the field of nuclear astrophysics (see section 5.1.2).

2.3.3.   The Out of Equilibrium Situations: a Major Difficulty

When one considers the present status of the stellar models, one is surprised by both their relative crudeness in some of their aspects (in particular those describing convection), and by their successes in spite of this shortcoming in describing a great part of the major stellar features. This is due mainly to the fact that stars, during most of their life, are systems in hydrostatic, thermal and, locally, in thermodynamic equilibrium. When these assumptions are no longer valid, the problem becomes much more difficult to handle and some simplifying hypotheses have to be made. Actually, out of equilibrium processes are far from being well understood and as a consequence are not numerically well described. Such processes occur both in the stellar interiors (e.g. convection, semi-convection) and at the surface (e.g. stellar winds) and are present in the stars during all their life.

2.3.4.   Mixing Processes

Many aspects of the stellar models depend on the way the chemical elements are mixed in their interior. Mixing may occur not only in convective zones, but also in more stable radiative regions where it may be driven by various instabilities, as for instance the turbulence induced by the friction of stellar layers rotating at different velocities (shear mixing). Among the numerous problems which remain to be solved in this field let us cite the extent of the convective penetration zone or overshoot region (cf. Zahn 1991) and the role of the gradient of the molecular weight in stabilizing turbulent flow (Grossman & Taam 1996; Maeder 1997; Meynet & Maeder 1997). The difficulty in solving these problems comes from our ignorance of important parameters such as the turbulent viscosity describing the effects of viscous dissipation. It is interesting to note that all these turbulent processes are at an intermediate stage between macroscopic and microscopic physics and involve many different length and time scales. In section 3, we shall present some observations indicating that some extra mixing processes are active in stars. By extra mixing processes, we mean here processes not yet accounted for in most (if not all) present stellar models.

2.3.5. Mass Inflows/Outflows

Hydrodynamic phenomena may also occur in the outer parts of the stars through accretion and/or ejection of matter. During the pre-main sequence phase of stellar evolution, accretion plays a very important role (cf. e.g. Palla & Stahler 1992 and the lectures of F. Palla in this volume). One of the main difficulty in this domain is to correctly predict the accretion rate. It depends not only on the nature of the accreting object but also on the physical conditions prevailing in the star forming cloud. This can have very interesting consequences on the way stars are distributed in masses in different environmental conditions (see Sect. 3).

Another process which deeply modifies the outputs of the stellar models is the mass loss by stellar winds. The loss, or at least the decrease in mass, of the original envelope of the stars by stellar winds has a strong impact on the evolution of surface abundances, the enrichment of the interstellar medium in heavy elements, the distribution of supergiants, the understanding of the Wolf-Rayet stars, of the Luminous Blue Variables, the nature of the stellar remnants (see the reviews by Maeder & Conti 1994). Very roughly one can say that in the mass domain greater than 30-40 $M_\odot$ the mass loss by stellar winds is clearly the dominant factor affecting almost all the outputs of the stellar models. On the other hand the rates of mass loss are still affected by great uncertainties. If on the main sequence, in the massive star range, one can with a certain confidence estimate the errors on the "observed" $\dot{M}$ to be of a factor of two, this is certainly not the case in the red supergiant phase for which the uncertainties are much greater (about one order of magnitude, see the review by Lamers and Cassinelli 1996).

The physical mechanisms by which mass is ejected from stars are still subject to debate. For luminous early type stars, the most developed of all the models available is the radiation-driven wind theory (Lucy & Solomon 1970; Castor et al. 1975; Abbott 1979; Pauldrach et al. 1990). In the framework of this theory, it is possible to show (see e.g. Kudritzki et al. 1988, p. 174 and followings) that the mass loss rate depends on the metallicity. This dependence comes out as a consequence of the increase of the opacity in the outer layers of the star, when the metallicity increases (more numerous bound-bound and bound-free transitions). Greater opacities favor a more efficient transfer of the momentum transported by the photons to the matter. Abbott (1982) finds an almost linear dependence between mass loss and metal abundance, while Kudritzki et al. (1988) find that the mass loss rate is proportional to approximately the square root of the metallicity. As we shall see in the next sections, this metallicity dependence of the mass loss rates has very important consequences on massive star evolution and therefore on the evolution of young starburst regions.

Stellar winds from late type giants and supergiants differ from their early type counterparts in that they have two components, gas and dust, which interact both thermally and dynamically. Present theories for this type of stars have not yet reached the degree of predictability of the radiation-driven wind theory for the hot stars.

At present, most evolutionary computations use empirical algorithms to take into account the mass loss effects; at every time step, a certain amount of mass, depending on the physical characteristics of the model, is removed from the star. An overview of the existing empirical formulae relating the mass loss rates to various physical quantities is presented in Chiosi & Maeder (1986).

3. TESTS OF STELLAR MODELS

Observations of individual stars give access to the following quantities: bolometric and/or "color" magnitudes, color indices, spectra, light curves, imagery, etc ... From these observed quantities it is possible to go back to quantities directly comparable with outputs of stellar models as the bolometric luminosity, the effective temperature, the surface abundances, the periods of pulsation, etc ... Of course the links between the observed quantities and the theoretical ones are not straightforward. To convert a visual apparent magnitude, m_v, to bolometric luminosity, first one must correct m_v for the reddening and distance effects, then a bolometric correction must be applied to take into account the non visible radiations. In the same manner, the color indices, as for instance $(B-V)$, must be corrected for reddening before a calibration relation between $(B-V)$ and $\log T_{eff}$ is applied (T_{eff} is the effective temperature). When comparing observations with predictions of stellar models one must keep in mind this process. In particular these examples show that the passage from the observed quantities to the theoretical ones necessitate the use of calibration relations. These relations can be either semi-empirical, or obtained from a model atmosphere. They depend on the luminosity class and on the initial metallicity and introduce some uncertainties of their own. An example of transformation of isochrones from the theoretical HR diagram [$\log L/L_\odot$ vs. $\log T_{eff}$] to the observational plane [M_v vs. $(B-V)$] is presented in Meynet et al. (1993).

Another kind of difficulty arises when one tries to compare the observed and theoretical positions in the HR diagram of a star surrounded by a dense stellar wind. Indeed, the wind can be so dense, that the photosphere is inside the wind. In that case, the observed $\log T_{eff}$ cannot be directly compared to the effective temperature at the surface of the hydrostatic core given by the stellar models. A correction must be applied (as e.g. in Schaller et al. 1992).

Complementary to the observations of individual stars, observations of stars in binary systems, clusters or of complete sample of field stars can give very important information on stellar features. In some favorable cases, from the observations of binaries one can deduce the masses of the components and this provides very tight constrains on the stellar models. In addition to the requirement that the evolutionary tracks for the observed masses must go through the observed positions of the stars in the HR diagram, since the two stars have the same age, a unique isochrone must link them. The width of the main sequence band, i.e.the distribution of the stars in the different phases can be deduced from the study of stars in clusters.

In the following we shall discuss some comparisons between stellar models and observations. We shall mainly concentrate on massive stars. The case of Wolf-Rayet stars will be discussed in section 4.

3.1. Pre-MS Evolution of Massive Stars

For a long time now, it has been found that massive stars seem to be absent from regions near the theoretical ZAMS (Garmany et al. 1982). More precisely, there seems to be a lack of stars more massive than about 40 $M_\odot$ and younger than ~ 2 million years. How to explain this fact ? One possibility was originally suggested by Garmany et al. (1982): the absence of the youngest O-type stars would be due to the fact that they are still immersed in their parental clouds, emerging only after 1-2 millions years. Wood & Churchwell (1989) have suggested that the first 10-20% of an O main-sequence lifetime is spent embedded in its parental cloud. In this scenario the upper ZAMS, as it is generally conceived, still exists but is buried inside thick clouds of interstellar matter.

A different picture emerges from a recent work by Bernasconi & Maeder (1996). These authors have computed the formation of stars in the framework of the accretion scenario (see Stahler et al. 1980; Palla & Stahler 1992; Beech & Mitalas 1994; see also the lectures by F. Palla in this volume) using realistic accretion rates derived from the observed line width in various molecular dark clouds. They computed the evolutionary track that a pre-main sequence (PMS) accreting object would follow starting from an initial value of 0.8 $M_\odot$ and finishing with 120 $M_\odot$. This evolutionary path is called a birthline since it is thought that, when the accretion stops on one point along this track, the star becomes visible and then continues its evolution without accreting matter. Two important features of this "accretion scenario" are the following ones:

1) Starting from the low mass stars and going up towards massive ones, the stars become optically visible at progressively later stages.

2) A given initial mass star becomes optically visible at later stages when the accretion rate decreases.

The consequences of these two facts are very interesting and deserve a brief comment here (see for more details the paper by Bernasconi & Maeder 1996; see also Bernasconi 1996 for grids of pre-MS stellar models):

• There is no longer a ZAMS. Indeed, in case of the most massive stars, for realistic accretion rates of 10^{-5}–10^{-4} $M_\odot$ per year, the accretion phase lasts some 2-2.5 millions years and is thus a significant portion of the H-burning lifetime (it is also much greater than the Kelvin-Helmholtz time scale which is of the order of 70 000 and 30 000 years for a 25 and 50 $M_\odot$ star respectively !). In that case, when the accretion stops the star has already burnt a part of its hydrogen in the central regions. Therefore in this scenario there is no longer a ZAMS at the upper end of the MS, in the sense that there is no longer a position in the HR diagram corresponding to a homogeneous state of the star with the corresponding initial mass, where the H-burning phase begins.

• The visible main sequence (MS) lifetime of massive stars is shortened. This is also true in the scenario of Garmany et al. (1982), but in the accretion scenario, the structure of the star when the star becomes visible is not strictly equivalent to that the star would have had if the star had evolved from a classical ZAMS model. Indeed as a result of previous evolution, the size of the convective core for a given central H content is reduced by about 5 to 10% still reducing the total main sequence lifetime.

• Since some massive young stars are still enshrouded in their parental clouds, the true number of massive stars in young association or starbursts is larger than observed. The "true" IMF is thus flatter than the one deduced from star counting. Moreover the observed initial mass function (IMF) depends on the accretion rate and then on the physical conditions prevailing in the molecular cloud out of which the stars have formed. Indeed if we postulate an intrinsic IMF, the observed one, as deduced from counting stars between theoretical mass bins will get steeper as the accretion rate is lowered. The reason for this is that, when the accretion rate decreases, massive stars will be more and more evolved at the time of their optical appearance. Therefore their visible MS phase will get correspondingly shorter, and they would turn up more rarely. The observations by Massey et al. (1995) who found that the observed IMF for massive field stars are systematically steeper than the observed IMF in asso ciations, could be interpreted as a higher accretion rate in associations than in the field.

• For realistic accretion rates applicable to ordinary star forming region in the Galaxy and Magellanic Clouds, the accretion process leads to a natural truncation of the IMF around 85-150 $M_\odot$ where the accretion time becomes comparable to the H-burning lifetime. Let us note that, according to the rate of accretion, one obtains different truncation of the IMF in its upper part, with the maximum mass which can be formed decreasing as the accretion rate decreases.

These few points show how the accretion process during the PMS phase can deeply modify our view of the H-burning phase of the most massive stars, typically more massive than about 40 $M_\odot$. The impact of such pre-MS scenarios on the output of very young starbursts remains largely to be explored.

3.2. Stars on the Main Sequence

Luminosities, effective temperatures, and surface abundances of main sequence stars should be accounted for by stellar models.

Very good fits of the sequences in the HR diagram for galactic clusters of all ages are achieved by models with modest overshooting (Meynet et al. 1993). To illustrate this point the observed sequence of the young cluster α Persei is represented in Fig. 1. Its age has been estimated equal to 52.5 Myr, by

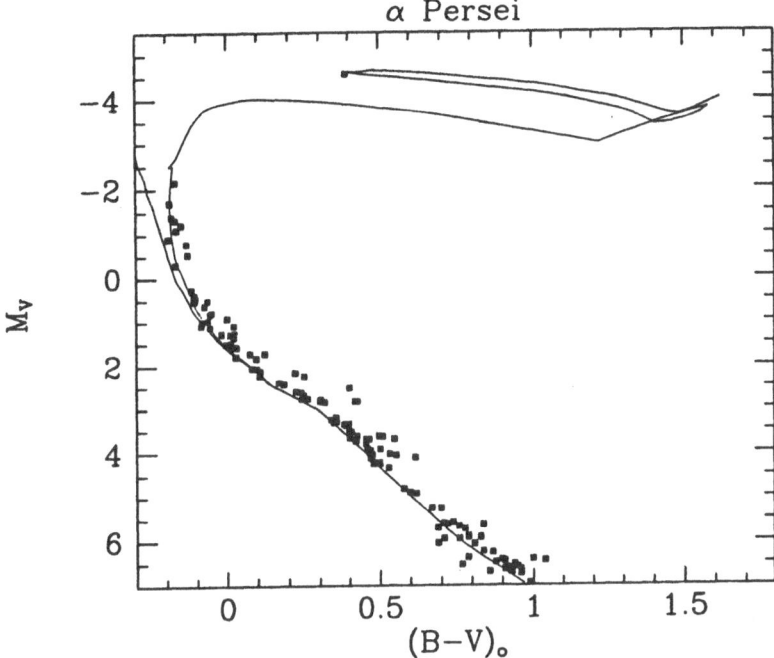

Fig. 1. — Fit of an isochrone computed from models with mass loss and overshooting on the color magnitude diagram for α Persei, $m - M = 6.36$, $E(B - V) = 0.09$, $\log$ age $= 7.72$. The isochrone perfectly fits simultaneously the MS shape and the position of the yellow supergiant α Per (Fig. from Meynet et al. 1993).

adjustment of an isochrone computed from the stellar models of Schaller et al. (1992). The terminal point on the MS is quite well reproduced (note the position of the yellow supergiant, α Per, which is also well reproduced). Similar good agreements have been found for younger and older galactic clusters. Other comparisons seem to indicate that the convective cores, either in the H-burning phase or in the He-burning phase must be greater than the size given by the Schwarzschild criterion (see Maeder & Meynet 1989; Andersen et al. 1990; Chiosi et al. 1992). It is however not clear what is the physical mechanism responsible for this increase. Is it convective penetration or mixing processes induced by rotation (Talon et al. 1997) ?

Generally main sequence stars do not present surface abundances which differ from their initial ones. However Herrero et al. (1992) found that all fast rotators among O-stars show surface He-enrichment. These observations show that rotationally induced mixing is strong enough to transport newly processed elements to the surface during a fraction of the main-sequence lifetime. For the moment these observational features are not yet well accounted for by stellar models. Let us emphasize here that these kinds of problems largely go beyond the resolution of some specific inadequacies between theory and observation. They have also an important impact on our understanding of much more general processes since all stellar properties are influenced by this effect: lifetimes, luminosities, nucleosynthesis, etc . . .

3.3. Evolved Massive Stars

If we consider theoretical evolutionary tracks in the HR diagram for stars more massive than about 2 $M_\odot$, we can recognize four different patterns after the end of the main sequence:

1) Stars with an initial mass $M_{ini} > M_{WR}$ become Wolf-Rayet stars. These stars may or may not go through a very rapid supergiant phase (see next section).

2) Stars with initial masses in between M_u and M_{WR} show no blue loops. Depending on the value for the mass loss rates and/or of the initial metallicity (both quantities are linked by the relation seen in section 2.3.4), the He-burning lifetime will be distributed differently between a blue and a red position in the HR diagram. At solar metallicity, models predict that the whole He-burning phase will be spent at the red location (the star is in a red supergiant stage), while at very low metallicity (or in case of no mass loss by stellar winds), a great portion of the He-burning phase will be spent in the blue location (the star is in a blue supergiant stage, see Chiosi & Maeder 1986).

3) In the mass range $M_l < M_{ini} < M_u$ the evolutionary tracks present extended blue loops. The presence and shape of blue loops depend on many parameters of the stellar models (see a discussion of this point in Maeder & Meynet 1989). The He-burning phase will be shared in consequence between a red and a blue location (at the extremity of the blue loop).

4) Below M_l, the loop never extends beyond 3.8 in $\log T_{eff}$. The whole He-burning phase occurs when the star is a red giant.

The mass limits M_{WR}, M_u and M_l at solar metallicity are about 30, 15 and 6 $M_\odot$ respectively (Schaller et al. 1992). From what precedes one can expect that at solar metallicity, in very young clusters with a mass at the turn off beyond 30 $M_\odot$, no red supergiants should be observed. Decreasing the mass at the turn off (or increasing the age of the cluster), one should first encounter clusters with red supergiants then clusters with blue and red supergiants and finally clusters with only red giants. Quantitative comparisons of this type have been performed by Meynet (1993) and have shown that, at least at solar metallicity, the same models which are able to fit the observed cluster sequences, can reproduce the relative number of red supergiants, with respect to the number of main sequence stars (more precisely of stars present in an interval of two visual magnitudes below the turn off).

What is the situation at other metallicities ? The number ratio of blue to red supergiants (B/R ratio) is a quantity which is known to depend extremely on the model parameters. Indeed supergiants are often close to a neutral state between a blue and a red location in the HR diagram. Even small changes in mass loss, in convection or other mixing processes greatly affect the evolution and the balance between the red and the blue. This is certainly the reason why, at a given metallicity, it is always possible, by slightly modifying the input ingredients of the models, to obtain a good correspondence between observation and theory. As an example, the number of blue to red supergiants in the SMC young cluster NGC 330 and in the LMC cluster NGC 2004 can be accounted for by increasing the mass loss rate by a factor of 1.4-1.6 (Meynet, 1993). Things become much more difficult when one attempts to reproduce the general behavior of this ratio with the metallicity. The main observed trend is that the B/R ratio increases steeply with Z. For M_{bol} between -7.5 and -8.5, the B/R ratio is up to 40 or more in inner Galactic regions and only about 4 in the SMC (Humphreys & McElroy 1984). A difference in the B/R ratio by an order of magnitude between the Galaxy and the SMC was also found on the basis of well selected clusters (Meylan & Maeder 1982).

Among the recent grids of massive star models one finds grids obtained with different assumptions regarding the mass loss rates, the processes of mixing and the binary nature (Claret & Gimenez 1992; Stothers & Chin 1992; Brocato & Castellani 1993; Bertelli et al. 1994; Langer et al. 1994; de Loore & Vanbeveren 1994; Meynet et al. 1994; Mowlawi & Forestini 1994). Recently Langer & Maeder (1995) confronted different stellar models with this observational feature and they concluded that all present day massive star models have problems in reproducing this observed trend.

In general the models with the Schwarzschild criterion (with or without overshooting) can reproduce the B/R ratio for solar composition. However, for what concerns the B/R supergiant ratio, at low Z, these models with current mass loss rates, predict no RSG for initial masses greater than 12 $M_{\odot}$ while many are observed.

At the opposite end, models with Ledoux criterion and models with semiconvection correctly reproduce the B/R ratio for the metallicities of the SMC or LMC but they fail at higher Z. As an example, the models by Arnett (1991) at Z=0.007 are able to simultaneously account for the observed distribution of the blue supergiants (BSG) and the presence of numerous red supergiants (RSG). But these same models predict only BSG at low Z and only RSG at high Z. A behavior which is in contradiction with the observed trend.

As stated by Langer & Maeder (1995), the restricted scheme for convection (Ledoux) looks better at low Z, while the extensive convection with overshooting is better at solar Z. These various results are undoubtedly the sign of something inadequate or missing in the present treatment of mixing in massive stars. The above authors conclude that a mixing efficiency in between the Schwarzschild and the Ledoux criterion might be the most appropriate to explain the B/R-trend with Z, but argue that rotational mixing may have some contribution as well.

As stated at the beginning of this section, the B/R ratio is a quantity which is known to depend extremely on the model parameters. In this sense this feature is a welcome amplifier which can be very useful to constrain the model physics very accurately. But this does not imply that the results of massive star theory have to be questioned altogether, especially the results concerning the WR stars for which the dominant effect is clearly the mass loss through stellar winds (see next section).

What is the situation for what concerns the surface abundances during the supergiant phase ? Walborn (1976, 1988) showed that ordinary OB supergiants have He– and N– enrichment as a result of CNO processing. Only the small group of peculiar OBC–stars have the normal cosmic abundance ratios (cf. also Howarth and Prinja 1989; Herrero et al. 1992; Gies and Lambert 1992). There are two alternatives to explain the He– and N–enrichment in supergiants and both bring about new problems:

1) The first alternative is that blue supergiants are on the blue loops after a first red supergiant stage where dredge–up has occurred, producing the observed surface enrichment. The problems related to this hypothesis are the following: a) At Z=0.02, current grids of models (Arnett 1991; Schaller et al. 1992; Alongi et al. 1993; Brocato and Castellani 1993) only predict blue loops for masses equal or lower than 15 $M_\odot$. At higher masses there are no blue loops and no predicted enrichment. b) Even on the blue loops the predicted enrichment does not seem high enough to account for the observations.

2) Another possibility is that the mixing responsible for the surface abundances of these supergiants occurred already on the main sequence. For instance we may assume that the effects producing enriched O–stars (as rotation, see above) are also active in lighter lower mass stars and that they lead to surface enrichment in supergiants resulting from a larger range of initial masses.

Finally let us mention the observations by Venn (1996), who has shown that most of the A-type supergiants present evidences for partial CN mixing, but at a level which is inferior to what is expected from a first dredge up episode. Does this mean that the convection zone during the red supergiant stage does not go so deep as the models predict or has this mixing another origin ? The question for the moment remains open.

4.   MASSIVE STAR POPULATIONS IN GALAXIES

If luminous matter represents only about one percent of the mass in the universe, massive O- and B- stars ($M_{ini} \geq 8 \ M_\odot$) contribute to about two thirds of the optical light of galaxies. One of the main properties of the massive stars is indeed their high luminosity ($\log L/L_\odot > \sim 3.5$ on the main sequence) which make them detectable, either as individual objects or by spectral features in the integrated light spectra of galaxies, at large distances. The observations by Pierce et al. (1994), Freedman et al. (1994), Ferrarese et al. (1996) of Cepheids in galaxies of the Virgo cluster at distances superior to 15 Mpc has considerably extended the region of the universe where individual massive stars have been detected. With the VLT the brightest blue supergiants ($M_V \sim -10$) will be seen up to distances of about 500 Mpc that means well beyond the Coma cluster. But already today it is possible to detect massive stars indirectly at such large distances. Indeed, massive stars may manifest themselves by the presence of some specific signature in the spectrum of the integrated light of galaxies. As an example, Wolf-Rayet (WR) stars may betray their presence through a small bump observed at 4686 Angström. This emission line has the particular property of being much broader than the other emission lines since it is generated in the expanding envelope of a WR star, while the other lines are nebular recombination lines. The ratio between the luminosity in the stellar broad line and that in one of the narrow nebular emission lines can be related to the number ratio of WR to O-type stars (Kunth & Sargent 1981; Vacca & Conti 1992). These data, as we shall see below, provide a very interesting diagnostic on what happened in these galaxies. At the present time, one of the most distant galaxies in which a "WR feature" has been detected is the IRAS galaxy 01003-2238 believed to be at a distance of 470 Mpc (Armus et al. 1988). Massive stars are thus privileged tools to study stellar evolution in various environments differing among other things by the metallicity and/or the intensity of the star formation rates.

Massive stars are not only the main sources of UV and ionizing radiation, they also power the far-IR luminosities of galaxies through the heating of dust. They also inject mass and momentum in the interstellar medium. As a spectacular illustration of this last property, let us mention the Homonculus nebula around η Carinae. This nebula is due to an eruption which occurred 150 years ago and expelled in less than 30 years, about 2 solar masses (Davidson 1989). Massive stars are very efficient nuclear reactors, processing great quantities of matter in short periods of time. Thus they contribute in an important manner to the enrichment of the interstellar medium in new synthesized heavy elements. They might also contribute to the chemical evolution of the intergalactic medium. Indeed some galaxies undergoing very powerful star formation events lose matter through galactic superwinds. For instance, M82 is believed to loose 1.3 $M_\odot$ per year, the ultraluminous galaxy Arp 220, 50 $M_\odot$ per year (Heckman et al. 1990). One of the mechanisms proposed for the ejection of this matter is the energy and momentum injections by massive stars through their stellar winds and their final explosion as a supernova. These ejections of matter may have strong influence on the way the galaxies evolve chemically. Indeed matter, which has been essentially enriched by massive stars, leaves the galaxy, contributing no longer to the enrichment of the galactic interstellar medium. Inversely, new synthesized heavy elements are injected in the intergalactic medium. This phenomenon is likely to be responsible for the high iron abundances observed in the X-rays in the intracluster medium (Arnaud et al. 1992).

To summarize, massive stars are great injectors of energy mass and momentum. As a consequence of this main property, they cannot live for long and thus trace the young, newly born stellar populations. They are the progenitors of type II, and possibly Ib/Ic supernovae, leaving after their nuclear life either a neutron star or a black hole.

In this section we shall study how the massive star populations are expected to evolve in environments differing by the metallicity and the star formation rate history. For this purpose we shall use the evolutionary population synthesis techniques, which enable us to follow the evolution as a function of time of the number of stars of different types, given a star formation rate history and an initial mass function (which may also evolve as a function of time). The most direct observational quantities to which outputs of such models could be compared are obviously the observed number of stars of various types in a given region. Of course, in most cases, the studied region is too far away for any individual star to be detected. In that case the observation provides access only to the integrated light of the stellar population and the observed quantities can only be compared with spectrophotometric models of evolving stellar populations.

4.1. The Wolf-Rayet Stars

Since populations of Wolf-Rayet stars will be very often referred in the next paragraphs, we recall here some basic facts about these stars.

4.1.1. Observed Properties

Wolf-Rayet stars exhibit, in addition to absorption lines corresponding to OB spectral types, numerous huge emission lines from essentially He, N, C, O Si (see the review by Abbott & Conti 1987, Willis 1996). These emission lines, formed in an expanding extended atmosphere, are indicative of strong stellar winds. Another feature testifying to the important mass loss undergone by these stars or their progenitors, is the presence around a significant number of WR stars of ring nebulae (Marston et al. 1994; Smith 1996).

Two main families of WR stars exist: the WN series in which the products of CNO-burning appear at the stellar surface, and the WC series which exhibit the products of partial He-burning (mainly ^4He, ^{12}C, ^{16}O, ^{22}Ne; see typical spectra of WN and WC stars in the review by Melnick 1992). Typically one has the following abundance ratios (in number):

	H/He	C/He	N/He	O/He
Solar System[1]	10	0.004	0.001	0.009
WN[2]	0-2.5	$\leq$ 0.004	0.003-0.008	
WC[2]	0	0.1-0.7		0.02-0.05

[1] given by Anders & Grevesse 1989
[2] given by Nugis 1991, Willis 1996

The actual WR masses are in the range $5 - 50$ $M_\odot$ (Massey 1982), with luminosities in the range of $\log L/L_\odot = 4.5 - 6.0$ and effective temperatures between 30 000 and 90 000 K (note that the observed effective temperature generally characterizes a layer in the dense wind surrounding the star, see the introduction of Sect. 3). Empirical positions of WR stars in the HR diagram are given in Hamann & Koesterke (1996).

WN stars have mass loss rates of the order of $10^{-4.5} - 10^{-4}$ $M_\odot$ per year (cf. Leitherer et al. 1995a). WNE and WC stars seem to have mass loss rates which depend on the actual mass of the star. Indeed Abbott et al (1986) proposed a relationship of the form $\dot{M} = 7 \ 10^{-8} \ (M_*/M_\odot)^{2.3}$ (in solar masses per year), where M_* is the actual mass of the WR star. They deduce such a relation from the determination of both the masses and the mass loss rates in detached binary systems containing a WR star. Langer (1989), from comparisons between outputs of stellar models computed with various prescriptions for $\dot{M}$ and observed characteristics of WR stars, deduced a relationship of the form $\dot{M} = (0.60 - 1.00) \ 10^{-7} (M_*/M_\odot)^{2.5}$ for WNE and WC stars respectively. According to Bandiera & Turolla (1990) this dependence of $\dot{M}$ on M_* seems to be an intrinsic feature of radiative wind acceleration in He-stars. Contrary to O-type stars, for WR stars, there is no indication yet about a dependence of $\dot{M}$-values on initial metallicity.

Let us end this brief description of the observed properties of WR stars by some comments on their number. At the present time, 201 WR stars have been observed in the Galaxy, 114 and 9 in the Large and Small Magellanic Cloud respectively (LMC and SMC, see van der Hucht 1996). The evaluation of their total number and distribution in the Galaxy is made uncertain by the fact that the inner galactic regions are obscured by dust, so that the WR catalogues are complete only at distances $R \sim 2.5$ kpc from the Sun, where $N(R) \approx 100$ WR stars have been observed up to now (van der Hucht 1995). From an extrapolation of the data concerning the WR density in low-metallicity regions like the Large and Small Magellanic Clouds and in the solar neighborhood, the number of galactic WR stars can be roughly estimated from $N_{\mathrm{WR}} \approx N(R)(R_\odot/R)^2$, where $R_\odot = 8.5$ kpc is the galactocentric distance of the Sun. This leads to $N_{\mathrm{WR}} \approx 1100$. Let us note that this number is likely a lower limit, since the star formation rate and the ratio of the number of WR to O-type stars increase significantly in the inner galactic regions (see section 4.2).

Some WR stars have been observed in other galaxies of the Local Group as M31 and M33 (see the review by Massey 1996), so that about 500 individual WR stars have been detected (5 per WR star's fan !).

4.1.2. Theoretical Models

A short historical account of the theoretical interpretations of the WR features is given in Maeder (1996). WR stars appear as a normal evolutionary stage of massive stars, with initial mass greater than $\sim 25 - 30$ $M_\odot$ for population I stars. This mass limit depends on the initial metallicity (Maeder 1991; Maeder & Meynet 1994). Due to strong stellar winds, the original stellar envelope is removed and nuclear matter which was initially in the CNO core (WN stars) and in the 3α-burning core (WC stars) appears at the surface. While the surface evolution is considerably influenced by mass loss, the evolution of central regions undergoes limited effects, so that most WR stars are likely to further explode as supernovae (Maeder & Lequeux 1982). The explosion must typically be of the same kind as Cas A and would likely be classified as Type Ib supernova. It was estimated that there is one supernova from a WR progenitor for, say, 5 − 7 supernova events in our Galaxy (van den Bergh and Tamman 1991).

As a consequence of their simple interior structure (nearly homogeneous stars), WR stars follow a mass-luminosity relationship (Maeder 1983). Moreover, since at least during the WNE-WC stages, $\dot{M}$ seems to be proportional to the actual mass of the star, other relations can be established between the mass, the effective temperature, the radius, the chemical composition (Schaerer & Maeder 1992). Maeder (1985) showed also that WR stars are vibrationally unstable with periods between 15 and 60 minutes. The high temperature dependence of the nuclear reaction rate at the center of the star is the driving mechanism of the pulsation ("ϵ−mechanism"). It is likely that this instability contributes to drive the huge mass losses undergone by these stars.

4.1.3. Theory Versus Observations

Many observed features of WR stars are well reproduced by theoretical models (see the review by Maeder 1995, 1996 and the references therein; see also the paragraphs 4.2 and 4.3 below). Let us emphasize however that this good correspondence is achieved under the assumption of higher than standard mass loss rates during the pre-WR and WNL phases (Maeder & Meynet 1994). Does this mean that the observed mass loss rates during these phases are too low ? This is not necessarily the case. In many respect an additional mixing process, as the one induced by rotation, would act as an increase of the mass loss rate. In the following, in the absence of grids of rotating stellar models (work now in progress), we shall discuss results obtained with these enhanced mass loss rate models (Meynet et al. 1994), keeping in mind that they likely represent a first approximation of more sophisticated future stellar models.

4.2. WR Star Populations in Galaxies of the Local Group

4.2.1. Populations of WC Stars

WC stars may be classified as WC9, WC8, ... down to WC4 star, the WO class corresponding to the class immediately following the WC4 class. Each WC subtype corresponds to different values of the (C+O)/He ratio (Smith & Hummer 1988; Smith & Maeder 1991). More precisely this ratio is found to increase going from the WC9 to the WC4-WO subtypes (see fig. 2 for a more quantitative comparison). This sequence thus corresponds to a sequence of more advanced chemical processing becoming visible at the stellar surface.

Some observed features of the WC populations are resumed below (cf. Smith & Maeder 1991):

1) WC stars are more numerous in the inner galactic regions.

2) The late WC subtypes WC9-WC8 are only found in inner metal-rich galactic regions, while in metal poor dwarf galaxies, only early WC4 and WO type stars are found.

3) The luminosity of a given WC subtype at low metallicity is smaller than the luminosity of the same WC subtype at high metallicity. This statement is important to bear in mind when one tries to deduce the number of WC stars from characteristics of the integrated light of young starburst regions; luminosity of WC stars of the correct metallicity should be used.

4) In a region of more or less constant metallicity, the luminosity of the early WC stars are smaller than the luminosity of the late WC stars.

Points 1 to 3 may be understood as consequences of the metallicity dependence of the mass loss rates (Maeder 1991). The higher $\dot{M}$ expected in metal-rich regions favor the appearance of the WC stage (point 1). From fig. 2, one can see that, for models presenting a WC phase, higher is the metallicity, lower is the (C+O)/He ratio at the entrance of the WC phase and thus later will be the WC subtype. The reason for this behavior was explained by Maeder (1991; cf. also Smith & Maeder 1991): when the masses (and/or the metallicities) are higher, the mass loss rates are more important. Thus the star enters the WC phase at an earlier stage of He processing. As a consequence, entry at WC9 only occurs at high metallicity (typically greater than $Z = 0.020$, point 2), Theoretical models predict also (see fig. 2) that stars of a given WC subtype are brighter in galaxies with low Z and that, at a given Z, late WC stars are brighter than early WC stars (points 3 and 4).

4.2.2. Number Ratios of WR to O-type Stars

When one compares the number of Wolf-Rayet stars with that of their progenitors, *i.e.* the O-type stars, one finds that this ratio increases when one moves towards the inner galactic regions. Despite the many claims that this increase is due to changes of the star formation rate and/or to changes of the initial mass function, it appears now that the main effect explaining this behavior is the metallicity-mass loss rate relationship (Maeder et al. 1980; Maeder 1991; Maeder & Meynet 1994). Higher the metallicity, stronger are the mass losses through stellar winds. Thus the earlier in the evolution that the star will become a WR star, longer will be the He-burning phase and thus the WR phase. Moreover when the metallicity increases the minimum initial mass which becomes WR is lowered, and due to the shift of the evolutionary tracks towards lower effective temperatures, the O type phase (defined by effective temperatures on the MS greater than 33 000 K) is shorter. All these facts contribute to an enhancement of the number ratio of WR to O type stars when Z increases.

Using the models by the Geneva group obtained with an enhanced mass loss rate during the main sequence and the WNL phase, and making the hypothesis of a constant star formation rate during at least the last 10 millions years, it is possible to reproduce the WR/O, WN/WC ratios observed in the Milky Way

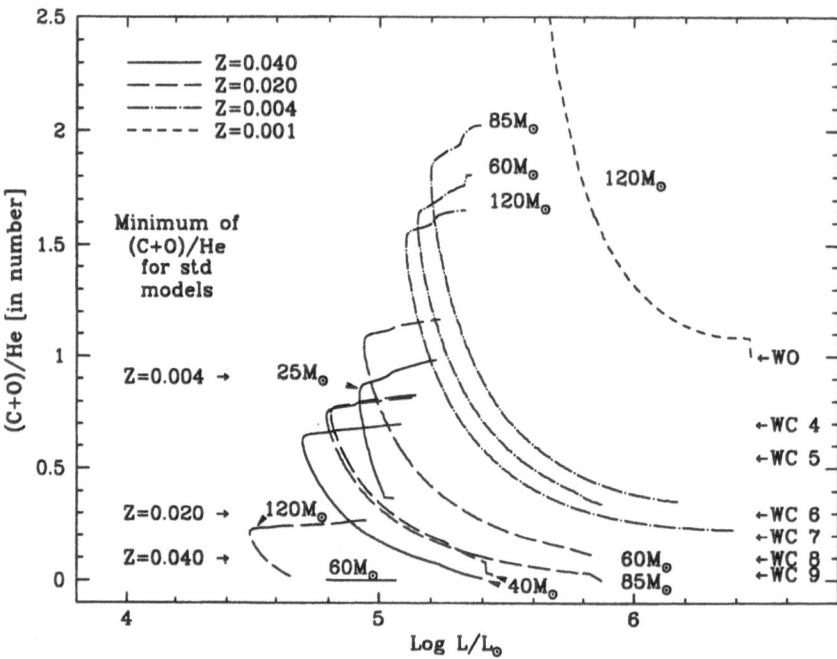

Fig. 2. — Theoretical evolution of the abundance ratio of carbon plus oxygen to helium (in number) at the surface of WC stars as a function of the luminosity. Evolution
proceeds from right to left. The initial masses and metallicities are indicated. On
the left, the minimum values of (C+O)/He predicted by models computed with lower
mass loss rates are given. On the right, the values of (C+O)/He at the entry point
in the different WC subtypes are indicated (cf Smith & Maeder 1991). Figure taken
from Maeder & Meynet (1994).

and in various galaxies of the Local Group (Maeder & Meynet 1994). Of course
these population syntheses models account only for the WR stars produce by
the evolution of single stars. However, many kinds of evidence show that mass
transfer through Roche Lobe overflows plays a limited role in massive star
evolution (see the discussion in Maeder 1996).

4.3.   Massive Star Populations in Starburst Galaxies

An important population of hot stars in a distant galaxy may affect its integrated light spectrum in many different ways:

1) If a great number of massive stars are present, great quantities of short

wavelength, ionizing photons are emitted: all stars having masses higher than 10 - 20 $M_\odot$ emit ionizing photons with wavelength of a few hundred Angströms; O and WR stars emit about 1/3 of their flux below 912 Angströms. Some of these photons may be absorbed and re-emitted in the stellar winds within about 10^{13}cm (about 10 stellar radii, Leitherer 1996). These processes contribute essentially to shape the UV part of the spectrum.

2) Since the stellar wind is optically thin to a significant fraction of the far UV photons, many of them are absorbed further away from the star, typically at radii of 10^{20} cm or tens of pc (Leitherer 1996). Indeed the opacity of the interstellar medium below 912 Å is very high and essentially all stellar far UV photons are absorbed and re-emitted in one or the other optical or IR emission lines.This process is responsible for the great number of strong emission lines superposed on to the optical continuum. This leads in fact to the formation of a classical H II region spectrum.

3) Radiation from massive stars can also be absorbed by dust and reemitted in the form of IR radiation (see e.g. Telesco 1993).

4) Free-free transitions in the ionized gas and synchrotron radiations from supernova remnants are responsible for the radio continuum emission (cf. Mass-Hesse & Kunth 1991).

4.3.1. Hot Star Populations from Optical Spectra

The optical spectrum of a starburst region is essentially composed of a continuum characteristic of late type B and early type A stars, superposed with numerous nebular recombination lines (see examples of starburst spectra from 1200 to 9000 Anströms in Kinney et al. 1996). In some cases, in addition to these nebular lines, a broad stellar He II $\lambda4686$ emission line is visible, indicating the presence of WR stars (see the introduction of this section). He 2-10 was the first object in which a feature resembling WR bands was detected (Allen et al. 1976). The first quantitative WR/O ratio inferred from the integrated spectra of a starburst galaxy was derived by Kunth & Sargent (1981). These authors found a strong He II $\lambda4686$ emission-line in the spectrum of the dwarf galaxy Tol 3, which they attributed to WR stars. Using the observed flux in this emission line to deduce the number of WR, and the H_β flux to obtain the number of O stars, they deduced a WR/O ratio of the order of 1/3 to 1. They interpreted such a high value for this ratio, roughly one order of magnitude above the value found in the solar neighborhood, as an evidence against a monotonic process of star formation. Since that time, this interpretation has been confirmed by many other works and the starburst nature of these objects seems now well established (see the recent review on massive star populations in nearby galaxies by Maeder & Conti 1994). At the present time, about 70 galaxies, or regions in galaxies, present a broad He II emission feature at 4686 Å attributed to WR stars (see the catalogue by Conti 1991).

Vacca & Conti (1992) have estimated, from long-slit optical spectra, the number ratio of WNL to O-type stars in 14 starburst galaxies. They used the same method as the one devised by Kunth & Sargent (1981). The most striking results of these observations are the following:

1) In general, the number ratios in these galaxies are larger than those expected in regions with a constant star formation rate. As already suspected by Kunth & Sargent (1981), this may well be the signature of a starburst episode and indeed, the population synthesis models for starbursts (see for instance Arnault et al. 1989; Mas-Hesse & Kunth 1991; Cerviño & Mas-Hesse 1994; Leitherer & Heckman 1995; Meynet 1995; Schaerer 1996; Schaerer & Vacca 1996) much better account for the observed ratios than the models for constant star formation regions. The slope of the IMF in starburst regions is quite a debated question. At the present there is no clear indication for the existence of a top heavy IMF in starburst regions. The observations by Vacca & Conti (1992) are compatible with a Salpeter's IMF (Schaerer 1996). What is the role of binaries in this context ? According to Cerviño et al. (1996), Schaerer & Vacca (1996), most of the observed starbursts are too young to have formed Roche Lobe Overflow WR stars.

2) The observed points do not draw a well defined relation, but instead cover a certain region in the WNL/O versus metallicity plane (see for instance figure 6 in Meynet 1995). This arises from the fact that in a starburst region, contrary to the case of a constant star formation zone, stellar populations have not reached an equilibrium state. Depending on the age of the starburst, *i.e.* on the time elapsed since the burst of star formation, on its intensity, duration, different values of WNL/O can be reached at a given time after the starburst. However one expects that the upper envelope of the observations in this plane be at or below the maximum values reached after an instantaneous burst of star formation.

3) Despite the scatter of the observed points, one notes however a tendency for increasing WNL/O ratios when the metallicity increases. This confirms the trend already observed in constant star formation rate regions. Let us note also that most starburst regions presenting a WR bump in their spectrum are metal weak. According to Conti (1994) this might be due to the fact that galaxies with high Z are likely to have a brighter underlying stellar population which could "drown out" the WR population.

4) As a natural consequence of stellar evolution, one expects that some starburst regions host WC stars. Depending on the mass loss rates used to compute the stellar models, quite different WC populations are expected (Meynet 1995). Using enhanced mass loss rate stellar models, Schaerer & Vacca (1996) estimate that, for metallicities $1/5 \leq Z/Z_\odot \leq 1$, 30% of WR galaxies should be dominated by WC stars if star formation occurred on time scale short compared to the lifetime of massive stars. Do we observe these WC stars ? Vacca & Conti (1992) report the presence of only WNL stars in nearly all the starburst galaxies of their sample. Indeed, from the absence or very weak strength of C III $\lambda4650$ and C IV $\lambda5808$ emission lines, characteristic of the spectra of WC stars, these authors suggest that the WC to WN number ratio, in most of the

galaxies of their sample, must be much smaller than the value found in the solar neighborhood (WC/WN$_\odot$=0.92, Conti & Vacca 1990), but perhaps similar to that found in the LMC and the SMC (WC/WN$_{LMC}$=0.26, WC/WN$_{SMC}$=0.13, see Breysacher 1981 and Azzopardi & Breysacher 1979, 1985). Only for He 2-10 A do they suggest the possible existence of a significant WC population. Presently very few "WR-galaxies" (about 10%) present a well established WC population (Schaerer & Vacca 1996), but no systematic searches have been performed. Moreover observations with signal/noise ratio superior to about 20 are required to derive the observed frequency of WC stars in WR galaxies (see e.g. Schaerer et al. 1997). Thus the apparent absence of WC stars in starburst regions might be due either to the fact that they have been overlooked in many previous observations or to some missing physical ingredient in the stellar models and/or in the population synthesis models. More observations are needed to settle on firmer ground this issue.

The procedure of deducing number ratios of WR to O-type stars from optical spectra suffers from several drawbacks. Two of these are :

1) The nebular emission lines, used to estimate the number of hot stars, originate from a much more spatially extended region than the stellar emission line, used to deduce the number of WR stars. Depending on the way the slit has been oriented, one will miss a more or less great part of the nebular light. According to Conti (1994), arbitrarily correcting for this would lower the ratio by a factor 2 (see also Conti 1993).

2) In Vacca & Conti (1992), the number of O-type stars is estimated by dividing the Lyman continuum luminosity, deduced from the H$_\beta$ line intensity, by the ionizing luminosity of an average O-type star on the ZAMS. However, as pointed out by Schaerer (1996), one should use in this estimate, not the ionizing luminosity of an average O-type star on the ZAMS, but the ionizing luminosity of an average O-type star having the age of the starburst (about 3-4 Myr for starbursts containing WR stars). Since the ionizing luminosity of an O star, after about 1 Myr, decreases with time, one needs more O-type stars to produce the observed ionizing power. Due to this effect, the true WR/O number ratios are systematically lower than those given in Vacca & Conti (1992) by a factor between 2 and 10.

From point 2 above, one sees that it is not possible to extract from the observed data the correct number of O-type stars without knowing the age of the starburst. Thus it is preferable to build population synthesis models predicting directly observable quantities, like as for equivalent widths of stellar and nebular emission lines (see for instance Schaerer 1996, Schaerer & Vacca 1996). Among the important quantities modelled in this manner, there is the equivalent width of the H$_\beta$ line, $W(H_\beta)$ (see e.g. Cerviño & Mas-Hesse 1994). This quantity measures the ratio of the ionizing flux to the continuum flux and is thus proportional to the number of massive stars (with initial mass superior to 20 $M_\odot$) divided by the total number of stars. After an instantaneous burst of star formation, $W(H_\beta)$ decreases as a function of time and thus constitutes a very good age indicator.

Up to now, we have examined the properties of starburst regions using characteristics of mainly the ionized gas. Let us now study the light coming directly from the hot stars, namely the UV part of the spectra.

4.3.2. Hot Star Populations from UV Spectra

Images in UV of the WR galaxy He 2-10 can be seen in Conti & Vacca (1994, HST faint object camera with a filter at 2200 Angströms). The central starburst has been resolved in several knots having diameters less than or equal to about 10 pc. Using the evolutionary population synthesis models of Bruzual & Charlot (1993), Conti & Vacca (1994) estimate, for an age around 3 Myr, a mass of $3 \ 10^6 \ M_\odot$ for the brightest knot. The masses, luminosities and diameters of these knots are consistent with those one would expect to observe for very young globular clusters (see also Meurer et al. 1995). Of course this interpretation would be strengthened by more direct measurement of the mass. This can be achieved through measurements of the radial velocity dispersion, which can be related to the total mass by the Virial theorem. Ho & Fillipenko (1996), with the Keck telescope, succeeded in measuring this radial velocity dispersion in one of the bright knot of the dwarf galaxy NGC 1569 (at 2.5 Mpc). They deduce from this observation a total mass for the knot of 3.3 ± 0.5 $10^5 \ M_\odot$ fully consistent with typical values of galactic globular clusters.

It is interesting to compare the number of O-type stars deduced from this observation at 2200 Angströms with that deduced from the H_β line intensity. On the base of the UV image, Conti & Vacca (1994) estimate the average number of O-type stars in a knot to be around 3400; this means 9 times the number of O-type stars in 30 Doradus and 6 times the number deduced from the H_β line intensity. As emphasized in the previous paragraph, the number of O-type stars deduced from the H_β line intensity may be underestimated due to a possible mismatch between the slit position and the nebular emission line region. Part of the discrepancy may also be due to the decrease of the ionizing luminosity of the O-type stars as the age increases.

Modeling of the spectrophotometric evolution of a starburst in the UV has revealed many interesting features: the ratios of the equivalent widths of the Si IV and C IV is a sensitive indicator of the spectral type of the dominant stellar population (Mas-Hesse & Kunth 1991). Combined with age informations deduced from $W(H_\beta)$, they can be used to disentangle age and IMF effects (Cerviño & Mas-Hesse 1994). The evolution of the profile of the CIV $\lambda1550$ line after an instantaneous starburst has been computed by Leitherer et al. (1995b, other lines and star formation rates are discussed in this paper). As long as stars with initial masses greater than about 50 $M_\odot$ are present on the main sequence, the profile keeps a well marked P-Cygni shape. As evolution proceeds, the profile reduces to an absorption line. This feature is thus a powerful age indicator. It has been used by Conti et al. (1996) to study one of the most luminous WR galaxies, NGC 1741 (merging system also known as Hickson 31, distance 51 Mpc, metallicity $\sim Z_\odot/4$). As He 2-10, this galaxy hosts discrete knots which are the main contributors to its UV luminosity. By fitting of the Si IV and C IV lines emitted by one of the knots (NGC 1741 B), Conti et al. (1996) obtain the following characteristics:

1) The population in this knot appears to result from a very short burst of star formation which occurred a few Myr ago. A good fit is obtained with a Salpeter's IMF and a maximum mass of $\sim 100\ M_\odot$.

2) From the WR bump at 1640 Angström (HeII line), they first obtain a confirmation of the age of a few Myr and deduce a number of WN stars equal to 500.

3) From the continuum luminosity in UV, they infer the presence of 10 000 O type stars.

These values are in good agreement with those obtained from the analysis of the optical spectra: 710 WN stars were deduced from the observation of the He II line at 4686 Angström and 17 000 O-type stars from the intensity of the H_α line. The slightly greater values deduced from the optics may be explained by the greater extent of the observed region in this wavelength range. The authors note also that the intensity of the optical continuum is much greater than the intensity predicted by models which fit the UV part of the spectrum, indicating the presence of an older underlying population. Thus, in this case, in contrast with the situation described above for He 2-10, one has a good agreement between the results obtained from respectively the optical and UV parts of the spectrum.

A very interesting point noted by Conti et al. (1996) is the striking resemblance between the spectrum of 1512-cB58 (Yee et al. 1996), a very remote galaxy (redshift equal to 2.72), with that of NGC 1741 B1. The galaxy 1512-cB58 is the most luminous non AGN galaxy observed so far ($M_V = -26$ for $H_0 = 75$ km s^{-1} kpc^{-1}). Its luminosity is two to three orders of magnitude greater than that of NGC 1741 (note however that the high luminosity might be an effect of gravitational lensing). From the similarity of the spectra in UV, one can wonder to what extent the burst mode of star formation, which is present in NGC 1741, could have some relevance in case of cB58 ? Are there knots as in NGC 1741 ? Whatever be the answers to these questions, there is some hope that the study of near starburst regions will shed new light on the processes active in far remote regions.

4.3.3. Hot Star Populations from FIR

The continuum emission from UV to FIR of a typical FIR luminous starburst galaxy is discussed in Telesco (1993). In these systems, the dust governs the energy transport. It absorbs stellar radiation, heats up to temperatures around 30-50 K, and reemits it in the wavelengths range from $10\mu m$ to a few $100\mu m$. Numerous emission line and absorption bands superpose over the continuum.

In order to link the emissivity in the IR to the stellar population, one must know which are the heating sources of the dust. Among the most important ones are (see Mas-Hesse & Kunth 1991):

1) The Lyman continuum radiation ($\lambda < 912$ Angström). A fraction of the ionizing photons emitted by the stars are directly absorbed by the dust.

2) The non-ionizing stellar continuum radiation.

3) The Ly α line radiation. As indicated in point 1 above, only a fraction of the ionizing photons are directly absorbed by dust. The other fraction will be absorbed by the interstellar hydrogen and produce nebular emission lines (Balmer series and Ly α lines). Photons emitted in the Ly α line are then absorbed with high probability by the dust.

4) Other sources contribute, like the nebular continuous emission, the rest of the nebular emission lines (other than the Ly α line) or the IR radiation emitted by dust. There are great uncertainties concerning the importance of the contribution of stars still embedded in their protostellar cloud (see section 3.1).

Evolution of the FIR luminosity, L_{FIR}, after a starburst episode can be seen in Mas-Hesse & Kunth (1991). The good correlation between the evolution of the FIR luminosity and the UV luminosity indicate that, in these models, the major contribution to the FIR luminosity comes from the stellar continuum. As a numerical example, four millions years after an instantaneous starburst, the Lyman continuum radiation contribute upto 6.5% to the L_{FIR}, the non ionizing radiation upto 65.5%, the Ly α nebular emission line upto 8% and the other sources upto 20%. Mas-Hesse & Kunth (1991) show also that the IR excess, *i.e.* the ratio between the FIR luminosity and the heating of the dust produced by absorption of the Ly α radiation (see Mezger et al. 1974), increases as a function of time after a starburst episode. This is a natural consequence of the fact that when the most massive stars (*i.e.* the ionizing sources responsible for the existence of the Ly α line) disappear, the contribution of the stellar continuum to L_{FIR} becomes more and more important.

4.3.4. Red Supergiants and Supernovae in Starbursts

The evolution of the V-K photometric index as a function of time correlates well with the number of red supergiants after a starburst episode (Cerviño & Mas-Hesse 1994). These authors show the great dependence of the expected number of red supergiants on the initial metallicity. In this respect it would be quite interesting to know if the observed metallicity dependence of the blue over red supergiant ratio found in constant star formation rate regions (see Langer & Maeder 1995) appears also in starburst regions.

In general the Geneva and Padova stellar models agree quite well, and, when used as ingredients of population synthesis models, give very similar outputs. However Charlot (1996) noticed that the Geneva tracks produce redder colors (greater B-V and V-K) during the period extending from about 5 to 16 Myr after an instantaneous starburst. For these ages, Geneva tracks predict red supergiants which are absent from the Padova ones. The physical origin of this difference is difficult to trace back, but whatever it is, observations of the number of red supergiants in young stellar clusters should permit to test the stellar models. The high number of red supergiants in the young open clusters NGC 884 and NGC 4755 seems to point towards the existence of the red supergiants predicted by the Geneva tracks (Meynet 1993). However more observations are needed to set this conclusion on firmer ground.

Once all the most massive stars end their nuclear life, the starburst should become a supernova burst. Predictions for the supernova rate after a starburst can be found for instance in Cerviño & Mas-Hesse (1994), Leitherer & Heckman (1995), Meynet (1995). These models show that the maximum of the supernova rate occurs typically 3 to 4 Myr after the burst when the most massive stars explode. Let us recall here that stars in the mass range from 40 to 120 $M_\odot$ have more or less the same lifetime and thus end their life approximately at the same time. Moreover, the first supernovae to explode will have WR stars as their progenitors and thus will likely be classified as type Ib supernova.

A rule of thumb suggests that the simultaneous birth of 13 500 O type stars gives rise to a rate of about 1 supernova per century at the maximum (Meynet 1995, case with a flat IMF). This means that for one of the youngest starbursts known (IRAS 01003-2238, Armus et al. 1988), the rate of supernovae could amount to ∼ 45 supernovae per century, slightly less than one every two years. The presence of numerous massive stars, which are great injectors of matter and energy through their stellar winds and their final supernova explosion,probably contribute to drive galactic superwinds (see the introduction of this section).

The supernovae leave neutron stars, sources of a non thermal radio emission produced by synchrotron radiation. Thus one can expect that the appearance of the supernovae will affect the radio continuum emission emitted by a starburst region. According to the models by Mas-Hesse & Kunth (1991), during the first 3 Myr after an instantaneous starburst, only thermal emission produced by free-free transition in the ionized gas contributes. During this period the radio continuum emission between 6 and 20 cm is expected to have a spectral index around -0.1. After 3 Myr, the spectral index decreases as an effect of both the decrease of the ionizing power of the stellar population and of the increase of the number of supernova remnants. After about 10 Myr, the spectral index stabilizes around -0.9, value characteristic of synchrotron radiation emission.

5.   CHEMICAL EVOLUTION

In this section we shall first briefly recall the main nucleosynthetic processes, sites and time scales. Then we shall discuss the dependence of the stellar yields on the metallicity. We shall end by discussing the effects of different star formation histories on the chemical evolution of galaxies.

5.1.   Nucleosynthetic Sites, Processes and Time Scales

One can distinguish essentially three main nucleosynthetic sites:

1) The nucleosynthesis which occurred during the Big Bang (primordial nucleosynthesis).

2) The stellar nucleosynthesis.

3) The nucleosynthesis induced by the interaction of the cosmic rays with the interstellar medium.

In the following we shall concentrate on the stellar nucleosynthesis which is responsible for the production of the bulk of the heavy elements.

Many books and reviews have been devoted to the presentation of the nucleosynthetic processes (Clayton 1968; 16th Saas-Fee Course, 1986, see the references to Chiosi or Woosley 1986 below; Rolfs & Rodney 1988; Arnett 1996). Here we shall simply recall some well known facts which are worthwhile to keep in mind to follow the discussions presented in sections 5.2 and 5.3.

5.1.1. CNO Cycle

The CNO cycle is the main H-burning process in main sequence stars with central temperatures superior to 18×10^6 K (initial mass > 1.5 $M_\odot$). CNO elements act as catalysers of the proton fusion reactions. The sum of their abundance remains constant during this process, but due to the very small rate for the reaction $^{14}N(p,\gamma)^{15}O$, most of the ^{12}C and ^{16}O transform into ^{14}N. As a numerical example, initially (*i.e.* before any CNO processing) number ratios of C/N and O/N amount to 4 and 9. When CNO has reached equilibrium, C/N and O/N take values as low as 0.03 and 0.02 respectively. Let us recall that in cosmic material, CNO represent about 80% of the heavy elements mass fraction, oxygen alone contributing to about the half of the heavy element mass fraction.

In addition to the CNO cycle, two cycles/chains are active: the Ne-Na cycle and the Mg-Al chain. These processes are responsible for the synthesis of some ^{23}Na and ^{26}Al (this last element is important for γ-ray line astronomy, see the review by Prantzos and Diehl 1995 on this subject).

The main products of H-burning are ^4He and ^{14}N. The ^{14}N formed by the CNO cycle is said to have a secondary origin or to be a secondary element, because it is formed from elements initially present in the star. Indeed the ^{12}C and ^{16}O from which ^{14}N forms were synthesized in previous generations of stars. Now one can imagine processes leading to the formation of primary nitrogen. Let us suppose that due to some mixing process, ^{12}C formed in a He-burning shell are injected into a H-burning zone. The protons can then react with the ^{12}C formed by the 3α process to form first ^{13}C and then ^{14}N. The ^{14}N formed in this way would be primary. Indeed it is formed from heavy elements synthesized in the star and not in previous generations of stars. Whether such a process is active in real stars remains an open question, but, as we shall see in the last paragraph of this section, there are some indications that it actually does occur.

5.1.2. He-burning Phase

At the beginning of the He-burning phase, ^{12}C is synthesized by the 3α reaction. At the end, part of the ^{12}C is transformed into ^{16}O by the reaction ^{12}C$(\alpha,\gamma)^{16}$O. This last reaction plays an important role in stellar evolution. It determines the chemical composition of the core at the end of the He-burning phase (typically at the end of the He-burning phase of a 20 $M_\odot$, one obtains 78% of ^{16}O and 19% of ^{12}C in mass fraction, see Schaller et al. 1992). It affects the end of the evolution of massive stars, by determining the quantity of available fuel for the carbon burning phase. The rate of this reaction is likely enhanced by a factor between 1.5 and 3 compared with the one given in Caughlan & Fowler (1988, see for instance the discussion in Schaller et al. 1992). The higher is this rate, the greater is the quantity of ^{16}O synthesized at the end of the He-burning phase, and less ^{12}C is available for the carbon burning phase. According to Woosley (1986), a short carbon burning phase favors the formation of big iron cores and thus the formation of black holes.

Other important chains of reactions accompany the He-burning process. In particular, the ^{14}N left by the previous H-burning phase is rapidly destroyed at the beginning of the He-burning phase and transformed, through a chain of reactions, into ^{22}Ne. The ^{13}C, produced by CNO processing in the previous nuclear phase, is also destroyed at the beginning of the He-burning phase by the reaction ^{13}C$(\alpha,n)^{16}$O. The neutrons emitted on this occasion can be captured by heavier elements and produce some "s-process" elements. A more efficient source of neutrons intervenes at the end of the He-burning phase when the ^{22}Ne is partly destroyed by the reaction ^{22}Ne$(\alpha,n)^{25}$Mg.

5.1.3. Advanced Evolutionary Phases in Massive Stars

Stars with initial mass greater than about 8-10 $M_\odot$ (see section 2.2) go through the whole sequence of nuclear burnings following the He-burning phase, namely the carbon burning phase, the neon-photodesintegration phase, the oxygen and silicon burning stages. The nuclei between neon and iron are synthesized during these advanced phases of the evolution of massive stars. Those nuclei formed from an entire number of α particles as ^{20}Ne, ^{24}Mg, ^{28}Si ..., are produced in greater quantities as a consequence of their great stability.

The explosion of the star as a supernova is accompanied by nucleosynthetic processes known under the generic name of explosive nucleosynthesis. The passage of the shock wave in the envelope of the star enhances the temperatures and densities and thus activates new nuclear reactions. The explosive nucleosynthesis will essentially affects the central layers of the star leaving unaltered most of its outer parts. Typically in a 20 $M_\odot$ model, the chemical composition of the envelope outside the 2 $M_\odot$ core remains unchanged (see e.g. Thielemann et al. 1996). Nucleosynthesis induced by neutron captures (r-process) as well as by neutrinos captures play an important role in the explosive nucleosynthesis (see e.g. Woosley & Weaver 1995).

5.1.4.   The Main Sites of Stellar Nucleosynthesis and Their Time Scales

Very schematically, one can say that there are three main sites of stellar nucle-osynthesis, whose time scales are quite different:

1) The massive stars: their very short lifetimes (a few times 10^6–10^7 years), and their ability to synthesize heavy elements, place these objects among the most important contributors to the chemical evolution of galaxies. At the beginning, only massive stars contribute, since the other sources have not yet had time to evolve. Thus these stars deeply affect the first phases of the chemical evolution of a galaxy or of a starburst region.

2) The Asymptotic Giant Branch stars (AGB): the nucleosynthetic active stars originate from the initial mass range between 1 and 8–10 $M_\odot$ (stars with initial masses below 1 $M_\odot$ have not yet had time enough to contribute to the enrichment of the interstellar medium; in new elements). These stars begin to enter the game of the chemical evolution only after a period of $\sim 10^8$–10^9 years. If the bulk of the heavy elements are produced by massive stars, asymptotic giant branch stars contribute to the synthesis of helium, carbon and nitrogen (primary and secondary). Numerous "s-process" elements are also produced by this type of stars.

3) The type Ia supernovae: these supernovae occur when a white dwarf in a binary system, as a result of accretion or of merging, acquires a mass greater than the Chandrasekhar mass. Carbon or helium ignition occurs then in highly degenerate conditions leading to the complete destruction of the star. Type Ia supernovae are the major contributors to the enrichment in iron group nuclei of the galactic matter. The evolutionary time scale of this source may vary in a large enough interval. Indeed to the lifetime, of a intermediate mass star (between 10^8 to 10^9 years), one must add the uncertain time needed for the accretion or merging process to occur.

5.1.5.   General Trends of the Chemical Evolution of our Galaxy

The different time scales of the various nucleosynthetic sites briefly sketched above play a key role in the interpretation of past chemical abundances. Let us for instance imagine how the ratio [O/Fe] $(=\log[(n_O/n_{Fe})_*/(n_O/n_{Fe})_\odot]$, where $(n_O/n_{Fe})_*$ and $(n_O/n_{Fe})_\odot$ are the number ratios of oxygen to iron atoms in a given star and in the sun respectively) varies as a function of [Fe/H].

First let us recall that this last quantity increases with time as more and more iron is produced by the stars and thus can be viewed as a kind of clock. As a numerical example, in the chemical evolution model proposed by Prantzos et al. (1994) for the halo and the disk of our Galaxy, [Fe/H] equals -3, -2 and -1, corresponding to galactic ages equal to 0.01 Gyr, 0.1 Gyr and 1–2 Gyr respectively. This means also that at [Fe/H] equal to -3, only stars more massive than about 20 $M_\odot$ have contributed to the chemical enrichment of the interstellar medium, at [Fe/H] equal to -2 and -1, only stars more massive than 6 and 2 $M_\odot$ respectively have contributed. Now, since AGB stars do not contribute to the enrichment in oxygen and iron, one can expect that the ratio [O/Fe] will keep a more or less constant value, characteristics of the yields of massive stars, for a great range of [Fe/H] values. At a given point however, type Ia supernovae intervene injecting great quantities of iron. From this point on, the [O/Fe] ratio will decrease as a function of [Fe/H]. This general trend agrees well with what is observed in the halo and the disk of our Galaxy (see Wheeler et al. 1989 for a review).

One sees that in this case, the short time scale with which oxygen is produced, and the much longer time scale for the iron production, are the main effects responsible for the evolution of the [O/Fe] ratio with [Fe/H] in our Galaxy. Other effects too influence the chemical evolution of stellar systems. Some of these are studied below.

5.2. Metallicity Dependence of the Stellar Yields

Maeder (1992) studied the metallicity dependence of the stellar yields. By stellar yields, we mean here the quantity of an element which has been synthesized and ejected into the interstellar medium by a given star during its whole life. Let us summarize here the main results of this study. As seen in paragraph 2.3.5, when the metallicity increases, the proportion of matter expelled by stellar winds increases. Thus at low metallicity, one expects that massive stars will eject most of their yields at the time of the supernova explosion, while at high Z, the ejection will occur both during the pre-supernova stage, through stellar winds, and during the supernova explosion itself. Now matter expelled by stellar winds will have a chemical composition characteristic of a less advanced evolutionary stage than matter expelled during the supernova explosion. Thus its chemical composition will be different.

To better understand this point, let us consider first an He-burning core just at the beginning of the He-burning phase. If the core is naked and suffers mass loss (typically a WC star, see section 4.1.2), the matter expelled by stellar winds will be rich in ^{4}He and ^{12}C (see section 5.1.2). At the end of the He-burning phase, a small O-rich core is formed. Now if we consider the evolution of the same core but with no mass loss, then the entire core matter will be transformed into essentially oxygen. Nearly no carbon and absolutely no helium will be ejected at the time of the supernova explosion. In this case, one expects that evolution with mass loss will produce ejecta richer in elements characteristics of the beginning of the He-burning phase (^{4}He and ^{12}C), while evolution without mass loss will produce ejecta richer in elements characteristics of the end of the He-burning phase (^{16}O). The stellar yields as well as the net yields (i.e. yields from a generation of stars) for metallicities Z = 0.001 and 0.020 are discussed in Maeder (1992, see also this reference for more precise definitions of the stellar and net yields).

It is interesting to mention that these metallicity dependent yields, when incorporated into models for the chemical evolution of our Galaxy, seem to provide a natural explanation for the evolution of the ratio [C/O] as a function

of [Fe/H] (Prantzos et al. 1994). Indeed the near constancy of this ratio in
the halo stars and its increase with [Fe/H] in the disk might be a consequence
of the fact that the yields of metal rich massive stars are much more rich in
carbon than the yields of metal poor massive stars.

According to Maeder (1992), yields might depend on the metallicity through
another mechanism. Let us suppose that when a black hole forms, all the
matter is swallowed by the black hole. In that case, the progenitors of black
holes contribute to the enrichment of the interstellar medium only through
their stellar winds. It is reasonable to believe that the initial mass range for
black hole formation depends on the metallicity. Indeed at low metallicity, the
most massive stars arrive at the end of their evolution still retaining a great
part of their initial mass and thus will certainly give birth to a black hole.
While at high metallicity, the star nearly evaporates under the effects of the
stellar winds and thus end its stellar life with very small final masses which will
produce neutron stars or possibly sometimes white dwarfs (see Meynet et al.
1994). If the scenario above is correct, one expects that at low Z, yields are poor
in heavy elements (locked into the black hole). This leads to values of $\Delta Y/\Delta Z$
(helium over heavy elements enrichment) as high as 4-5 in agreement with
the determinations by Pagel et al. (1992). Let us recall here that also while
he is produced in great quantities by intermediate mass stars, massive stars
through their winds. Helium is thus much less affected than heavy elements by
black hole formation. At high Z, lower values of $\Delta Y/\Delta Z$ are expected, since
strong winds carry away a lot of heavy elements and the process of black hole
formation is expected to be less frequent as a result of smaller final masses.

5.3. Chemical Evolution in Starbursts: the Puzzling Case of Nitrogen

The chemical evolution of a galaxy is a combination of various time scales for
the formation of different elements and of different star formation histories.
When the star formation rate is quite high during a period, as it is believed
to have been at the birth of elliptical galaxies for instance, many massive stars
can contribute to the enrichment of the interstellar medium before the type Ia
supernovae intervene. In such systems, one expects that the [O/Fe] ratio will
begin to decline for higher [Fe/H] values than in a system with a lower star
formation rate (see *e.g.* Matteucci & Brocato 1990). This example illustrates
the effect of the star formation rate history on the chemical evolution of a
system.

Let us now concentrate on less evolved regions, namely the blue compact
galaxies. Blue compact galaxies contain young populations, have low metallicities (0.02 to 0.1 $Z_\odot$) and significant gas contents (10^7–10^8 $M_\odot$). They are thus
poorly evolved objects. There are two possibilities for their past star formation
histories: either these systems are galaxies formed recently and we see in them
the first generations of stars, or they have undergone discontinuous star formation events. This second possibility is favored by the detection of an underlying

old stellar population in most cases. Moreover it seems that fluctuations of the star formation rate seem to decrease as the size of the galaxy increases (see Marconi et al. 1994 and references therein): if blue compact galaxies undergo global, short and intense bursts of star formation, the dwarf irregulars appear to have a more or less constant star formation rate, interrupted by short periods of quiescence, while giant irregular galaxies seem to present a constant star formation rate. These facts give support to the view according to which the probability for a galactic region to start forming stars is proportional to the number of adjacent regions where this process is already active. In that case, it can be shown that fluctuations in stellar births vary as the inverse of the dimension of the galaxy (self-propagating star formation model, see Gerola et al. 1980).

Marconi et al. (1994) have studied the chemical evolution of blue compact dwarf galaxies, considering that the chemical composition observed today is the result of a certain number of bursts of star formation. In order to reproduce the observations plotted in the plane He/H versus O/H:

1) they must allow for the existence of differential galactic winds, *i.e.* of winds bringing away the elements produced by massive stars (like oxygen) and leaving in the galaxy the elements essentially produced by low and intermediate mass stars (as helium).

2) they must also suppose that a great number of bursts of star formation have occurred, typically between 10 and 15.

However, using these hypotheses they cannot account for the observed positions of the galaxies in the N/O versus O/H diagram: too much nitrogen is produced, leading to too high N/O ratios. Thus models that are in better agreement with the He constraints are inconsistent with the nitrogen data. This might be an indication that the nitrogen nucleosynthesis is still not well understood. In particular, one can wonder what would be the results of such chemical evolution models, if a significant proportion of nitrogen is produced by massive stars and is of primary origin ? It is likely that the N/O ratio would remain more or less constant for a great range of O/H values, since the nitrogen would more or less follow the same chemical history as oxygen. Other observations point towards a primary origin for nitrogen. Particularly relevant are the recent determinations of N/O ratios in damped Lyman α systems at redshifts equal to 1.775 and 3.39 (Green et al. 1995; Molaro et al. 1996). These quite young and metal deficient systems ($12+\log(O/H) =\sim 6$, while in the sun this quantity is equal to 9) show N/O ratios which are higher than solar. Such observations are well accounted for by chemical evolution models making the hypotheses that nitrogen is partially primary and is produced on short time scale, *i.e.* by massive stars in a starburst episode (see Matteucci 1996 and references therein). Let us add here that not all damped Lyman α systems present high N/O ratios (see for instance the system observed by Pettini et al. 1995). In that case, it might be that we see a system with a much lower star formation rate, may be a protospiral galaxy.

One sees here that observations of starbursts can provide important clues for the existence of primary nitrogen. Moreover the determination of the nitrogen abundance in damped Ly α systems might be used to trace the history of star formation in galaxies at high redshifts. Of course, this makes sense only if the different N/O ratios are due to differences in the star formation history, and are not caused for instance by oxygen depletion resulting from the formation of grains.

6. CONCLUSION

In these lessons we have tried to put in evidence some links between stellar physics and the evolving appearance of young starburst regions. We emphasized in particular the determining role of the mass loss by stellar winds and of its dependence on the metallicity.

In the future, it is likely that more unified models of stars and of starburst regions will be built. By more unified stellar models, I mean here models accounting, in addition to the evolution of the interior of the star,

1) for the evolution of the atmosphere and of the stellar wind, as Schaerer et al. (1996ab) have done for O-type stars,

2) for the evolution of the wind interactions with the surrounding interstellar medium, as has been studied for instance by Garcia-Segura et al. (1996ab), and

3) for the interaction of the supernova events on the circumstellar medium (see Yepes et al. 1997).

Such a unified picture is particularly important in order to better understand the process of star formation in a starburst episode. How the processes of the star formation depends on the strength of the formation rate itself (non linearity of the process), in particular how the accretion processes so important in the pre-main sequence are realized in starbursts ? What are the effects of the huge amounts of energy, mass and momentum ejected by the new born massive stars on the processes of star formation ? How such mechanisms would change the initial mass function ? Not only the way stars form but also the way they evolve could be different. In very dense star clusters, collisions between stars, stars and protostellar clouds could be frequent. Massive stars, and a fortiori clusters of massive stars are not isolated objects. They must be understood as part of a bigger ensemble of which the interstellar medium is a major constituent.

Acknowledgments

I express my warmest thanks to A. Maeder and D. Schaerer for many fruitful discussions and collaborations in the past years.

References

[1] Abbott D.C. (1979) in IAU Symp. 83, eds. Conti P.S. & de Loore C.W.H.,
     p. 237.
[2] Abbott D.C. *ApJ* **259** (1982) 282.
[3] Abbott D.C., Conti P.S. *ARAA* **25** (1987) 113.
[4] Abbott D.C., Bieging J.H., Churchwell E., Torres A.V. *ApJ* **303** (1986)
     239.
[5] Allen D.A., Wright A.E., Goss W.M. *MNRAS* **177** (1976) 91.
[6] Alongi M., Bertelli G., Bressan A., Chiosi C., Fagotto F. *A&AS* **97** (1993)
     851.
[7] Anders E., Grevesse N. *Geochim. Cosmochim. Acta* **53** (1989) 197.
[8] Andersen J., Nordström B., Clausen J.V. *ApJL* **363** (1990) L33
[9] Armus L., Heckman T., Miley G. *ApJ* **326** (1988) L45.
[10] Arnaud M., Rothenflug R., Boulade O., Vigroux L., Vangioni-Flam E.
     A&A **254** (1992) 49.
[11] Arnault Ph., Kunth D., Schild H. *A&A* **224** (1989) 73.
[12] Arnett D. *ApJ* **383** (1991) 295.
[13] Arnett D. (1996),"Supernovae and Nucleosynthesis", Princeton Univer-
     sity Press.
[14] Azzopardi M., Breysacher J. *A&A* **75** (1979) 243.
[15] Azzopardi M., Breysacher J. *A&A* **149** (1985) 213.
[16] Bandiera R., Turolla R. *A&A* **231** (1990) 85.
[17] Barkat Z., Rakavy G., Sack N. *Phys. Rev. Lett.* **19** (1967) 379.
[18] Becker S.A., Iben I. *ApJ* **232** (1979) 831.
[19] Beech M., Mitalas R. *ApJS* **95** (1994) 517.
[20] Bernasconi P.A. *A&AS* **120** (1996) 57.
[21] Bernasconi P.A., Maeder A. *A&A* **307** (1996) 829.
[22] Bertelli G., Bressan A., Chiosi C., Fagotto F., Nasi E. *A&AS* **106** (1994)
     275.
[23] Bowen G., Willson L.A. *ApJ* **375** (1991) 53.
[24] Breysacher J. *A&AS* **43** (1981) 203.
[25] Brocato E., Castellani V. *ApJ* **410** (1993) 99.
[26] Bruzual G., Charlot S. *ApJ* **405** (1993) 538.
[27] Burrows A., Hubbard W.B., Lumine J.I. *ApJ* **345** (1989) 939.
[28] Castor J., Abbott D.C., Klein R. *ApJ* **195** (1975) 157.
[29] Caughlan G.R., Fowler W.A. *Atom. Data Nucl. Data Tables* **40** (1988)
     283.
[30] Cerviño M., Mas-Hesse J.M. *A&A* **284** (1994) 749.
[31] Cerviño M., Mas-Hesse J.M., Kunth D. (1996), in "Wolf-Rayet Stars
     In The Framework Of Stellar Evolution", 33rd Liège Intern. Astrophys.
     Coll., eds. Vreux J.M. et al., Université de Liège, p. 613.
[32] Chandrasekhar S. *MNRAS* **96** (1936) 644.
[33] Chandrasekhar S. *Rev. of Mod. Phys.* **56** (1984) 137.

[34] Charlot S. (1996), in "From Stars to Galaxies: the Impact of Stellar physics on Galaxy Evolution", ASP Conf. Series 98, eds. Leitherer C., Fritze-von Alvensleben U., Huchra J., p. 275.

[35] Chiosi C. (1986) in "Nucleosynthesis and Chemical Evolution", 16th Saas-Fee Course, eds. Hauck B., Maeder A., Meynet G., Geneva Observatory, p. 199.

[36] Chiosi C., Bertelli G., Bressan A. *ARAA* **30** (1992) 235.

[37] Chiosi C., Maeder A. *ARAA* **24** (1986) 329.

[38] Claret A., Gimenez A. *A&AS* **96** (1992) 255.

[39] Clayton D.D. (1968), "Principles of Stellar Evolution and Nucleosynthesis", Mc Graw-Hill Book Company, New-York.

[40] Conti P.S. *ApJ* **377** (1991) 115.

[41] Conti P.S. (1993), in "Massive Stars: Their Lives in the Interstellar Medium", eds. Cassinelli J.P., Churchwell E.B., ASP Conf. Ser. 35, p. 449.

[42] Conti P.S. *Space Science Reviews* **66** (1994) 37.

[43] Conti P.S., Vacca W.D. *AJ* **100** (1990) 431.

[44] Conti P.S., Vacca W.D. *ApJ* **423** (1994) L97.

[45] Conti P.S., Leitherer C., Vacca W.D. *ApJ* **461** (1996) L87.

[46] D'Antona F., Mazzitelli I. *ApJ* **296** (1985) 502.

[47] Davidson, K. (1989), in "Physics of Luminous Blue Variables", IAU Colloq. 113, eds. Davidson K., Moffat A.F.J., Lamers H.J.G.L.M., Kluwer, Dordrecht, p. 101.

[48] Dorman B., Nelson L.A., Chau W.Y. *ApJ* **342** (1989) 1003.

[49] Dziembowski W.A., Pamyatnykh A.A. *MNRAS* **262** (1993) 204.

[50] Eddington A.S. (1926) "The Internal Constitution of The Stars", Cambridge University, Cambridge.

[51] Ferrarese L. et al. *ApJ* **464** (1996) 568.

[52] Fowler W.A., Hoyle F. *ApJSS* **9** (1964) 201.

[53] Freedman W.L., Madore B.F., Mould J.R., Hill R., Ferrarese L., Kennicutt Jr R.C., Saha A., Stetson P.B., Graham J.A., Ford H., Hoessel J.G., Huchra J., Hughes S.M., Illingworth G.D. *Nature* **371** (1994) 757.

[54] Frost C.A., Lattanzio J.C. (1995), in "Stellar Evolution: What Should be Done", 32nd Liège Intern. Astrophys. Coll., eds. Noels A. et al., Université de Liège, p. 307.

[55] Garcia-Segura G., Langer N., Mac Low M.M. *A&A* **316** (1996a) 133.

[56] Garcia-Segura G., Mac Low M.M., Langer N. *A&A* **305** (1996b) 229.

[57] Garmany C.D., Conti P.S., Chiosi C. *ApJ* **263** (1982) 777.

[58] Gerola H., Sneiden P.E., Schulman L.S. *ApJ* **242** (1980) 517.

[59] Gies D.R., Lambert D.L. *ApJ* **387** (1992) 673.

[60] Green R.F., York D., Huang K., Bechtold J., Welty D., Carlson M., Khare P., Kulkarni V. (1995), in "QSO Absorption Lines", ESO Workshop, ed. G. Meylan, Springer Verlag, 85.

[61] Grossman S.A., Taam R.E. *MNRAS* **283** (1996) 1165.

[62] Hamann W.-R., Koesterke L. (1996), in "Wolf-Rayet Stars In The Frame-work Of Stellar Evolution", 33rd Liège Intern. Astrophys. Coll., eds. Vreux J.M. et al., Université de Liège, p. 491.

[63] Heckman T.M., Armus L., Miley G.K. *ApJS* **74** (1990) 833.

[64] Herrero A., Kudritzki R.P., Vilchez J.M., Kunze D., Butler K., Haser S. *A&A* **261** (1992) 209.

[65] Ho L.C., Filippenko A.V. *ApJ* **472** (1996) 600.

[66] Howarth I.D., Prinja R.K. *ApJS* **69** (1989) 527.

[67] Huebner W.F., Merts A.L., Magee N.H., Argo M.F. (1977), Astrophysical Opacity Library, UC-34b.

[68] Humphreys R.M., McElroy D.B. *ApJ* **284** (1984) 565.

[69] Iben I., Renzini A. *ARAA* **21** (1983) 271.

[70] Iglesias C.A., Rogers F.J. *ApJ* **412** (1993) 752.

[71] Kawaler S. (1997), in "Stellar Remnants", eds. Meynet G. & Schaerer D., Springer-Verlag, Berlin, p. 1.

[72] Kinney A.L., Calzetti D., Bohlin R.C., McQuade K., Storchi-Bergmann. T., Schmitt H.R. *ApJ* **467** (1996) 38.

[73] Kippenhahn R., Weigert A. (1990) "Stellar Structure and Evolution", Springer-Verlag, Berlin.

[74] Krauss A., Becker H.W., Trautvetter H.P., Rolfs H.P. *Nucl. Phys.* **A467** (1987) 273.

[75] Kudritzki R.P, Pauldrach A., Puls J. (1988), in "O-stars and WR stars", NASA SP-497, eds. Conti P.S. & Underhill A.B., p. 173.

[76] Kunth D., Sargent W.L.W. *A&A* **101** (1981) L5.

[77] Lamers H.J.G.L.M., Cassinelli J.P. (1996), in "From Stars to Galaxies: The Impact of Stellar Physics on Galactic Evolution", eds. Leitherer C., Fritze-V. Alvensleben U., Huchra J., ASP, San Francisco, p. 162.

[78] Langer N. *A&A* **220** (1989) 135.

[79] Langer N., Maeder A. *A&A* **295** (1995) 685.

[80] Langer N., Hammann W.-R., Lennon M., Najarro F., Pauldrach A.W.A., Puls J. *A&A* **290** (1994) 819.

[81] Lattanzio J.C. *ApJ* **311** (1986) 708.

[82] Leitherer C. (1996), in "Wolf-Rayet Stars In The Framework Of Stellar Evolution", 33rd Liège Intern. Astrophys. Coll., eds. Vreux J.M. et al., Université de Liège, p. 591.

[83] Leitherer C., Heckman T.M. *ApJS* **96** (1995) 9.

[84] Leitherer C., Chapman J.M., Koribalski B. *ApJ* **450** (1995a) 289.

[85] Leitherer C., Robert C., Heckman T.M. *ApJS* **99** (1995b) 173.

[86] Leitherer C. et al. *PASP* **108** (1996) 996.

[87] de Loore C., Vanbeveren D. *A&AS* **103** (1994) 67.

[88] Lucy L.B., Solomon P. *ApJ* **159** (1970) 879.

[89] Maeder A. *A&A* **120** (1983) 113.

[90] Maeder A. *A&A* **147** (1985) 300.

[91] Maeder A. *A&A* **242** (1991) 93.

[92] Maeder A. *A&A* **264** (1992) 105.

[93] Maeder A. (1995), in "Wolf-Rayet Stars: Binaries, Colliding Winds, Evolution", IAU SYMP. 163, eds. van der Hucht K.A., Williams P.M., Kluwer, Dordrecht, p. 280.

[94] Maeder A. (1996), in "Wolf-Rayet Stars In The Framework Of Stellar Evolution", 33rd Liège Intern. Astrophys. Coll., eds. Vreux J.M. et al., Université de Liège, p. 39.

[95] Maeder A. (1997), *A&A*, in press.

[96] Maeder A., Conti P. *ARAA* **32** (1994) 227.

[97] Maeder A., Lequeux J. *A&A* **114** (1982) 409.

[98] Maeder A., Meynet G. *A&A* **210** (1989) 155.

[99] Maeder A., Meynet G. *A&A* **287** (1994) 803.

[100] Maeder A., Lequeux J., Azzopardi M. *A&A* **90** (1980) L17.

[101] Marston A.P., Yocum D.R., Garcia-Segura G., Chu Y.-H. *ApJS* **95** (1994) 151.

[102] Marconi G., Matteucci F., Tosi M. *MNRAS* **270** (1994) 35.

[103] Mas-Hesse J.M., Kunth D. *A&AS* **88** (1991) 399.

[104] Massey P. (1982), in "Wolf-Rayet Stars: Observations, Physics, Evolution", IAU Symp. 99, eds. de Loore C. & Willis A.J., Reidel, Dordrecht, p. 251.

[105] Massey P. (1996), in "Wolf-Rayet Stars in The Framework of Stellar Evolution", 33rd Liège Intern. Astrophys. Coll., eds. Vreux J.M. et al., Université de Liège, p. 361.

[106] Massey P., Lang C.C., DeGioia-Eastwood K., Garmany C.D. *ApJ* **438** (1995) 188.

[107] Matteucci F. (1996), in "From Stars to Galaxies: the Impact of Stellar physics on Galaxy Evolution", ASP Conf. Series 98, eds. Leitherer C., Fritze-von Alvensleben U., Huchra J., p. 529.

[108] Matteucci F., Brocato E. *ApJ* **365** (1990) 539.

[109] Mazzitelli I., D'Antona F. *ApJ* **308** (1986) 706.

[110] Melnick J. (1992), in "Star Formation in Stellar Systems", eds. Tenorio-Tagle G., Prieto M., Sanchez F., Cambridge University Press, Cambridge, p. 253.

[111] Meurer G.R., Heckman T.M., Leitherer C., Kinney A., Robert C., Garnett D.R. *AJ* **110** (1995) 2665.

[112] Meylan, G., Maeder, A. *A&A* **108** (1982) 148.

[113] Meynet G. (1993) in "The Feedback of Chemical Evolution on the Stellar Content of Galaxies", eds. Alloin D. & Stasinska G., Observatoire de Paris, p. 40.

[114] Meynet G. *A&A* **298** (1995) 767.

[115] Meynet G., Maeder A (1997), *A&A*, in press.

[116] Meynet G., Mermilliod J.C., Maeder A. *A&AS* **98** (1993) 477.

[117] Meynet G., Maeder A., Schaller G., Schaerer D., Charbonnel C. *A&AS* **103** (1994) 97.

[118] Mezger P.G., Smith L.F., Churchwell E. *A&A* **32** (1974) 269.

[119] Molaro P., D'Odorico S., Fontana A., Savaglio S., Vladilo G. *A&A* **308** (1996) 1.

[120] Mowlawi N., Forestini M. *A&A* **282** (1994) 843.

[121] Nomoto K. (1984) in "Stellar Nucleosynthesis", eds. Chiosi C. & Renzini A., Reidel, Dordrecht, p. 238.

[122] Novikov I.. (1997), in "Stellar Remnants", eds. Meynet G. & Schaerer D., Springer-Verlag, Berlin, p. 237.

[123] Nugis T. (1991) in "Evolution of Stars: The Photospheric Abundance Connection", IAU Symp. 145, eds. Michaud G. & Tutukov A., Kluwer, Dordrecht, p. 209.

[124] Pagel B.E.J., Simonson E.A., Terlevich R.J., Edmunds M.G. *MNRAS* **255** (1992) 325.

[125] Palla F., Stahler S.W. *ApJ* **392** (1992) 667.

[126] Pauldrach A.W.A, Kudritzki R.P, Puls J., Butler K. *A&A* **228** (1990) 125.

[127] Pettini M., Lipman K., Hunstead R.W. *ApJ* **451** (1995) 100.

[128] Pierce M.J., Welch D.L., McClure R.D., van den Bergh S., Racine R., Stetson P.B. *Nature* **371** (1994) 385.

[129] Prantzos N., Diehl R. *Physics Reports* **267** (1995) 1.

[130] Prantzos N., Vangioni-Flam E., Chauveau S. *A&A* **285** (1994) 132.

[131] Pringle J.E., Wade R.A. (1985), "Interacting Binary Stars", Cambridge University Press, Cambridge.

[132] Rolfs C.E., Rodney W. (1988), "Cauldrons in The Cosmos: Nuclear Astrophysics", University of Chicago press, Chicago.

[133] Schaerer D. *ApJ* **476** (1996) L17.

[134] Schaerer D., Maeder A. *A&A* **263** (1992) 129.

[135] Schaerer D., Vacca W.D. (1996), in "Wolf-Rayet Stars In The Framework Of Stellar Evolution", 33rd Liège Intern. Astrophys. Coll., eds. Vreux J.M. et al., Université de Liège, p. 641.

[136] Schaerer D., Contini T., Kunth D., Meynet G. (1997), *ApJ Letters*, submitted.

[137] Schaerer D., de Koter A., Schmutz W., Maeder A. *A&A* **310** (1996a) 837.

[138] Schaerer D., de Koter A., Schmutz W., Maeder A. *A&A* **312** (1996b) 47.

[139] Schaller G., Schaerer D., Meynet G., Maeder A. *A&AS* **96** (1992) 269.

[140] Seaton M.J., Yan Y., Mihalas D., Pradhan A.K. *MNRAS* **266** (1994) 805.

[141] Simon N.R. *ApJL* **260** (1982) 87.

[142] Smith L.F., Hummer D.G. *MNRAS* **230** (1988) 511.

[143] Smith L.F., Maeder A. *A&A* **241** (1991) 77.

[144] Smith L.J. (1996), in "Wolf-Rayet Stars In The Framework Of Stellar Evolution", 33rd Liège Intern. Astrophys. Coll., eds. Vreux J.M. et al., Université de Liège, p. 381.

[145] Srinivasan G.. (1997), in "Stellar Remnants", eds. Meynet G. & Schaerer D., Springer-Verlag, Berlin, p. 97.

[146] Stahler S.W., Shu F.H., Taam R.E. *ApJ* **241** (1980) 637.

[147] Stothers R., Chin C.-W. *ApJ* **390** (1992) 136.

[148] Talon S., Zahn J.P., Maeder A., Meynet G. (1997), *A&A*, in press
[149] Telesco C.M. (1993), in "Infrared Astronomy", eds. Mampaso A., Prieto M., Sánchez F., Cambridge University Press, Cambridge, p. 173.
[150] Thielemann F.K., Nomoto K., Hashimoto M. *ApJ* **460** (1996) 408.
[151] Trautvetter H.P. (1993), in "Nuclei in The Cosmos", eds. Käppeler F. & Wisshak K., Institute of Physics Publishing, Bristol, p. 139.
[152] Vacca W.D., Conti P.S. *ApJ* **401** (1992) 543.
[153] van den Bergh S., Tamman G.A. *ARAA* **29** (1991) 363.
[154] van den Heuvel, E.P.J. (1994) in "Interacting Binaries", 22nd Saas-Fee Course, eds. Nussbaumer H. & Orr A., Springer-Verlag, Berlin, p. 263.
[155] van der Hucht K.A. (1995), in "Wolf-Rayet Stars: Binaries, Colliding Winds, Evolution", IAU Symp. 163, eds. van der Hucht K.A. & Williams P.M., Kluwer, Dordrecht, p. 7.
[156] van der Hucht K.A. (1996), in "Wolf-Rayet Stars In The Framework Of Stellar Evolution", 33rd Liège Intern. Astrophys. Coll., eds. Vreux J.M. et al., Université de Liège, p. 1.
[157] Venn K.A. *PASP* **108** (1996) 309.
[158] Walborn N.R. *ApJ* **205** (1976) 419.
[159] Walborn N.R. (1988), in "Atmospheric Diagnostics of Stellar Evolution", IAU Colloq. 108, ed. Nomoto K., Springer-Verlag, Berlin, p. 70.
[160] Weidemann V. *ARAA* **28** (1990) 103.
[161] Weiler K.W., Sramek R.A. *ARAA* **26** (1988) 295.
[162] Wheeler J.C., Sneden C., Truran J.W. *ARAA* **27** (1989) 279.
[163] Willis A.J. *Astrophys. and Space Science* **237** (1996) 145.
[164] Wood, D.O.S., Churchwell, E. *ApJ* **340** (1989) 265.
[165] Woosley S.E. (1986) in "Nucleosynthesis and Chemical Evolution", 16th Saas-Fee Course, eds. Hauck B., Maeder A., Meynet G., Geneva Observatory, p. 1.
[166] Woosley S.E., Weaver T.A. *ApJS* **101** (1995) 181.
[167] Yee H.K.C., Ellingson E., Bechtold J., Carlberg R.G., Cuillandre J.C. *AJ* **111** (1996) 1783.
[168] Yepes G., Kates R., Khokhlov A., Klypin A. *MNRAS* **284** (1997) 235.
[169] Zahn J.P. *A&A* **252** (1991) 179.

Starburst Triggering and Environmental Effects

F. Combes

Observatoire de Paris, DEMIRM
61 Av. de l'Observatoire, F-75014, Paris, France

---

1. INTRODUCTION

Observations of star-forming galaxies reveal an obvious relation between the star-formation rate and dynamical features. Spiral arms are mostly a tracer of young stars, giant HII regions often accumulate at the extremities of bars, or in conspicuous hot spots delineating nuclear rings; the most spectacular starburst occur in tidally interacting galaxies and mergers. The problem is to understand in details the large-scale mechanisms that trigger star-formation coherently over such scales (larger than one kpc), and their relation with the small-scale physics: cloud gravitational collapse, and feedback processes provided by the newly formed stars such as stellar winds, jets, supernovae, etc...

The triggering of cloud collapse followed by star-formation is to be searched in gravitational instabilities, but the problem of the stability of two-fluid media, where a non-dissipative component (stars) is coupled to a dissipative one (the gas) is complex (cf Jog & Solomon 1984; Jog 1992). The stability of the stellar component alone and its possible structures have been widely studied: even when a stellar disk is stable against axisymmetric perturbations (i.e. the velocity dispersion is enough for the equivalent pressure to prevent Jeans instability at small scales, while the large scales are stabilised by rotation, Toomre 1964), it can develop spiral and bar waves, or z-instabilities. But these gravitational instabilities saturate rapidly, since they heat considerably the stellar disk, that can then no longer sustain spiral structure, or thicken so that z-instabilities are suppressed. Any instability can be totally suppressed by increasing the velocity dispersion, or by the addition of a hot spherical component (bulge) which both reduce the self-gravity of the stellar disk. The relevant scale of the self-gravity is $\lambda_c = 4 \pi^2 G \Sigma(r)/\kappa^2(r)$, where $\Sigma(r)$ is the disk surface density and $\kappa(r)$ the epicyclic frequency (λ_c typically varies from 1 to 15 kpc over galaxy disks).

The behaviour of the gas component is completely different. Alone, submitted only to its self-gravity, it will never be stable on a galactic scale. Its dissipation, and propensity to radiate quickly its kinetic energy away, keeps its velocity dispersion always lower than the critical one. It will be Jeans unstable at any scale, which results in a patchy and clumpy structure, leading in some cases to star formation. Only when coupled to a non-dissipative component such the stellar or dark component, can its self-gravity be reduced, and the critical velocity brought down below its velocity dispersion. In that case, it can stay as stable as a diffuse component. However, the coupling between the various components cannot be neglected, and separate stability analysis is misleading.

2.   STABILITY OF A TWO-FLUID MEDIUM

For a single component, the standard stability criterion against local axisymmetric perturbations, developed first by Safronov (1960) is

$$Q = \frac{\kappa c}{\pi G \Sigma} > 1$$

where c is the one-dimensional random velocity or sound speed. The numerical coefficient π corresponds to a fluid approach, while the kinetic theory would give 3.36 (Toomre 1964). This criterion can be applied separately to the stellar and gas component, where the corresponding values of Σ and c are used, leading to Q_s and Q_g. Jog & Solomon (1984) showed however that, since $c_g << c_s$, only a small percentage of mass in gas can destabilize the whole disk, even when $Q_s > 1$ and $Q_g > 1$. They give a condition of neutral stability in the form of the dispersion relation:

$$\frac{2\pi G k \Sigma_s}{\kappa^2 + k^2 c_s^2} + \frac{2\pi G k \Sigma_g}{\kappa^2 + k^2 c_g^2} = 1$$

where k is the wave number. This formula has the advantage of giving directly an idea of the relative weight of gas and stars in the instabilities. However, there is no simple criterion of local stability, and the study can only be done numerically (Jog 1996). Indeed the equation above is only the condition $\omega^2=0$, where ω is the angular frequency of the radial perturbation of wave number k. But the neutral equilibrium requires the simultaneous solution of $d\omega^2(k)/dk = 0$; the system have several solutions, and the analysis is cumbersome. Fig. 1 gives a global idea of the numerical solutions, in terms of a two-fluid Q_{s-g} value, which is always lower than the Q_s or Q_g values. The stability problem depends strongly on the gas mass fraction ϵ (between 5 and 25%). The fastest growing two-fluid modes are characterized by a dimensionless wavelength $l_{s-g}\lambda/\lambda_c =$

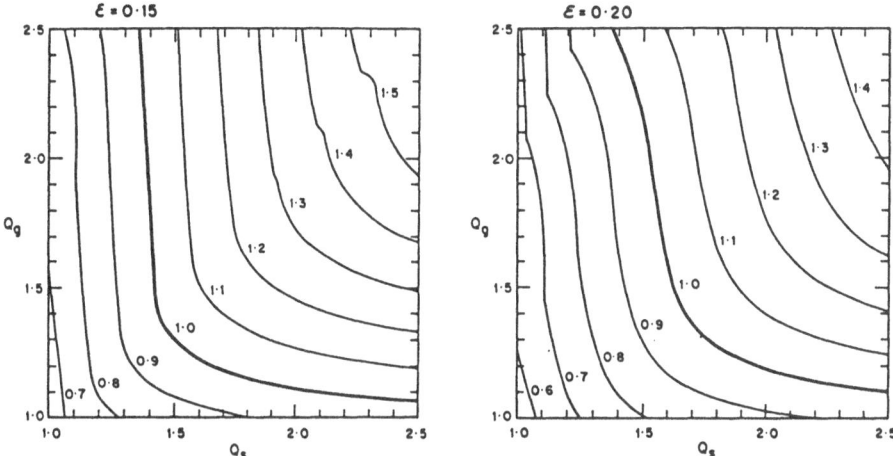

Fig. 1. — A contour plot of Q_{s-g} the local stability parameter for a two-fluid star-gas disc, as a function of Q_s and Q_g. The gas mass fraction is ϵ (from Jog 1996).

$\kappa^2/2\pi Gk(\Sigma_s + \Sigma_g)$, which is displayed in fig 2. What is striking is how sharp is the transition from high to low values of l_{s-g}, as the mass fraction increases from $\epsilon = 0.1$ to 0.15 for instance. Also, l_{s-g} can be very low, and equal to the gas l_g when Q_s is high and Q_g is low. These results imply several interesting effects: a gas-rich galaxy (with $\epsilon > 0.25$) is only stable at very low surface densities (this explains the Malin 1-type galaxies, Impey & Bothun 1989); the center of early-type galaxies, where Q_s is high, and Q_g might be low (low gas velocity dispersion), could be dominated by the gas wavelength, even at very low gas mass fraction: this can explain spiral arms in galaxies such as NGC 2841 (Block et al 1996). Also interaction of galaxies, by bringing in a high amount of gas, may change abruptly the fastest growing wavelength from l_s to l_g, and trigger star-formation.

It is important to note that, even beyond the neutral stability criterion, when $Q_{s-g} > 1$, a galaxy disk can be unstable with respect to non-axisymmetric perturbations, such as bars or spirals; in this case also, the two-fluid coupling increases the instability, i.e. the disk will form a bar, even if the stars or the gas alone are stable with respect to such perturbations (Jog 1992).

Also important are the stabilising effects of the vertical thickness of the galactic planes (e.g. Romeo 1994). These can be represented by reduction

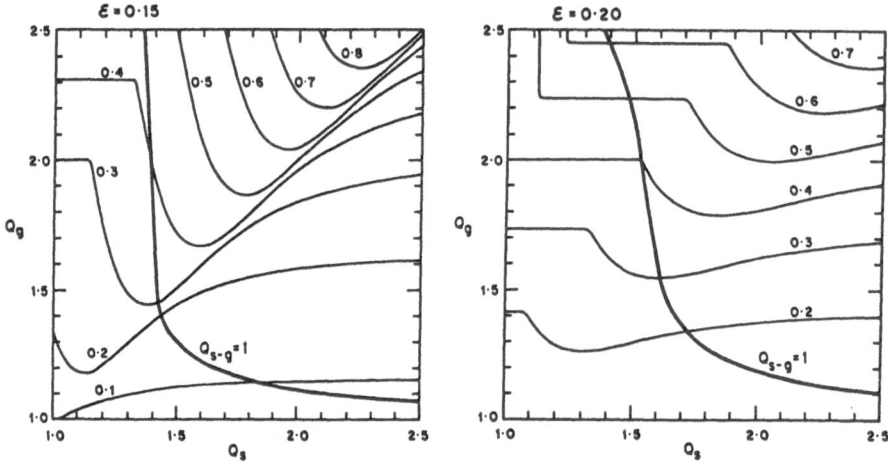

Fig. 2. — A contour plot of l_{s-g} the dimensionless wavelength corresponding to the fastest growing mode of the two-fluid star-gas disc, as a function of Q_s and Q_g, for ϵ the gas fraction equal to 0.15 and 0.20 (from Jog 1996).

factors for stars and gas $f_s = (1+kh_s)^{-1}$ and $f_g = (1+kh_g)^{-1}$ of scale-heights h_s and h_g respectively. Note that the stabilising effects depend on wavelengths $2\pi/k$. The vertical and planar heating mechanisms, and related self-regulation processes, are almost decoupled, as are the stability properties (Romeo 1992).

3. MECHANISMS TO TRIGGER STARBURSTS

Several mechanisms to explain enhanced star formation in colliding galaxies have been proposed in the literature. Since Giant Molecular Clouds (GMCs) are in our Galaxy the sites of massive-star formation, and that they are just in virial equilibrium, with a rather low star formation efficiency, it was proposed that the direct collision between them at high velocities could trigger their instability. Harwit et al (1987) have proposed such a model based on direct collisions between GMC of the two galaxies, at a relative velocity of about 300km/s. But since the filling factor of molecular clouds is so low, they required that the collision between the two galaxies be edge-on. Jog & Solomon (1992) have taken along this idea, but only the HI gas with a large filling factor was able to collide directly, form a hot ionized, high pressure medium. This overpressure caused a radiative shock compression of the outer layers of the GMCs and triggered the burst of star formation. This occurred in the region of overlap of the two galaxies ($\approx$ 1kpc in size), and not in the galaxy nuclei.

But the energy deposited by a direct collision, given the very restricted overlap zone, and the small mass involved (the diffuse HI gas), is negligible relative of the energy available in a galaxy interaction. Tidally interacting galaxies reveal active star-forming zones without any contact, and the most frequent starbursts occur at the nuclei. The fundamental reason for star bursting activity in interacting galaxies is more the dissipative character of the collision: the energy to dissipate is the total relative orbital energy of the two colliders. This is at least 5 orders of magnitude larger than the energy dissipated in the direct collision of diffuse clouds at the overlapping zone. Through tidal interaction and dynamical friction, all the energy ends up in deformation and heating of the galactic disks (velocity dispersions, waves, etc..), which triggers the star formation. In particular, the large gas inflows driven by the waves produce huge gas surface densities in spiral arms and particularly in the nuclei, where gravitational instabilities at all scales are quickly growing. Other possibilities have been suggested, like an enhanced rate of cloud-cloud collisions (Noguchi & Ishibashi 1986, Sanders et al 1988). In any case, the rate of star formation ends up as a non-linear function of the gas density, as is observed empirically as the Schmidt law (e.g. Larson 1987); see lectures by Kennicutt in this volume

4.   DYNAMICAL MECHANISMS: NON-AXISYMMETRY AND TORQUES

4.1.   Angular Momentum Transfer for the Stellar Component

In a pioneering paper, Lynden-Bell & Kalnajs (1972) have shown how spiral waves can carry angular momentum (A.M.), and that only trailing spirals can carry it outwards, which explains the predominance of trailing waves in observations. For a steady wave, stars can exchange angular momentum at resonances only: they emit A.M. at inner Lindblad resonance, while they absorb at corotation and outer resonance. The authors also show that, while stars do not gain or lose angular momentum on average away from resonances, act as vehicles to transport angular momentum during their orbiting around the galactic center. When they are at large radii, they gain A.M., while they lose some at small radii: even if the net balance is zero, they carry A.M. radially, and the sense of this transport is opposite to that of the spiral wave. This phenomenon can then damp the wave, if the amplitude is strong enough. Lynden-Bell & Kalnajs (1972) noticed that this damping phenomenon became negligible for small wavelengths, i.e. when $kr \gg 1$.

This A.M. transport has been investigated recently by Zhang (1996), who puts forward another point of view: due to the phase-shift between the stellar density and the potential of the spiral wave, gravity torques are exerted by the wave on the basic state stars, and stars gain or lose angular momentum, even away from resonances. This is the consequence of a kind of dissipational process, corresponding to small-angle scattering of neighboring stars in the spiral arm. This process transforms ordered motions into disordered ones (resulting in increased velocity dispersion, or large epicycle amplitudes), and when collective effects are taken into account, secular modifications of the stellar orbits can result: they can lose energy and angular momentum inside corotation.

4.1.1. Expressions for the Torque and Phase-shift

It is obvious that the torque exerted by the wave on particles is a second order term, since the tangential force $\frac{\partial V_1}{\partial \theta}$ is first order, and has a net effect only on the non-axisymmetric term of the surface density $\Sigma_1(r, \theta)$. The torque $T(r)$ applied by the wave on the disk matter in an annular ring of width dr, can be expressed by (Zhang, 1996):

$$T(r) = rdr \int -\Sigma_1(r, \theta) \frac{\partial V_1}{\partial \theta} d\theta$$

With this formula, it is easy to see that the torque vanishes if the density and potential spiral perturbations are in phase. But there must be a phase-shift in the general case, according to the Poisson equation. The potential is non-local, and is influenced by the distant spiral arms. So the sign and amplitude of the phase-shift depends on the radial density law of the perturbation. It has been shown by Kalnajs (1971) that the peculiar radial law of $r^{-3/2}$ for an infinite spiral perturbation provides exactly no phase shift. The phase-shift is such that the spiral density leads the potential if the radial falloff is slower than $r^{-3/2}$, and the reverse if it is steeper.

Now in a self-consistent disk, the phase-shift given by the Poisson equation must agree with that given by the equations of motion. Through the computation of linear periodic orbits in the rotating frame, Zhang (1996) has found that the forcing consists of two terms in quadrature, and that the phase shift δ of the orbit orientation with respect to forcing potential has the expression:

$$\tan(m\delta) \approx \frac{-2\Omega}{(\Omega - \Omega_p)kr}$$

This shows that in the WKBJ approximation, the phase shift is negligible ($kr \gg 1$), and that it changes sign at corotation. The fact that the torque is a non-linear effect, and vanishes in the WKBJ approximation, may explain why it was neglected before, but it has a quite important effect in real spiral galaxies. Inside the corotation radius, we expect that the density leads the potential, and the contrary outside the corotation. These predictions have been confirmed through N-body simulations by Zhang (1996), as far as the sign of the phase-shift on each side of the corotation is concerned. The amplitude of the phase-shift appears quite high, up to $20°$. The existence of the phase-shift is no surprise of course, but the amplitude of δ was not suspected to be so high.

I have reproduced a comparable N-body simulation (2D polar grid PM with 10^5 particles), and reported the Fourier analysis results in Figure 3. In this simulation, initial parameters were chosen such as to stabilise the stellar disk with respect to bar formation, through the presence of an analytical bulge component of mass equal to the mass of the self-gravitating stellar disk. A bar developed still, but was delayed until 2×10^9 yr, i.e. ≈ 20 dynamical times. In the mean time, a spiral wave developed, and remained for several dynamical times. Its power spectrum revealed a well-defined pattern speed, at least up to corotation (≈ 7 kpc). We can see in fig. 3 that the phase-shift between stellar density and potential is indeed quite high, up to 28° inside corotation, where it is the most meaningful.

4.1.2. *Exchange of Angular Momentum between Particles and Wave*

Previous studies had neglected the exchange of energy and angular momentum between the stars and the wave outside resonances (Lynden-Bell & Kalnajs 1972; Goldreich & Tremaine 1979). This interaction is mediated by the graininess of the particles, which under the local gravitational instabilities in the spiral arms, can scatter particles and produce dissipation (Zhang 1996).

During all its growth, the unstable spiral mode is able to deposit negative angular momentum in every annulus inside the corotation radius, and positive angular momentum outside the corotation radius. As the wave reaches the nonlinear regime, an increasing fraction of the deposited angular momentum by the spiral wave is channeled to the basic state, since by the very definition of the quasi-steady state the wave should not grow any further. We can note that this stellar collective dissipation process and angular momentum transport, induced by the phase-shift between density and potential, is nonexistent for a bar. The latter can therefore be robust and long-lived in a collisionless ensemble of particles (galaxy disk without gas).

4.2. Angular Momentum Transfer for the Gas Component

The mechanism of the A.M. transfer for the gas is the same as was previously described: the gas settles in a spiral structure which is not in phase with the potential. Gravity torques are exerted by the wave on the gas. The dissipation here is of course different, since the gas radiates away its energy. The gas component is then maintained cool and responsive to new gravitational instabilities. This is the source of more drastic secular evolution, with the possibility of the whole gas component inflowing towards the center.

The gas response in a barred potential has been tackled through many simulations by several authors (e.g. Sanders & Huntley 1976, Schwarz 1981) and can be understood in terms of periodic orbits families in a bar potential (e.g. Combes 1988). The sign of gravity torques also changes at corotation. They can change also at ILR, according to the shape of the $\Omega - \kappa/2$ precession-rate curve.

Figure 4 shows the relative phases of the gas and stellar component with

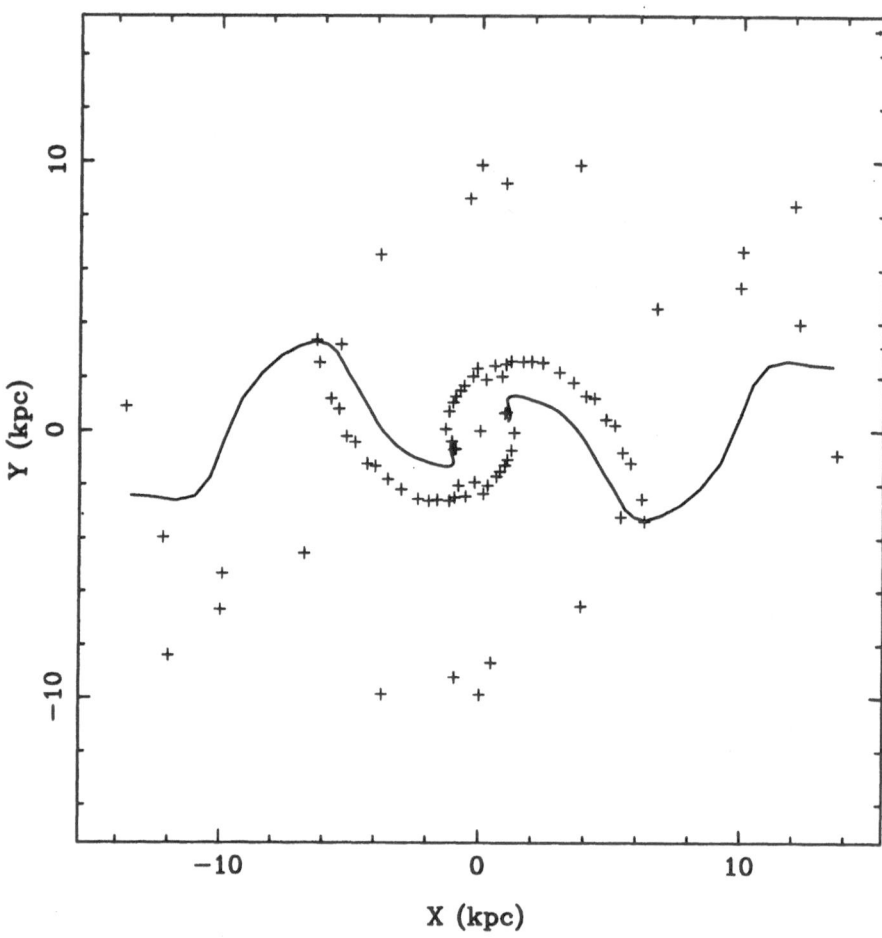

Fig. 3. — Results of the Fourier analysis of the stellar density (crosses) and potential
(full lines) in a purely stellar N-body simulation, while a spiral structure rotates in the
disk. Plotted here are the phases of the pattern at each radius. The density leads the
potential almost everywhere inside corotation. The perturbation was weaker outside
corotation.

respect to the total potential, in an N-body simulation, while a bar was developing, with its external spiral structure at the same pattern speed. There is
now a negligible phase-shift of the stellar density with respect to the potential
in the bar, while there is a slight gas phase-shift. In the arms, outside corotation, the potential leads both components, but the phase-shift is always larger
for the gas.

What is the actual role of dissipation? In an axi-symmetric galaxy, viscous torques would also transfer the A.M. of the gas towards the outer parts,

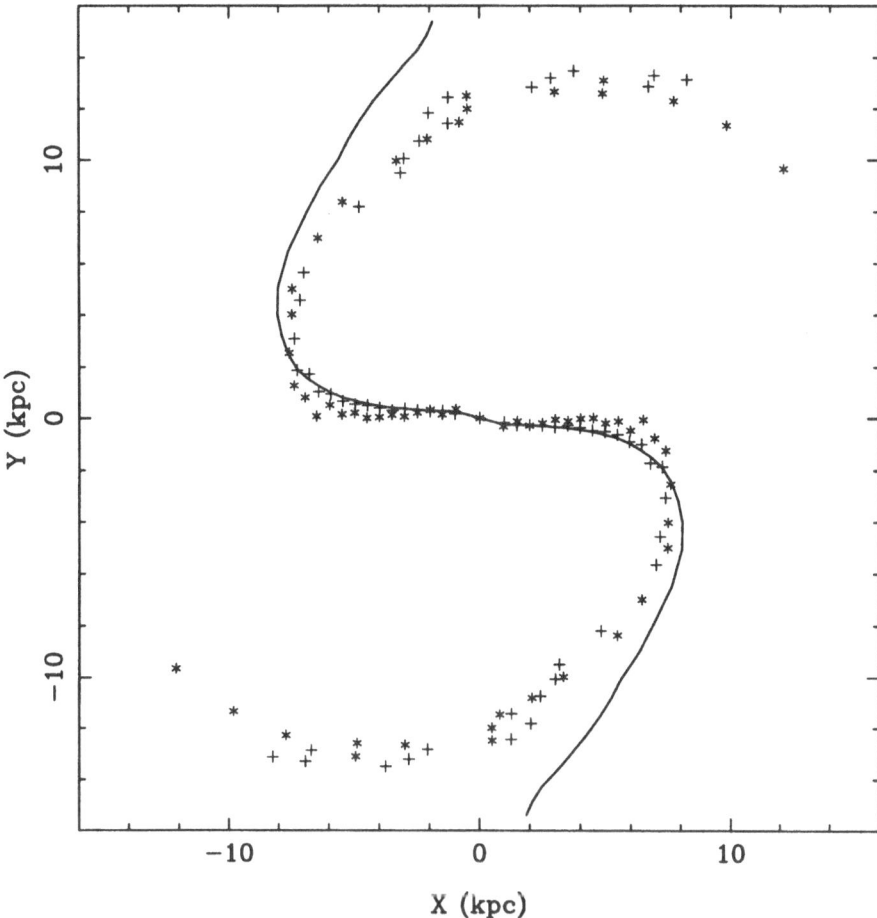

Fig. 4. — Same as fig. 3 for an N-body simulation taking into account gas and stars, and where a bar develops. The phases of the gas pattern are indicated by asterisks, while crosses indicate the stars, and the full lines the potential. Now the potential is in phase with the stellar density in the bar, and leads the density in the spiral outside corotation.

but the time-scale would be longer than a Hubble time at large radii (Lin & Pringle 1987a). Gravity torques are then directly responsible for the gas inflow, and dissipation prevents the gas heating, and maintains the non-axisymmetric features required for gravity torques (Combes et al 1990a). That is why the global gas behaviour should not depend much on the gas hydrodynamics, and the point of view adopted: gas as a fluid or as ballistic colliding clouds. In details this is not quite true, and the two approaches lead to different results: the characteristic shocks along the dust lanes in a barred galaxy are reproduced only with the fluid approach, and all the variety of resonant rings are better obtained with the sticky particles approach. The reality must be in between, when the multi-phase property of the gas is taken into account.

4.3. Feedback and Self-regulation

Given the importance of gravitational instabilities in cloud and star formation, it is interesting to summarize here observational data about Q values. The stellar dispersion has been studied in our own galaxy and in a few external galaxies by Lewis & Freeman (1989) and Bottema (1993). The velocity dispersion decreases exponentially with radius, in parallel to the stellar surface density, and Q_s appears to be constant in function of radius, at least for large galaxies, and vary between 2 and 3 from galaxy to galaxy.

If the stellar disk can be considered as a self-gravitating infinite slab, which is locally isothermal (σ_z independent of z), its density obeys:

$$\rho = \rho_0 \, sech^2(z/z_0) = \frac{\rho_0}{\cosh^2(z/z_0)}$$

where z_0 is the characteristic scale-height of the stellar disk, given by

$$z_0 = \frac{\sigma_z^2}{2\pi G \Sigma_s(r)}$$

where σ_z is the vertical velocity dispersion, and $\Sigma_s(r) = z_0 \rho_0$ is the surface density. The latter has a general exponential behaviour (e.g. Freeman 1970), with a radial scale-length h. The scale-height z_0 has been observed to be independent of radius (van der Kruit & Searle 1981, 82), and there are only small departures from isothermality in z (van der Kruit 1988). Then, if the mass to light ratio is constant with radius for the whole stellar disk, and there is no or little dark matter within the optical disk, we expect to find a velocity dispersion varying as $e^{-r/2h}$. This is exactly what is found, within large uncertainties (Bottema 1993). Since some galaxies of the sample are edge-on and others face-on, the comparison requires to know the relation between z and r projection of the dispersion. In the solar neighbourhood, $\sigma_z = 0.6\sigma_r$, and this ratio is assumed to be valid in external galaxies too.

As for the gaseous component, the vertical velocity dispersion is constant with radius in the outer parts of galaxies, where the rotation curve is flat (Dickey et al 1990): $c_g \approx 6km/s$. The behaviour of $\kappa \propto 1/r$ (for a flat rotation curve) is exactly parallel to the gas surface density $\Sigma_g \propto 1/r$ (e.g. Bosma 1981), and therefore $Q_g \propto c_g \kappa / \Sigma_g$ is constant with radius in the outer parts. This strongly suggests a regulation mechanism, that could maintain the values of Q about constant for both stars and gas.

The mechanism could be simply gravitational instabilities coupled with gas dissipation. When the medium is cool enough, so that the Q value is too low, gravitational instabilities quickly provide heating. The stellar component cannot cool down and keeps hot, although this is somewhat moderated by the gravitational action of the gas, and the young stars born in the cool component. The gas is even more sensitive to the heating, but it can dissipate its disordered motions. The key point is that cooling encourages dynamical instabilities, and therefore produces heating, which is how the regulation works (cf Bertin & Romeo 1988).

Kennicutt (1989) remarks that the extreme radius of HII regions in galaxies coincides with the radius where Q_g is about 1, in a one-fluid analysis of a thin plane; he then suggests that there is a gas surface density threshold for star formation in galaxy disks, and that this threshold corresponds to the onset of gravitational instabilities. In fact, the existence of a star-formation threshold is unavoidable, but could be slightly different than the cloud-formation threshold constrained by gravitational stability criteria. In the outer parts of galaxies, the HI gas that extends much further than the last radius of star-formation, appears patchy, clumpy, and following some kind of spiral structure. The outer gas is unstable at any scale. This suggests that instabilities are present, at the origin of cloud formation, and are the regulators of the constant c_g and Q_g. Sellwood & Carlberg (1984) have shown by simulations that such a feedback does occur, and Lin & Pringle (1987b)have proposed that the self-gravitational instabilities act as a local effective viscosity, giving rise to a local heating time-scale $\propto Q^2/\Omega$ for $Q < 1$.

It is interesting to note that there are several ways to maintain Q constant, and that the stars and gas have chosen two different ways: the stellar component keeps its scale-height constant, while its velocity dispersion is exponentially decreasing with radius; the gas keeps its velocity dispersion constant, while its scale-height increases steadily with radius (linearly, when the rotation curve is flat). This might be related to the different radial distribution. The gas does not display an exponential radial profile, may be due to continued infall or accretion.

5.   FUELING ACTIVITY BY BARS

Activity here refers to either starbursts in the central region, or nuclear activity. These are often related, since they are both triggered by central massive gas inflow in a spiral galaxy. The issue has been addressed by many authors, observationally as well as theoretically. I refer to the proceedings of "Mass-transfer induced activity in galaxies", ed. by Shlosman (1994). The main problem to solve in this matter is that of the angular momentum. We have already seen above the efficiency of non-axisymmetric perturbations and gravity torques. Since they are efficient on a time-scale of the order of the rotation time-scale, their efficiency is increasing towards the center, where the dynamical time is short. This is specially true in mass-concentrated galaxies, where the rotation curve has a steep central gradient (typically $V_{rot} \approx 200km/s$ in nuclear rings of 100pc radius, corresponding to rotation periods of 3 Myr). Even at those

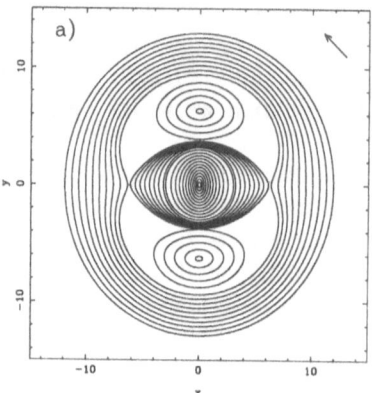 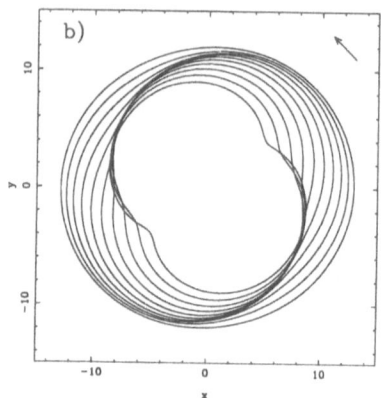

Fig. 5. — a) Periodic orbits in a $\cos 2\theta$ potential. Their orientation rotates by $\pi/2$ at each resonance. b) The gas tends to follow these orbits, but is forced to precess more rapidly while losing energy and angular momentum.

scales, the viscous torques are less efficient. Only dynamical friction exerted on the GMCs could be considered: the Chandrasekhar time-scale is about $3\ 10^7$ yr at r=200pc for a cloud of mass $10^7\ M_\odot$ in our Galaxy, and it varies as the square of the distance from the center.

5.1. The Inner Lindblad Resonance

Radial gas flows are driven the most efficiently by a bar, or a companion producing an $m = 2$ non-axisymmetric potential. Simkin et al (1980) show that bars are also the main agent to bring gas towards active nuclei and to fuel their strong energetic activity. However, as emphasized by Combes and Gerin (1985), the gas is in general efficiently driven to the inner Lindblad resonance, where it accumulates, but not to the very center. We can try to understand schematically, in terms of orbits, the sense of winding of gaseous spiral arms in the center, and consequently the sense of gravity torques.

In a $\cos 2\theta$ bar potential, the main families of periodic orbits are aligned or anti-aligned to the bar, according to the position with respect to Lindblad resonances (see Figure 5). Due to dissipation or cloud collisions, the gas component cannot follow these periodic orbits; instead, upon losing energy, the gas streams in elliptical trajectories at lower and lower radii, with their major axes leading more and more the periodic orbit, since the precession rate (estimated by $\Omega - \kappa/2$ in the axisymmetric limit, for orbits near ILR, and by $\Omega + \kappa/2$ near OLR) increases with decreasing radii in most of the disk. This regular shift forces the gas into a trailing spiral structure, from which the sense of the grav-

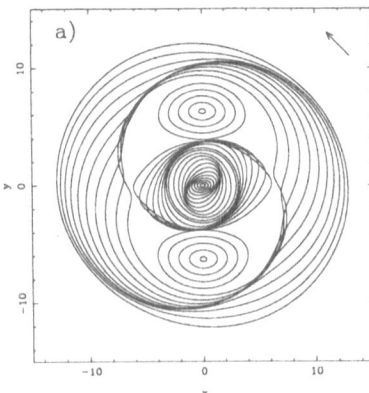 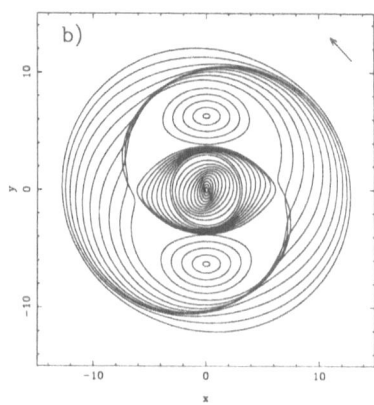

Fig. 6. — a) Without a central mass concentration, the gas winds up in a leading spiral inside the ILR ring; b) with a central mass concentration, the gas follows a trailing spiral structure.

ity torques can be easily derived. Inside the corotation radius the torques are negative, and the gas is driven inwards towards the inner Lindblad resonance (ILR).

Inside the inner ILR, and from the center, the precessing rate is increases with radius, so that the gas pattern due to collisions will be a leading spiral, instead of a trailing one (see Figure 6). The gravity torques are positive, which also contributes to the accumulation of gas at the ILR ring. This situation is only inverted in the case of a central mass concentration, like a black hole, for which the precession rate $\Omega - \kappa/2$ is monotonically increasing towards infinity with decreasing radii. Only then, the gravity torques will pull the gas towards the very center, and "fuel" the nucleus.

Note that inside the two ILRs, it is not easy to predict whether the gas spiral will be leading or trailing. This depends on the gas density radial distribution, which will determine the dominant contribution. Gas simulations do tell us that the gas accumulates then at the inner ILR (e.g. Combes & Gerin 1985). The torque is therefore negative inside the two ILRs, and the dominant spiral must be leading.

If there is no central mass concentration, the torque is positive inside the inner ILR (leading spiral) and the gas is not driven towards the very nucleus. To fuel the nucleus, we are therefore back to the formation of the active nucleus or mass concentration in the first place. This might be a formation in several steps, since bars can be weakened or destroyed by a sudden mass concentration, and reformed afterwards. The first gas accumulation, followed by star formation, can steepen the rotation curve; this will generate a large gradient in dynamical time-scales, and produce a decoupling of a second bar inside the primary bar. This nuclear bar, and possibly other ones nested inside like Russian dolls, can take over the action of gravity torques to drive the gas to the nucleus, as first proposed by Shlosman et al. (1989).

5.2.   Nuclear Disks and Nuclear Bars

This is a phenomenon progressively revealed by increasingly high spatial resolution: either optically through HST imaging (Barth et al 1995), and high-resolution spectroscopy, or by observation of the molecular gas with millimetric interferometers (Ishizuki et al 1990). Recently, Rubin et al (1996) have studied a sample of 80 Virgo cluster galaxies, and have identified 14 ($\approx$ 20%) with a decoupled nuclear disk or ring, of at most 500pc in radius, and revealing large velocity gradients. When face-on these decoupled central disks show frequently a nuclear ring of star formation (e.g. NGC 1068, 1097, 7469, Genzel et al 1995). When edge-on, these systems are peculiar from their large velocity gradients, often showing "forbidden" velocities, like in the Milky Way (Dame et al 1987).

In these decoupled nuclear disks, the gas surface density is often very high, explaining the star formation bursts traced by the $H\alpha$ hot spots. This high density makes the gas more self-gravitating and unstable against bar formation in the center. A secondary bar can form inside the first one, with a common resonance with the primary one. The more general case is where the ILR of the primary is the corotation radius of the secondary one (Friedli & Martinet 1993).

Depending on the physical parameters, and essentially the physics of the gas (dissipation, star formation), there can form a secondary component perpendicular to the primary, but with the same pattern speed, or a secondary bar with a higher speed (Combes 1994). In these two cases, the torques exerted on the gas are quite different. The phase-shift between the gas bar and the potential is very large in the second case, and the amount of fuel arriving on the nucleus is much larger. The high-speed nuclear bar can live for several rotations, until the non-axisymmetry of the potential is destroyed.

Asymmetries and lopsidedness is often observed in these nuclear disks. Any $m = 1$ wave is not expected to amplify in the stellar component, but in the presence of a sufficiently high gaseous mass, there exist amplification mechanisms (Shu et al 1990). However, the development of such a wave does not favour gas inflow (Junqueira & Combes 1996).

The phenomenon of nuclear disk decoupling can occur on very small scales, as is observed in our neighbour M31, thanks to its proximity. The nucleus of M31 appears like a compact stellar disk completely decoupled morphologically and kinematically from the surrounding bulge and large-scale disk. Its radius is less than 10pc. The rapid rotation is known from the early work of Lallemand et al (1960), who showed that the rotation curve is very compact in the nucleus, returning to zero at about 2" from the center. With higher resolution, the rotation peak has increased to 200km/s, which implies the presence of a mass concentration of the order of a few 10^7 $M_\odot$ (Kormendy & Richstone 1995). The dynamics is complicated by the appearance of a double nucleus (Lauer et al 1993), and an $m = 1$ asymmetry.

5.3. Bar Destruction through Mass Concentration

When gas is not present, a bar instability can be very robust in the stellar component, as demonstrated by N-body simulations. The spiral structure that develops at the beginning of the instability, and which transfers the angular momentum outwards to allow the bar growth, quickly disappears through heating. The bar then stays until the end of the simulations, i.e. of the order of a Hubble time. But as soon as a small fraction of gas is present, this is not true anymore. The bar can trigger its own destruction, through the gravity torques exerted on the gas. When enough mass is driven into the center (at least 5% of the mass), the bar begins to dissolve. Hasan & Norman (1990) have shown that the central mass concentration prevents the regular $x1$ orbits from building the bar. In deflecting the central particles, the central mass transforms the inner orbits into chaotic ones. In other words, the central mass deepens the potential, and modifies the characteristic frequencies Ω and $\Omega - \kappa/2$: the resonance structure is lost, and the orbital distribution as well (Pfenniger & Norman 1990, Hasan et al 1993).

Although the central mass considered in models is often a black hole, in real galaxies it is more likely a massive enough decoupled nuclear disk or compact bulge. In that case, the growth of the central mass concentration is slow, and the bar destruction is likely to be a self-regulated process, based on gas accretion. Several bar episodes can successively occur in the same galaxy.

Let us follow such a process: a first strong bar instability develops in the disk; the torques due to this strong initial bar drive the gas inwards in a short time-scale. This increases the central mass concentration, up to the point where the bar weakens; the bar begins to dissolve as soon as there exists two well defined ILRs in the center (due to the perpendicular $x2$ orbits between the ILRs, not supporting the bar). The bar torques weaken and the gas infall is slowed down or halted; the gas coming from the outer parts of galaxies has now time to form stars in the disk, which re-establishes the mass balance between the nucleus and the disk. The central mass concentration loses its dynamical efficiency, and a new bar can form through gravitational instability, which closes the cycle. To have several such cycles in a Hubble time however requires however a substantial mass accretion, i.e. important gas reservoirs in the outer parts of galaxies (cf Pfenniger, Combes & Martinet 1994). Another possible loop of the cycle is provided by the decoupling of a second bar, as developed in the previous section.

One consequence of this feedback cycle is that in evolved bars, there are always just about 2 ILRs. The destruction of the bar first takes place at its intermediate radii. In the center, orbits remain regular, due to the axisymmetry of the dominant mass concentration (figure 7). Then comes a region of chaotic orbits, where the central mass is in competition with the bar influence: this region of the bar is destroyed. Near the end of the bar is a region still dominated by the x_1 elongated orbits. This development of chaotic orbits during the central mass growth could produce the characteristic lenses, often observed associated with barred galaxies. These orbits are bounded only by their limiting energy curve, in the rotating frame, where the potential is $\Phi(r) - 1/2\Omega^2 r^2$, so that outside corotation, there is no bound. Moreover, most orbits between CR and -4/1 resonance are stochastic (Contopoulos & Grosbol 1989). For the ensemble of chaotic orbits inside CR, the corotation region acts as a boundary. This could explain the sharp cut-off at the border of observed lenses.

5.4. Gas-dominated Central Disk

When the gas mass is a large fraction of the dynamical mass in the center, the violent Jeans instability prevents any bar or coherent wave formation (Shlosman & Noguchi 1993). The gas gathers into big clumps; these clumps are first trapped in almost radial orbits, lose energy through collisions and shocks and form finally a binary system, that evolves into a single object at the nucleus. Inhomogeneities in the gas heat up the stellar component, which cannot sustain any bar either. Heller & Shlosman (1994) have simulated the gas flows with or without the presence of a central black hole. The infall is accelerated by dynamical friction against the stellar component, very efficient for such big gas clumps. A mass of 10^9 $M_\odot$ can be gathered inside 100pc within 6×10^8 yr. However, the rate of infall is uncertain, since including star-formation acts as a very efficient feed-back in depositing energy, and also stops the infall in locking part of the gas in a non-dissipative component. When a black hole already sits in the center, it helps the disruption of the big gas clumps through its tidal forces.

6. ENVIRONMENTAL EFFECTS

Mergers are without doubt the most efficient starburst triggers. Observationally, the cause of the starburst appears to be the concentration of a huge gas mass in the center of the merged system. CO emission suggests that the molecular gas represents a significant part of the dynamical mass in the center (Scoville et al 1991). Non-nuclear starbursts are very rare (cf the Antennae, Stanford et al 1990, Whitmore & Schweizer 1995; Arp299, Sargent & Scoville 1991; VV114 Yun et al 1994). They occur in less advanced mergers, where some star formation is enhanced both in giant complexes in the disks, and in the center.

Numerical simulations have brought much of our understanding in those systems, although many uncertainties remain about the physics of gas and star formation (see the review by Barnes & Hernquist 1992a). To trigger such starbursts, gas must be brought towards the center on a time-scale short enough with respect to the feedback time-scale of star-formation (a few times 10^7 yr), that will blow the gas back outwards (e.g. Larson 1987). This requires extremely strong gravity torques. We have seen that viscous torques are not able to drive the gas on such short time-scales.

6.1. Numerical Codes and Gas Modelling

Thanks to the increased power of modern computers, completely self-consistent simulations are now current, including the stellar and gas components. The dark matter halos are also included self-consistently, and this is very important since the halos participate actively in the angular momentum transfer. Barnes (1993) has shown that the dark matter component receives all the angular momentum from the central parts, allowing a rapid merger of the luminous component. The deformation of the halos also accelerates dynamical friction and the sinking of the satellites. There are however fundamental uncertainties in any simulation, since the physics and spatial distribution of most of the mass (the dark component) is unknown, and the physics of the dissipative component, the gas, is not well known either.

Gas dissipation is a key factor in the formation of density waves and non-axisymmetric structures, although viscous torques are negligible versus the gravity torques (Combes et al. 1990a). Dissipation essentially cools down the gas clouds, and cloud collisions deviate their trajectory to form the spiral structure. But only a few collisions per rotation are necessary, and that alone will be insufficient to drive the gas inwards in a Hubble time. Once the non-axisymmetric structure is built up, gravity torques produce the gas flows. We do not know precisely the actual viscosity of the ISM, but given its very complex multi-phase and small-scale structure, it is not relevant to model it in any accurate manner. Any large-scale hydrodynamical simulation can reproduce the main characteristics of gas flow in galaxies, provided that viscous torques are negligible. Two families of gas models are currently used, one based on a continuous diffuse fluid, essentially governed by pressure forces. Such models include artificial viscosity to spread shock waves over a few resolution cells. The physics of the gas is assumed to be isothermal at 10^4K (case of SPH or finite difference codes). The other model used in galaxy hydrodynamics is the sticky particles approach, where an ensemble of gas clouds move on ballistic orbits and collide, without extra pressure and viscosity terms. This model represents more closely the fragmented structure of the molecular component.

6.2. Star-formation Processes

The large-scale star-formation rate and a IMF are not known in every dynamical circumstance, and present simulations taking into account star formation, and the gas released by young stars, can be at best exploratory. Note that the star formation is itself fundamental for the dynamics, since it can lock the gas into the non-dissipative medium and halt for a while the mass transfer towards the center. Since stars are observed to be formed inside giant molecular clouds

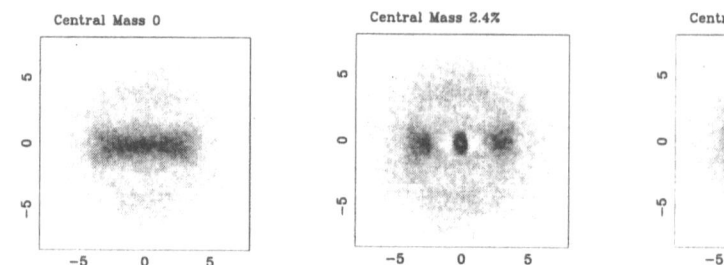

Fig. 7. — Steps in bar destruction, according to the amount of central mass concentration. The destruction of the bar begins in its middle regions, due to the existence of chaotic orbits.

in our Galaxy, the latter being the result of agglomeration of smaller entities, one process could be to relate star formation to cloud-cloud collisions, in the sticky particles approach (Noguchi & Ishibashi 1986). Another more widely used, method is to adopt a Schmidt law for the SFR, i.e. the rate is taken to be proportional to a power n of the gas volume density, n being between 1 and 2 (Mihos et al 1992). In both cases, it was shown that interacting galaxies were the site of strong starbursts, that could be explained both by the orbit crowding in density waves triggered by the tidal interactions, and by the gas inflow and central concentrations, accumulating the gas in small and very dense regions. This depends of course on the non- linearity of the Schmidt law, and SF-efficiency strongly depends on the power n (see e.g. Mihos et al 1992). Mihos & Hernquist (1994a) use a hybrid-particles technique, within SPH, to describe the effects of gas depletion and formation of a young star population. The SPH/young star particles are converted from gaseous to collisionless form, as soon as their gas mass fraction drops below 5%. Although this method has many computational advantages (number of particles fixed), it does not decouple the behaviour of young stars and gas, especially in the shocks. This also inhibits contagious star formation, or any feedback mechanisms (Struck-Marcell & Scalo 1987, Parravano 1996).

6.3. Formation of Large Complexes

It has long been suggested that instabilities in the long tidal tails could form condensations of gas of masses and sizes characteristic of dwarf galaxies. Star formation has been observed in such condensations in the antennae and super-antennae (e.g. Mirabel et al 1992). Hernquist & Barnes (1994), and Elmegreen et al (1993) have shown that such condensations easily occur in N-body simulations. More generally, the ISM of interacting galaxies is likely to condense in large gaseous complexes, of say 10^8 $M_\odot$.

One of the most striking features in interacting galaxies is the enhanced velocity dispersions in the gaseous component. This comes from the gravitational perturbations, that produce streaming motions, asymmetries, spiral arms, strong bars, etc.. In particular, in the molecular gas of M51, the 500pc-scale average of the dispersion is 25km/s, while it is more around 6km/s in a quiet population (Garcia-Burillo et al 1993). This has the consequence to increase the critical Jeans scale for gravitational instabilities, and to create giant complexes (e.g. Rand 1993). The global Jeans length is $\lambda \propto \sigma^2/\Sigma_g$, where σ is the velocity dispersion and Σ_g is the gas surface density of the galactic disk. The corresponding growth time is $\tau_{ff} \propto \sigma/\Sigma_g$, and the instabilities will occur as soon as $Q \propto \sigma\kappa/\Sigma_g$ becomes lower than 1. For the same ratio σ/Σ_g, a perturbed system with elevated σ and Σ_g will see the condensations of larger complexes of mass M $\propto \sigma^4/\Sigma_g$, on the same time-scale τ. These complexes with larger internal dispersions, and larger gravitational support will be less easy to disrupt through star-formation, which enhances the star-formation efficiency (Elmegreen et al 1993). The thermal Jeans length is also larger, due to the hotter gas temperature induced by the larger number of stars in the clouds, and high mass stars are favored.

This can explain the formation of giant and dense star clusters, and the presence of young globular clusters in mergers (e.g. Schweizer 1993). This interesting issue is a clue to the formation of ellipticals through mergers, since present-day ellipticals are clearly richer in globular clusters than their presumed spiral progenitors. Ashman & Zepf (1992) and Zepf & Ashman (1993) have developed a semi-analytical model where globular clusters (GC) form in the merger of two spirals, and predict the existence of several populations of GC, more centrally concentrated and metal-enriched for the more recently formed ones. This receives support from the observations of metal-enriched GC in the cD galaxies M87 and NGC1399 (Mould et al 1990, Geisler & Forte 1990), and recent discovery of young GC with HST in NGC 7252 (Schweizer & Seitzer 1993, see also Meurer, 1996), and NGC1275 (Holtzman et al 1992): the latter detected 50 bright and blue clusters in the cD of the Perseus cluster, and estimated from population synthesis models their age to 3 10^8 yr and their mass to 10^7 M$_\odot$. These would evolve into the well-known halo globular clusters in about 15 Gyr. The uniformity of colors and derived ages supports their formation in one violent merging event. Note that the 'turbulent' conditions giving rise in a merger to new GC must be very similar to the pregalactic collapse conditions, where the first GC are supposed to be formed.

6.4. Lessons from Mergers

In major mergers, where the two interacting galaxies are of comparable mass, strong non-axisymmetric forces are exerted on the gas. But contrary to what could be expected, the main torques responsible for the gas inflow are not directly due to the companion. The tidal perturbations destabilise the primary disk, and the non-axisymmetric structures generated in the primary disk (bars, spirals) are responsible for the torques. The self-gravity of the primary disk, and its consequent gravitational instabilities, play the fundamental role. The internal structure therefore takes over from the tidal perturbations on the outer parts. The gas is provided by the primary disk itself. Again it is not the gas accreted from the companion which is the main trigger for activity, until the final merger of course. This is why the first parameter determining the characteristics of the merger event is the initial mass distribution in the two interacting galaxies (Mihos & Hernquist 1996). The mass ratio between the bulge and the disk is a more fundamental parameter than the geometry of the encounter.

The central bulge stabilises the disk with respect to external perturbations. If the bulge is sufficiently massive, the apparition of a strong bar is delayed until the final merging stages, and so is the gas inflow, and the consequent star-formation activity. But the starburst can then be stronger. When the primary disk is of very late type, without any bulge, the gravitational instability settles in as soon as the interaction begins of the interaction, there is then continuous activity during the interaction, but at the end the starburst is less violent, since most of the gas has already been progressively consumed (see fig 8).

The effects of the geometry are more visible on the direct manifestations of the tidal interaction, i.e. on the tails and debris. For the galaxy experiencing a retrograde collision, no extended tidal tail is formed. There are no resonance effects in the target galaxy disk, and less violent material perturbations; a transient leading tidal arm is instead developed. There is however a tidally-induced two-arm density wave, and the subsequent torques produce gas inflow towards the center, as for the prograde encounter. During the interaction, it happens that the retrograde disk has more star formation than the prograde one, due to the larger gas content. For coplanar merger, the retrograde disk accretes a significant fraction of the gas from the prograde disk (Mihos & Hernquist 1996).

In the simulations, about 75% of the gas is consumed during the merger, whatever the internal structure of galaxies, or the geometry of the encounter (Mihos & Hernquist 1996). This is the most uncertain parameter, however. What is the fate of the rest of the gas? In general long tidal tails are entrained in the outer parts, especially in neutral hydrogen, since this is the most abundant component in external parts of galaxies. But most of the material of the tails is still bound to the system, and will rain down progressively onto the merger remnant (Hibbard 1995).

During minor mergers, where the mass ratio between the two colliding galax-ies is at least 3, the same features can be noticed: the first relevant parameter is the mass concentration in the galaxies before the interaction (Hernquist & Mihos 1995). The torques responsible for the gas mass inflow are exerted by the non-axisymmetric (essentially $m = 2$) potential developed in the disk. The torques are much stronger towards the center; the inflow time is of the order of the rotation time-scale at each radius.

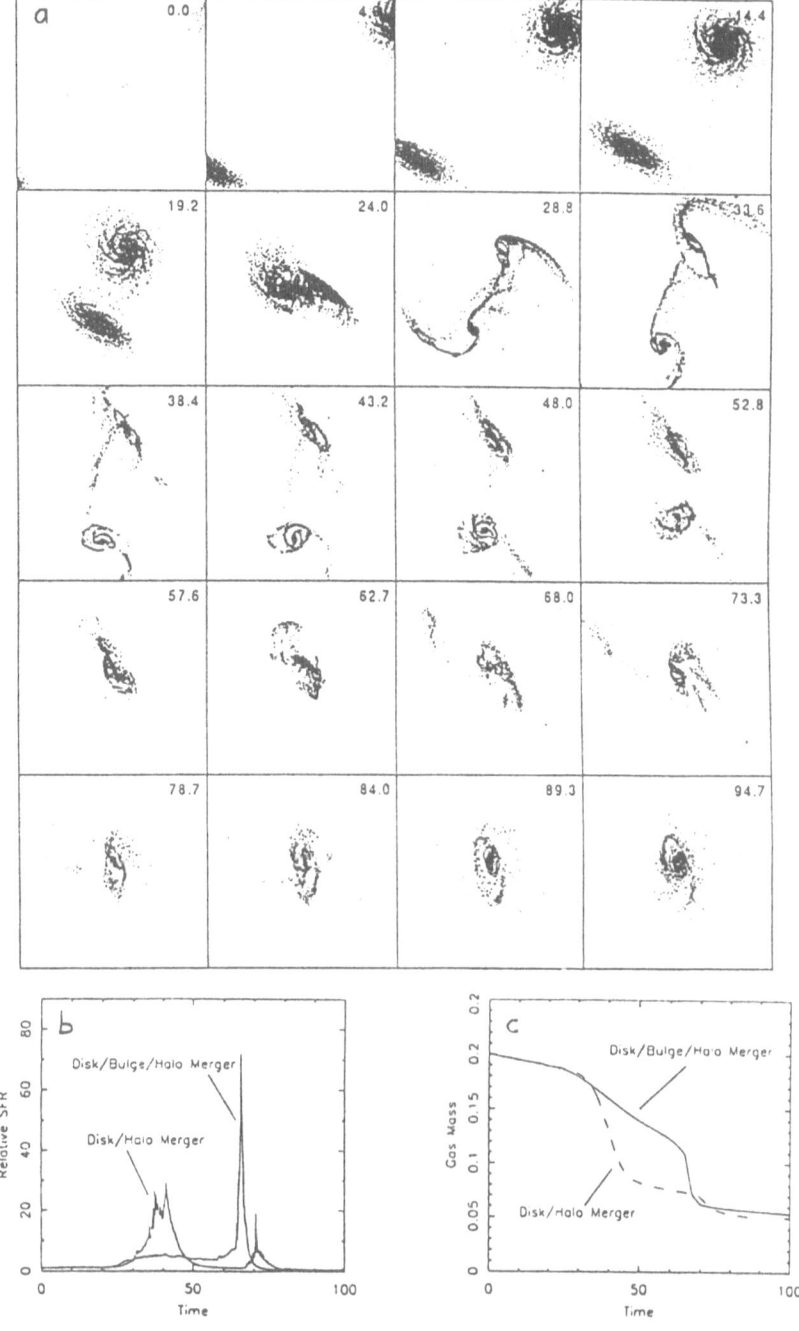

Fig. 8. — a) Evolution of the gas and young stellar components in a merger between two spiral galaxies with bulges. b) evolution of the global star formation rate as a function of time for the mergers of spirals with and without bulges; c) evolution of the total gas mass, for the same runs, from Mihos & Hernquist (1996).

6.5.   Gas Morphology in Mergers

The HI gas is present in the outer parts of galaxies, well beyond the optical disk. There the matter is less bound to the galaxy, and is very sensitive to tidal forces, which grow as r^2. Spectacular HI tidal tails and bridges are observed in mergers. Several systems have recently been observed with high sensitivity and resolution with the VLA: M81/82 (Yun et al 1993), Arp 295, NGC 520 and the Mice (Hibbard et al 1993), NGC 7252 (Hibbard et al 1994).

It appears that the percentage of the total HI mass that is contained in the tidal tails increases during evolution towards the final merger stage. Only 20% of the HI is in the tidal tails in the M81 system, it amounts to 30% in Arp295, 60 to 65% in the Mice and NGC 520, and 75 % for NGC 7252 and NGC 3921, both advanced mergers. The total HI content, however, remains normal, between 10^9 and 10^{10} $M_\odot$.

Detailed simulations of the gas behaviour in mergers help to interpret these observations (Barnes & Hernquist 1992b, Hibbard & Mihos 1995). Only a fraction of the disk gas can be ejected into the tails; due to the symmetric nature of the tidal force, this fraction is at most 50%. Once launched, the tidal tails are not in free expansion, but more than 99% of the visible matter remains bound to the remnant. The bases of the tail, that are more tightly bound than the ends, turn around quickly in their orbits, and fall back towards the remnant (see fig. 9). The phenomenon is a phase wrapping of the material in the central potential that becomes more and more time independent, and approaches an elliptical. Progressively the rest of the tail will rain down, but slower and slower; for may be 20% of the tail material, this will take more than a Hubble time. This phase wrapping creates loops, shells and ripples, such as those frequently observed in merger remnants. As the HI rains down into the merger remnant, it must be transformed into other phases, since no atomic gas is observed within the inner regions of merger (e.g. Hibbard et al 1994). These forms can be molecular gas, stars, and/or hot ionised medium (Dupraz et al 1990, Wang et al 1992). The observation of merger remnants in X-rays have not however disclosed large amounts of hot gas, the L_X/L_B ratio is much lower than in elliptical galaxies (Fabbiano & Schweizer 1995, Read et al 1995).

It appears therefore more likely that the HI gas is transformed into the molecular phase. This corresponds to the large column densities of H_2 observed towards the nuclei of merging galaxies, with millimetric interferometers. Masses up to 10^{11} $M_\odot$, i.e. 10 to 50 times the normal molecular content of a spiral galaxy, are detected in the ultra-luminous infrared galaxies (Sanders et al 1991). These high concentrations of molecular matter in the nuclei explain easily the violent nuclear starbursts: the gas density is well above the critical one for global gravitational instabilities. In the center, the molecular component can represent up to 45% of the dynamical mass (Scoville et al 1994). It is very rare to find CO extensions in maps of merging galaxies, since more than 50% of the molecular component is concentrated to the central spot (about 1.7" in Arp 220, i.e. 300pc).

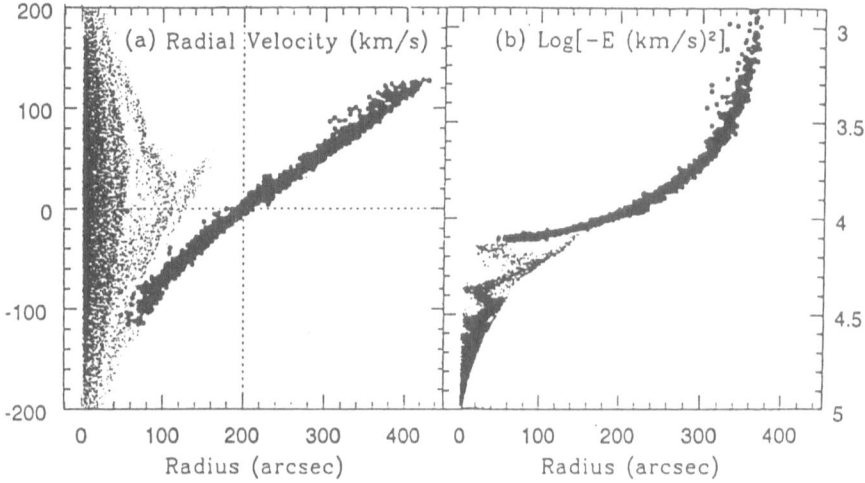

Fig. 9. — The orbital structure of the tidal tails in the NGC 7252 merger remnant, from the simulations of Hibbard & Mihos (1995). a) Radial velocity as a function of radius (dash lines indicate turn-around). b) Energy as a function of radius.

Such high gas concentrations in nuclei, and the violent starburst that results, solve a long-debated problem about the elliptical formation through mergers: it was objected that the central phase-space density in the cores of ellipticals is much larger than in the spiral precursors, and that in a non-dissipative merging, the phase-space density could only decrease (e.g. Carlberg 1986). In fact, we know now that the gas density becomes a large fraction of the total density in merger nuclei, and the process is far from non-dissipative. Mihos & Hernquist (1994b) find even too strong central concentrations in simulations of disk galaxies mergers. The driving of the gas towards the center is so efficient, that subsequent star formation produces dense stellar cores in the merger remnant that are not consistent with the $r^{1/4}$ radial profile of elliptical galaxies. This is however strongly model dependent, i.e. the viscosity of the gas could be artificially enhanced in the simulations, or the star-formation law too strongly non-linearly dependent on local density, etc...

6.6. Tidal Tails and Dark Matter

The extent of tidal tails can help to constrain the amount of dark matter around galaxies. Simulations have shown that, as the dark-to-luminous mass ratio increases, the length of the tails and the mass involved in them is considerably reduced (Dubinski et al 1996). Larger masses imply higher speed encounters, that will detune resonances between the angular frequency of the orbital and internal motions, required to form tails. The deeper potential wells to climb induce shorter tails. Simulations are compatible with observations for dark-to-luminous mass ratios between 0 and 8, but not higher. This corresponds to the dark matter detected from HI rotation curves, but rules out more massive estimates, such as those obtained from timing argument for the Milky Way (10^{12} $M_\odot$ from the orbit of our companion M31).

6.7. Ring Galaxies

6.7.1. Dynamics

Ring galaxies, which have the Cartwheel as prototype, are believed to be formed during a head-on encounter between a disk galaxy and a companion crossing its plane (e.g. Lynds & Toomre 1976, Theys & Spiegel 1976). Although they are rare objects, ring galaxies are an ideal laboratory to study the dynamical triggering of starbursts. An impulse is given by the crossing of the plane by the companion, which starts the dynamical perturbation in the gas and stellar disk, that will eventually trigger the starburst. This school-case triggering event is not perturbed by multiple tidal perturbations. The disk of the target remains in general intact, the star formation is enhanced outside the nuclear regions, in the disk. The geometry of the phenomenon is very simple, relative to messy mergers, and it is relatively easy to follow the chronology of the star formation, in the expanding ring, involving more and more gas (see e.g. Appleton & Struck-Marcell 1996).

The dynamical perturbation is a density wave that propagates outward in the target disk. In a first approximation justified if the collision occurs at high velocity, the positions of the particles are assumed the same just before and just after the collision with each particle having acquired an inward velocity. This is the impulse approximation. The orbits of the stars can then be followed as if they performed radial epicyclic oscillation; most of the effect can be obtained through the kinematic approximation, neglecting self-gravity. Since the period of oscillation is an increasing function of radius (increasing linearly for a flat rotation curve for instance), the central particles have time to reach the center and begin to move outwards, while particles from a larger radius are still moving inwards. This will therefore build a ring of higher density where the orbits crowd, and this ring will expand as a density wave. When the central particles oscillate inward again, the same phenomenon can form a second ring, however expanding much more slowly, and phase mixing soon washes out any structure (see fig. 10). The radii of the rings, their width and spacing, and their expansion rates depend only on the mass distribution of the primary, i.e. on the rotation curve. The amplitude of the perturbation determines the amplitude of the epicyclic motions, and therefore the radial extents of the rings

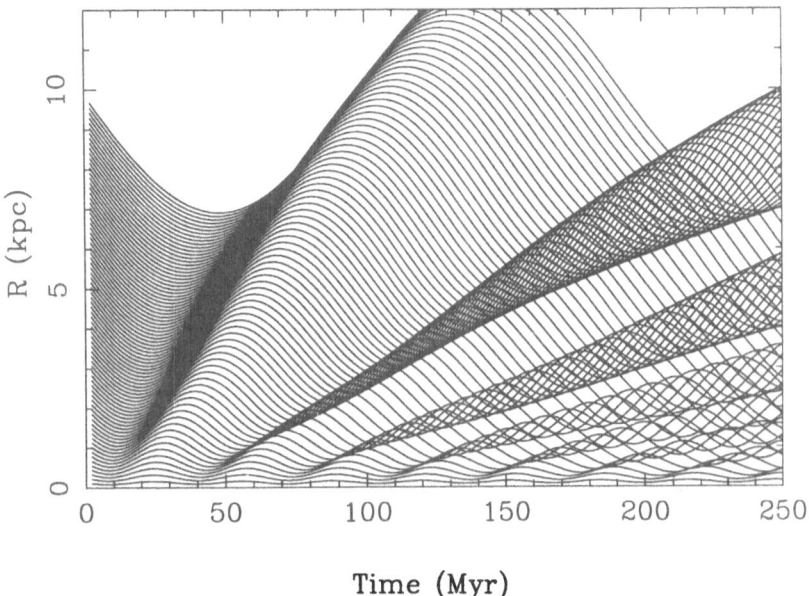

Time (Myr)

Fig. 10. — Kinematic model of ring wave formation, during the head-on collision of two galaxies. The lines are radial trajectories of particles as a function of time.

(or caustics) zone.

In that context, we can expect that gas will experience shocks, or in any case enhanced collisions in the crowded region of the ring-like density wave, enhancing star formation. The behaviour of gas depends strongly on its clumpy or diffuse nature; if highly fragmented with a low volume filling factor, its behaviour should approach that of the stars, and the gas will rebound outside of the ring. On the contrary, the diffuse gas is braked, and by dissipation falls inwards.

6.7.2. "Wheels of Fire"

Ring galaxies are the site of enhanced star formation. Even if they are not among the most violent starburst (> 100 $M_\odot$/yr), they reveal about 10 times the star formation activity of normal galaxies. The dynamical triggering is thus very effective. They are among the rare objects where the starburst is non-nuclear, and occurs in the disk. In the Cartwheel, the Hα comes exclusively from the outer ring (Higdon 1995), and 80% of it comes from just one quadrant, quite asymmetrically. The recent star formation rate (Hα) is about 10 times that over the last 15 Gyr (B-band), and the consumption time-scale of the HI gas (although abundant $1.3\ 10^{10}\ M_\odot$) is only 430 Myr, as the same order as the ring expansion time-scale.

This prototypical behaviour is reproduced in every ring observed in the Hα survey of Marston & Appleton (1995) (except in a Seyfert galaxy): star-formation is exclusively found in the outer ring, and not in the central ring or nucleus. Within the ring, the star formation is distributed unevenly, with some privileged quarters, and in strong discrete knots. The whole ring is also traced by diffuse, broadly distributed, Hα emission. These asymmetries, and strongly preferential regions for star formation, suggest the existence of a threshold dependence on ring gas density to trigger the burst. However, this simple view does not seem to be supported by observations. The Cartwheel is in this respect exemplary: the quarter of the maximum HI density in the ring does not coincide with the quarter of maximum star formation (Higdon 1996). ([1])

The history of star formation has been studied within different hypotheses about the gas structure by Appleton & Struck-Marcell (1987): interesting phenomena are predicted if the ring expansion time-scale commensurates with the collision time τ_c between clouds in the galaxy disk. Star formation is strongly enhanced in the leading edge of the density wave. If a third time-scale is introduced, corresponding to the cloud collapse time, or the time-scale of the feedback cycle (τ_d), then a series of quasi periodic secondary peaks are expected, as a secondary inner ring. However, this is not observed, and predictions of the gas infall towards the nucleus are not supported by observations as well.

Completely self-gravitating models, with stars and gas, have been widely developed in the 1990s, and are quite instructive about the details of gas dynamics and star formation triggering. Gravitational instabilities, mainly in the more sensitive gas component, arise in the disks in the spacing between two rings, and are wound by differential rotation: these might be the explanations of the spokes, although no gas is observed in the spokes of the Cartwheel, nor recently formed stars. Dissipation appears necessary for spoke formation in the models. The number of spokes is predicted to be $R/a = 6$, the ratio of the radius to thickness of the ring.

The VLA HI map by Higdon (1996) does not reveal any correlation of the HI emission with the spokes. The central region including the inner ring appears devoid of gas, since there is no CO detection in this galaxy either (Horellou et al 1995). This appears in contradiction with the models, that predicts a large gas concentration in the center through dissipation (e.g. Hernquist & Weil 1993). One explanation is that there was no gas in the center at the time of the impulse. This is not likely, however, since the lack of gas in the center is an universal feature of ring galaxies. Self-consistent models show however that the gas and stars are decoupled in their ring-like response. There exist some epochs of the simulation, where the center is completely deprived of gas, confined into

---

([1]) A solution to this problem is to invoke large amounts of invisible H_2 gas, although no CO emission has been detected in this galaxy (cf Horellou et al 1995). The CO-to-H_2 conversion ratio could be much higher than standard in this low-metallicity object

an outer ring (somewhat doubled, see fig. 11), while the stars have already fallen back to the center. This could correspond to the time where rings are more conspicuous on the sky, and explain the observations. At that stage, the spokes are not well developed, but this could be an artifact of the limitations of the model (not enough particles, excess heating of the stellar component cutting off gravitational instabilities, etc...). This stage corresponds to the higher compression in the gaseous rings, and must imply the most spectacular starburst. Self-gravity in the compressed rings trigger fragmentation, and the formation of star-forming complexes (knots).

The third dimension of the simulation reveal the characteristic z-perturbations suffered by the disk (Hernquist & Weil 1993, Horellou & Combes 1994). The stellar disk is heated and thickens, while the dissipation in the gas maintains a very small thickness. After new stars have formed in the disk, this could explain why the galactic disks remain relatively unperturbed, in spite of the frequent interactions since their formation.

6.8. Groups and Clusters

6.8.1. Limited Star-forming Efficiency

If there is a large amount of data in favor of enhanced star-formation in interacting galaxies (see Roberto Terlevich, this summer school), the evidence in groups and clusters of triggered activity is much less abundant. Tidal interaction is however effective: distorted galaxies with asymmetries, tails and plumes are often observed in the stellar component (which cannot be due to ram-pressure), but stellar activity requires a large abundance of gas in the outer parts of galaxies, available for gas inflow, and this gas is likely to be removed by tidal stripping. It is heated and then forms the coronal gas observed in X-rays in compact groups and clusters. In the case of strong gas deficiency, the interaction frequency can even be anti-correlated with star-formation rate.

6.8.2. Compact Groups

Compact groups are small systems composed of 4 to 6 galaxies tightly packed on the sky, isolated from other galaxy clusters. Since their internal velocity dispersion is of the same order as the rotational velocity inside an individual galaxy (≈ 200km/s), tidal interactions are expected to be quite efficient. Their small size (≈ 40kpc) and their density ensure frequent galaxy encounters, and they appear as ideal laboratories to study the effect of environment on galaxy evolution. The sample of Hickson compact groups (HCG) has been widely studied at all wavelengths. Their reality as physical groups was questioned (e.g. Mamon 1986), since their merging time should be much shorter than a Hubble time. But the observations of tidal tails and deformations of the galaxies in the group has been advanced as an evidence for the existence, instead of a chance alignment (Mendes de Oliveira & Hickson 1994). There is some (controversial) evidence of excess far-infrared emission (Zepf 1993, Sulentic et

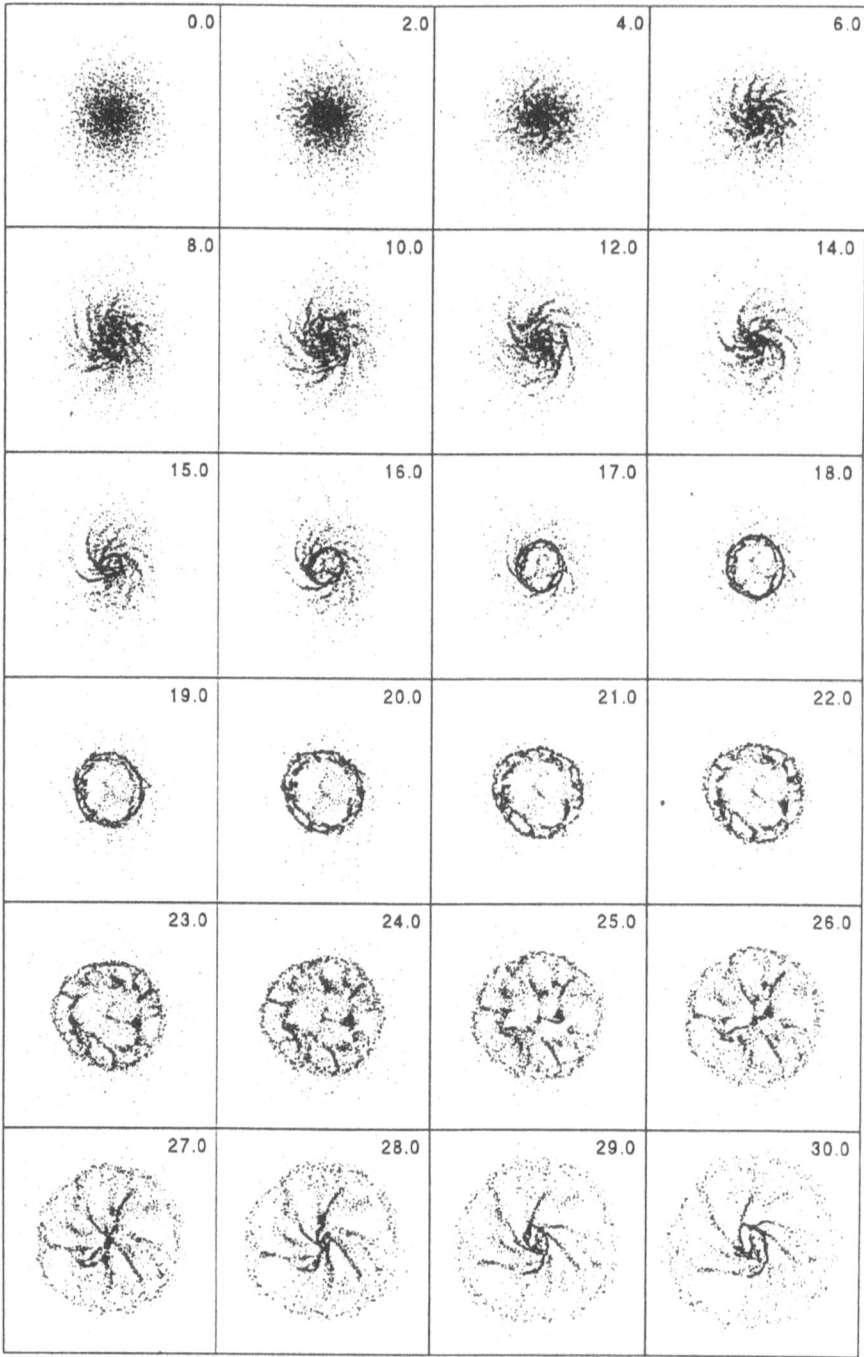

Fig. 11. — Time evolution of the gas component of the primary galaxy in a head-on collision, from Hernquist & Weil (1993).

al 1993) indicating more star-formation than isolated galaxies (especially the $60/100\mu$ colors). There is also excess radio emission (Menon 1995) and HI deficiency (Williams & Rood 1987, Williams et al 1991), which confirm their peculiarity with respect to field galaxies. A recent ROSAT survey of 22 HCGs (Ponman et al 1996) detected the hot coronal gas through X-ray radiation in 75% of them. Pildis et al (1995) show that there is a correlation between extended X-ray emission and low spiral fraction in compact groups. Contrary to binary galaxies, where CO emission is enhanced (Combes et al 1994), there is no evidence of higher molecular gas content in most HCG galaxies (Leon et al 1996, Boselli et al 1996). Therefore, although there exist some obvious signs of interaction, these indicate limited efficiency (there is no merging), and limited star-formation triggering. One interpretation could be that the main consequence of frequent and repeated close tidal interactions, on a time-scale short with respect to the usual dynamical friction time (~ 1 Gyr) is to strip galaxies of their gas, which is heated to the virial temperature of the group system. The gas, extended and diffuse, and hot enough to emit in X-rays, does not participate in the dissipation and braking of galaxies, nor in the star-formation in individual galaxies.

Another possibility is that compact groups are compact only in projection, and they are filamentary structures seen edge-on. Pildis et al (1996) propose a scenario of formation of compact groups, within the CDM model ($\Omega = 1$, $\Omega_b = 0.1$). Groups form along filaments of matter, and become compact only in the final 20% of the simulation. They contain X-ray emitting gas well before becoming compact. However, their end-states at $z = 0$ are more gas-rich, and have an X-ray temperature 3 times too low compared with observations.

6.9. Rich Clusters

6.9.1. HI-deficiency

It has been claimed about two decades ago (Davies & Lewis 1973), and further established (Chamaraux et al 1986) that spiral galaxies in Virgo are actually HI deficient with respect to field galaxies. Cayatte et al (1990) show that the deficiency is a function of distance from the center (M87). CO emission does not seem to follow this trend, and is not depleted in Virgo spirals (Kenney & Young 1989). This can be understood since CO emission comes only from the central parts of galaxies (the optical disk), probably due to radial metallicity gradients. Tidal stripping occurs only in the outer parts, outside the optical disk. The tidal forces increase as r^2 with the distance to the galaxy center, and the gas in the outer parts is much less gravitationally bound to the disk.

As far as star-formation is concerned, rich clusters are not very peculiar, and the star-forming activity of their remaining spirals follows the CO content (Boselli et al 1995). In Virgo, there is even evidence that the global star formation rates have been reduced (Kennicutt 1983). There exists however some distorted galaxies with evidence of enhanced star formation (Cayatte et al 1990): the triggering is likely to be due to normal tidal interaction, since these galaxies are binaries. Here the effects of stripping are much more important, and the remaining spirals could only be due to galaxies that have entered the cluster recently.

Since the HI deficiency in clusters is related to the amount and density of hot X-ray gas detected, it has been proposed that the main cause of stripping is ram-pressure or thermal evaporation. The geometry of the HI deficiency, where all gas-rich galaxies are confined to the border of the cluster, supports this idea. However, the competing effects of tidal stripping and ram-pressure cannot be disentangled, since they both occur preferentially in the cluster center. More-over, the existence of the hot gas itself could come from the tidal stripping of neutral gas from galaxies: the two processes are intimately related. Some clues to the non-predominance of ram-pressure are the stripping of the clumpy molecular gas (Combes et al 1988), and the existence of many gas rich dwarf galaxies, only moderately HI-deficient, in Virgo (Hoffman et al 1988). Gavazzi & Jaffe (1987) find that the Coma Supercluster spirals tend to have stronger radio-continuum emission, and they identified three cases of distorted galax-ies, good candidates for an ICM-ISM interaction. However, the red and NIR images of these galaxies are also distorted. It is difficult to know whether this comes from the old stellar disk (in that case the galaxies are tidally distorted), or whether recent star formation in the ICM-distorted gas could account for the data (Boselli et al 1994). It has been proposed that the first encounter of the intra-cluster gas by a galaxy entering the cluster could enhance star-formation in its disk, by the shock-waves produced. The supposed over-pressure exerted on the HI gas could compress them to form molecular clouds, which would be the site of star formation. However the ICM pressure, even in the center of the Coma cluster, is only comparable to the internal pressure of a normal ISM (see e.g. Kenney 1990).

6.9.2. Harassment and Evolution

The accumulation of frequent high-speed close encounters, dubbed "harass-ment" by Moore et al (1996a), has peculiar consequences, quite different from the normal galaxy-galaxy binary interactions and merging. Simulations have shown that these close encounters, although they are not resonant with the internal motions, can produce the tidal damage observed in most perturbed galaxies in clusters (e.g. Combes et al 1988). One issue is to distinguish be-tween the influence of the global tidal action of the cluster, and the two-body encounters between individual galaxies. The former global effect is expected to be significant only in the central part, while the best cases of tidal deformations are observed on the periphery. But this is not a strong argument, given the morphological segregation: only early-type galaxies (ellipticals, or lenticulars with large bulges) survive in the center, and they are not sensitive to tidal perturbations.

It is instructive to compute the order of magnitude of the two effects: tidal interactions between two galaxies are important, as soon as their relative distance is of the order of their diameters. For galaxies located at the periphery, the ratios of tidal forces from the nearby companions and from the cluster is proportional to the ratio of masses ($\sim$ 100), divided by the cube of the distances ($\sim$ 1000). Two-galaxies interactions will dominate at the periphery, but not in the center. This is only an indication, since the number of penetrating collisions, more frequent in the center, has not been estimated. The collision time is of the order of $t = (n\sigma v)^{-1} \sim 1 Gyr$, with a density of galaxies of $n=$ 100 Mpc^{-3}, and a velocity dispersion of $v=1000$km/s.

Stripping due to the global tides of the cluster can be estimated through the tidal radius r_T (cf Merritt 1984): near the core radius of the cluster r_c, $r_T/r_c \approx 0.5 v_g/v_c$ where v_g and v_c are respectively the velocity dispersions of a galaxy and the cluster. This gives an order of magnitude of 15kpc for r_T in a typical rich cluster of $v_c = 1000$km/s.

Harassment is so efficient that it is possible to account for the rapid galaxy evolution in rich clusters (Moore et al 1996a): within 4-5 Gyr, the morphology of galaxies in clusters can be transformed from late-type spirals to early-type, lenticulars and ellipticals. Only 4-5 high-speed collisions are necessary for this evolution. However, the whole picture is made uncertain by our missing knowledge of the gas dynamics, and the repartition of the dark matter: either bound to each galaxy, or diffuse and bound to the cluster globally. The "overmerging" problem faced in dissipationless N-body simulations could be cured by the consideration of the gas component (Moore et al 1996b).

7. GALAXY EVOLUTION

7.1. Evolution along the Hubble Sequence

The Hubble sequence is morphological and is determined by some main physical parameters:

1. The bulge-to-disk luminosity ratio increases from late to early type spiral galaxies. Given the similar mass-to-light ratios for both components, this means that the mass concentration increases from Sc to S0. If there is an evolution along the sequence, it can only be from late to early, since the mass concentration is irreversible.

2. The total visible mass increases also from late to early types. This confirms the sense of evolution.

3. The percentage of gas mass decreases from Irr and Sc to S0 and E. This had been already noticed by Hubble (1926) and Sandage (1961), due to the patchy aspect of late-types, with a lot of dust. HI observations demonstrate clearly the progression of gas mass fraction along the sequence, while the situation is not so clear for the molecular gas component. The CO emission does not vary as much as the HI emission along the sequence (Young & Scoville 1991), but this could be a metallicity effect, the CO emission being depleted in late-type galaxies.

4. The percentage of dark matter decreases from Sc to Sa. This is known from recent results of HI rotation curves in early-type galaxies. These galaxies possess much less gas, and more sensitive studies are necessary to actually sample the rotation curves in the outer parts. Casertano & van Gorkom (1991) and Broeils (1992) have established that the rotation curves of early-type galaxies tend to fall at large distance from the center, while rotation curves are still rising for very late galaxies (see also the compilation by Persic et al 1996). There is clear evidence of decreasing dark-to-visible mass ratio along the Hubble sequence.

5. A secondary morphological parameter noted by Hubble is the pitch angle of the spiral: early-types tend to be more wound, and late-types more open. This is only a consequence of mass concentration, and bulge-to-disk mass ratio. Early-types have a much larger epicyclic frequency for a given disk, and their Q parameter is higher. They are less violently unstable with respect to spirals, and their pitch angle is smaller. The critical wavelength is $\lambda_c \propto \Sigma/\kappa^2$ (cf section 1), which also points towards smaller wavelengths for early types.

We know that bars and spiral waves can transfer angular momentum, and cause mass concentration on a time-scale much shorter than the Hubble time. Since most spiral galaxies are observed with these features, this implies rapid evolution along the Hubble sequence, from late to early types. But along the sequence, galaxies gain visible mass, and their dark-to-luminous mass ratio decreases: this suggests that some of their dark matter has been transformed in visible stars (Pfenniger et al 1994, Pfenniger & Combes 1994).

Recent large galaxy surveys in the near-infrared band have slightly modified and clarified the criteria of the Hubble sequence. These surveys are less affected by dust extinction, which blurs the classification. de Jong (1996) and Courteau et al (1996) claim that the bulge mass is a better criterion than the bulge-to-disk ratio to define the sequence. The total mass is also a fundamental parameter (e.g. Gavazzi et al 1996). Courteau et al (1996) see moreover a constant ratio between the characteristic scales of bulge and disks ($r_b/r_d \approx 0.1$); according to these authors, this supports theories of bulge formation through a secular evolution from the disks (e.g. Pfenniger & Norman 1990). Andredakis et al (1995) emphasize that the bulge morphology evolves continuously from late to early types, from an exponential radial distribution to a de Vaucouleurs $r^{1/4}$ profile. The smooth sequence observed for bulges along the Hubble sequence suggests a single mechanism for bulge formation.

Alternatively, there could be two main processes contributing to bulge formation. First, strong bar instabilities in late-type systems can be the cause of box- or peanut-shaped bulges, due to vertical Lindblad resonances (Combes et al 1990b). This secular evolution can form small bulges, of exponential profiles essentially. Second, minor mergers, or the accretion of a small companion may form bigger bulges, of de Vaucouleurs profiles (Pfenniger 1991). These processes are cumulative along the evolution.

7.2. Fragility of Disks

The main argument against frequent interactions in a galaxy life, or formation of the bulge through minor mergers, is that galaxy interactions can easily
thicken or even destroy a stellar disk (e.g. Gunn 1987). Vestiges of interactions, such as ripples, plumes, shells (e.g. Schweizer 1990) do witness some
interactions, but those must have occurred in the past, before the formation of
the disks. The fragility of stellar disks with respect to thickening has been used
by Toth & Ostriker (1992) to constrain the frequency of merging and the value
of the cosmological parameter Ω. They claim for instance that the Milky Way
disk has accreted less than 4% of its mass within the last $5\ 10^9$ yr. Numerical simulations have tried to quantify the thickening effect (Quinn et al 1993,
Walker et al 1995). They show that the stellar disk thickening can be large and
sudden, but it is strongly moderated by gas hydrodynamics and star-formation
processes, since the thin disk can be reformed continuously through gas infall.

Since the efficiency of disk thickening is not yet settled, it is interesting to
check observationally the influence of tidal interactions on the disk thickness of
galaxies. In a sample of 31 edge-on galaxies, out of which 24 were interacting,
B, V and I photometry has been analysed, to tackle particularly this problem
(Reshetnikov & Combes 1996). It was found that the ratio h/z_0 of the radial
disk scale length h to the scale height z_0 is indeed 1.5 to 2 times lower for
interacting galaxies than for a normal sample. While the h/z_0 ratio is 11 for
late-type galaxies, and 5.2 at minimum for large galaxies in a control sample
(Bottema 1993), it is 2.9 on average in tidally interacting galaxies. This comes
both from an increase of z_0 and a slight decrease of h through stripping. This
is in good agreement with simulations.

However, since galaxies have experienced many interactions in the past, including the presently isolated galaxies, all these perturbations, thickening of
the planes and radial stripping, must be transient, and disappear after an interaction time-scale, i.e. one Gyr. Present galaxies are thought to be the result
of merging of smaller units, according to theories of bottom-up galaxy formation; a typical galaxy has accreted several tens of percents of its mass, and the
existence of shells and ripples attests of the frequency of interactions (Schweizer
& Seitzer 1992). This implies that the global thickness of galaxy planes can
recover their small values after galaxy interactions. Since the stellar component
alone cannot cool down, this can only be the result of gas accretion. Such an
accretion is possible from the huge HI reservoirs that late-type galaxies possess in their outer parts. This gas is driven inwards by gravity torques during
a tidal interaction (Braine & Combes 1993). The accretion also produces a
deeper potential well in the plane, and the stellar component will react with a
slightly reduced thickness; the main effect will be the formation of young stars,
in a much thinner disk. The signature of past interacting events can thus be
found in the existence of thick disks (cf Burstein 1979). In the Milky Way, the
thick disk has been widely studied (e.g. Robin et al 1996). We can note that
the low value of h/z_0 in our Galaxy is a sign that it is still presently interacting.

7.3. Evolution at High Redshift

High resolution imaging with HST has recently allowed the study of galaxy morphology at intermediate redshift ($0.1 \leq z \leq 1.0$), and often revealed tidal features and distorted morphologies indicative of interactions and mergers. To interpret these data, we must be aware that the detectability of tidal features is reduced at high redshift (Mihos 1995). Tidal tails and bridges rapidly disperse, and fall below the surface density limit for detection. The debris and signatures of mergers become invisible after 1 Gyr for $z = 0.4$ and after 0.2 Gyr only at $z = 1$. This leads to underestimating the influence of tidal interactions in objects displaying spectroscopic signatures of starbursts. For instance, E+A galaxies could be in fact merger remnants.

From high-resolution images of a cluster at $z = 0.4$, Dressler et al (1994) suggest that the excess blue galaxies seen in distant clusters are predominantly normal late-type spirals, undergoing tidal interactions or mergers. The Butcher-Oemler effect, associated with enhanced star-forming activity appears now convincingly to be due to enhanced interactions, arising from the hierarchical merging of clusters. Clusters at $z = 0.4$, i.e. 3-4 Gyr ago, appear to possess a much larger fraction of disk-dominated galaxies than present day clusters (Couch et al 1994), and galaxy interactions were undoubtedly more frequent at high z.

At very high redshift ($z \approx$ 2-3), HST images have revealed a population of faint, small, compact objects, that are identified to the sub-galactic clumps, or building blocks of present day galaxies through multiple merging processes (Pascarelle et al 1996, Steidel et al 1996). However, caution must be used to interpret high redshift data, since fading of low surface brightness features (as $(1+z)^{-4}$), and k-correction effects transform galaxies into later types and more irregular objects, even without evolution (Giavalisco et al 1996). For instance, the fact that barred and grand design galaxies appear under-represented in distant deep field samples, while early-types are not (van den Bergh et al 1996) could be only an artifact of the non-detectability of disks. At very high redshifts, only the high surface density objects, such as bulges and ellipticals, can be detected.

In spite of all these biases underestimating the influence of interactions, many studies have put forward evidence for an increasing merger rate with redshift, as $(1+z)^{m}$ with $m = 2-3$ (e.g. Yee & Ellingson 1995). Lavery et al (1996) claim that ring galaxies are even more rapidly evolving, with $m = 4-5$, although statistics are still insufficient. Evolution at high redshift is a rapidly growing field, but results are not yet settled on very firm ground.

8. GAS AND DARK MATTER

The possibility that there exists large amounts of gas in the Universe, up to the closure density, has been considered mainly in the 1960's (see the review by Peebles, 1971). The hypothesis was rejected, since the gas was assumed homogeneous, and self-gravity was ignored. For an homogeneous gas, the Gunn-Peterson test applies, i.e. absorption lines accumulate along the line of sight in front of a remote continuum source such as a quasar, and we should detect a trough of absorption for redshifts lower than the quasar redshift. The results of the tests at many wavelengths constrain the density of any intergalactic gas several orders of magnitude below the closure density. However, the test is not valid for a clumpy medium, where the filling factor is very small. Most of the time, the line of sight will be empty, except some lines of sight where the absorption will be highly optically thick.

The hypothesis that cold gas could in fact be the dark matter has been reconsidered only recently, when it was realised that the fundamental state of the cold self-gravitating gas is not homogeneous, but a hierarchical structure of dense clouds, permanently gravitationally unstable, very similar to the fractal structure of the detected interstellar medium (Pfenniger et al 1994). Several models were developed, considering the cold gas either in the halo, associated with clusters of Machos (de Paolis et al 1995, Gerhard & Silk 1996), or confined to a self-gravitating disk (Pfenniger & Combes 1994). In the latter model, dark matter around galaxies is made of basic "clumpuscules" of molecular hydrogen, of Jupiter mass (10^{-3} $M_\odot$), but with much smaller density than brown dwarfs, 10^{9-10} cm^{-3}, and radius of ≈ 20 AU. All along the hierarchy, structures are unstable gravitationally, permanently coalescing and fragmenting in the isothermal regime. The clumpuscules are self-gravitating, statistically in equilibrium with pressure forces, at the limit of the adiabatic regime, since then the radiation transfer time equals the dynamical time. They compose the smallest scale of a hierarchical structure, that ranges over six orders of magnitude, up to giant molecular clouds of 100 pc; above this scale, bigger gaseous complexes are torn apart by galactic shear. The ensemble of clumpuscules is in quasi isothermal equilibrium with the bath of photons of the cosmic background radiation at $T = 2.73$ K$(1 + z)$. Due to the fractal structure (of dimension $D < 2$), the clumpuscules collide together frequently, with, for such fractal dimension, a rate of the order of the cooling-heating and dynamical times. The condensed structure resembles closely the well-studied ISM gas, already known as a $\dot{D} < 2$ fractal-like structure over several orders of magnitude in scale (Larson 1981; Scalo 1985; Falgarone et al. 1992). The main difference is that star-formation in the visible disk has metal-enriched the medium that can then cool down much faster, and the heating sources have partly destroyed the condensed fractal and formed a diffuse intercloud medium.

In galaxy outskirts, the condensed H$_2$ phase is almost only in contact with the intergalactic radiation field, which photodissociates a small fraction of it into HI gas, because at the envisaged column densities ($\sim 10^{25}$ cm^{-2}) H$_2$ can self-shield easily. From this point of view, the atomic gas in the outer parts serves as a tracer of dark matter. This easily explains why, observationally, the gas and dark matter appears to be intimately related. From the flat rotation curves, the surface density of dark matter Σ_{DM} varies asymptotically as r^{-1}, and as well the HI surface density Σ_{HI}. Bosma (1981) was the first to notice a constant ratio of Σ_{DM}/Σ_{HI} as a function of radius in spirals, which has been confirmed by many authors (Sancisi & van Albada 1987; Puche et al. 1990), and varies between 10 and 20 according to the morphological type (Broeils 1992, Carignan 1996).

The hypothesis of cold gas as dark matter helps also to explain the evolution of the Hubble sequence (and in particular the decreasing dark-to-luminous mass ratio along the evolution, cf above), the overabundance of gas in interacting galaxies (Braine & Combes 1993), and also the presence of huge amounts of hot gas in clusters.

8.1. Hot Gas in Rich Clusters

The larger sensitivity of the ROSAT satellite, that allowed the detection of the hot gas in the outer parts of clusters, and its spatial resolution, from which the temperature as a function of radius $T(r)$ was obtained, have considerably changed our physical view of clusters. David, Jones and Forman (1995) summarize the results, and emphasize that some of the previously dubbed dark matter is now seen as hot gas. The mass of the hot gas is increasing with scale, and reaches 30% of the total mass, including the remaining dark matter. This leads to the so-called baryon catastrophe, since the baryon fraction in rich clusters is representative of the baryon fraction in the Universe; in other words, the dark matter cannot be segregated outside of clusters (White et al 1993). This high baryon fraction (30%) is incompatible with $\Omega = 1$ given the constraints of primordial nucleosynthesis ($\Omega_b < 0.1$).

David et al (1995) show that the mass-to-light ratio is not increasing with scale, as was previously believed, but remains of the order of 100-150 on scale 10^{12} $M_\odot$ to 10^{15} $M_\odot$. They compare the mass of hot gas around a giant elliptical, in loose groups, clusters, and rich clusters. The gas mass fraction increases from 2 to 30%. The radial distribution of hot gas reveals a flatter distribution than for the other visible mass, and even the dark matter. The hot gas could be only a tracer of a multiphase medium, with even more mass in a cold gas phase (Ferland et al 1994, Pfenniger & Combes 1994).

These new results reveal a striking phenomenon: the dark matter is more important relatively on galaxy scale than on cluster scale (see figure 12). In other words, some of the dark matter becomes visible as hot gas when galaxies belong to rich clusters, or when large-scale structures virialize and relax gravitationally. It has been proposed that feedback from galaxy formation can reheat the intergalactic medium, and especially galactic winds can produce the flatter gas distribution relative to the gravitating mass observed. Alternatively, galaxy interactions and tidal stripping could produce the hot gas from the cold gas reservoirs present around spiral galaxies in large amounts (Pfenniger et al 1994).

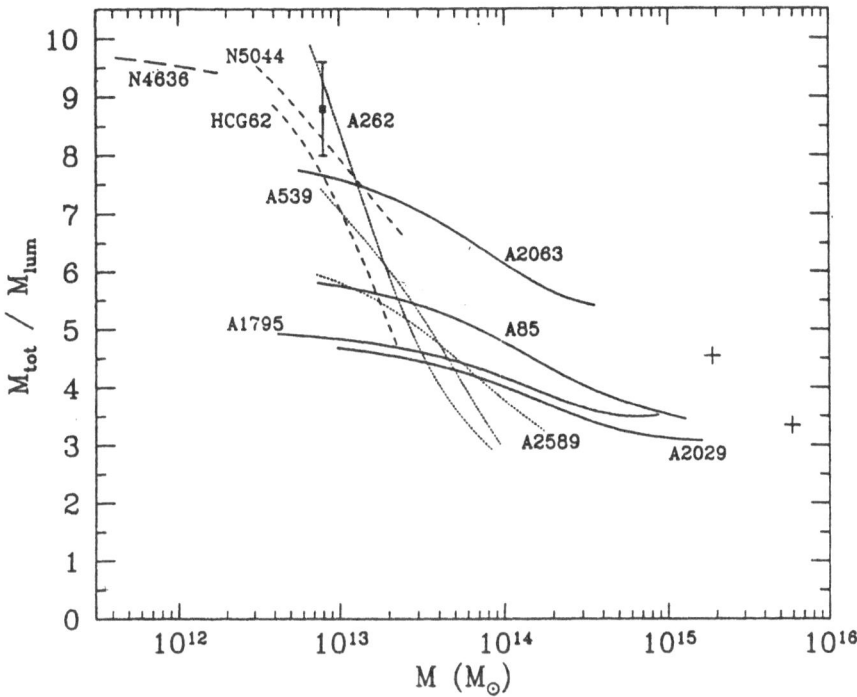

·Fig. 12. — Ratio of total gravitating mass-to-luminous mass (gas plus stars) as a function of the total gravitating mass, for various systems and scales, from a single galaxy to rich clusters (David et al 1995).

8.2.   Self-gravity and Fractal Structure of the ISM

The hypothesis of cold gas as dark matter is based on the existence of a fractal structure for the molecular clouds, in isothermal equilibrium with the cosmic background radiation. If we consider the outer parts of galaxies, devoid of any heating and star formation, we can consider the gas only submitted to its own gravity. The theory of such a medium has not yet been established. We are presently trying to built such a theory, which will explain why the medium has to form a fractal structure, and to predict what is the Haussdorf dimension d_H of the fractal (de Vega, Sanchez & Combes, 1996). The underlying idea is that the fractal is build through recursive Jeans instability and fragmentation. In an isothermal context, the gas has the tendency to collapse and follow an isothermal sphere, but as soon as the density contrast reaches a critical value, fragmentation occurs: the large-scale structure fragments in clouds, for which the same collapse instability applies. This recursive fragmentation proceeds until the density is high enough to reach the adiabatic regime. Therefore, self-gravity alone could be the origin of the fractal, with generated turbulent motions in virial equilibrium at each scale.

We have developed the grand partition function of an ensemble of particles, interacting only through their own gravity, and in equilibrium at a given temperature T. The number of particles N is allowed to vary, keeping the gravito-chemical potential μ constant. In transforming the partition function through a functional integral, it can be shown that the system is exactly equivalent to a scalar field theory. The theory does not diverge, since the system is considered only between two scale limits: the short-scale cut-off (the size of the smallest fragments, or clumpuscules), and the large-scale limit (the largest self-gravitating entities, or GMCs). We demonstrate through a perturbative approach that the system has a critical behaviour, for any parameter (effective temperature and density). That is, we can consider the self-gravitating gaseous medium as correlated at any scale, as for the critical points phenomena in phase transitions.

Since scaling behaviors are the best studied through renormalisation group theory, we use these methods to derive by analogy the critical exponents of the system. It is shown that the critical exponent ν of the correlation length can be identified with the inverse Haussdorf dimension $d_H = 1/\nu$. The long distance critical behaviour must then be governed either by the Ising fixed point, for which $\nu = 0.63$, or the mean field point ($\nu = 0.5$). Both fixed points give a fractal dimension compatible with observations. It is well known that there exists only two independent critical exponents: the second one concerns the potential, and this theory makes some specific predictions, that could be checked through gravitational lenses observations in front of remote sources.

This study shows that self-gravity can account alone for the fractal structure of a self-gravitating isothermal medium. It is well known that fixed points are quite robust with respect to perturbations, and we can hope that this generalizes to the interstellar medium, provided that it is not too much perturbed by star formation. The latter heats the medium and destroys progressively the fractal and condensed structure to the benefit of the diffuse phase.

9. CONCLUSION

The high level of star formation found in starbursts (10-100 times the normal rate), is almost always due to a dynamical triggering. Galaxy interactions are the main agent to de-stabilise gravitationally a galaxy, and trigger the development of spiral and bar waves. These non-axisymmetric perturbations produce gravity torques, that transfer the gas inwards, and agitate the medium which elevates the velocity dispersion. When the interacting galaxies are bound, they merge rapidly through dynamical friction: galaxy collisions are highly inelastic, and the relative orbital energy is dissipated in galaxy deformations and heating. Star formation is then more efficient, since these perturbations produce gas accumulations and piling up towards the center, and the star formation rate is non-linear with density (global Schmidt law, due to cloud-cloud collisions, gravitational instabilities, etc..). A starburst is obtained when the gas inflow is provided on time-scales shorter than the feedback processes of star formation (supernovae, stellar winds, bipolar flows..). This occurs for strong enough perturbations, tidal interactions and subsequent bars.

Processes triggering star-formation are not yet known in details: criteria for gravitational instabilities just control the cloud formation, but interstellar clouds can be quiescent even when unstable to fragmentation and small-scale structure formation. Certainly the final criterion is dynamic (and not static) and must take into account the energy liberated in the first stars formed.

Galaxy evolution therefore depends strongly on the environment. Dynamical interactions and subsequent evolution make galaxies more concentrated, with higher surface densities. Low surface density galaxies (such as Malin I types) are always found in poor environments; they have not suffered many galaxy interactions, which has slowed down their evolution. They might represent a state close to primeval galaxies.

It is well known that non-axisymmetric structures like spiral structures and bars transfer angular momentum, and produce central mass concentrations. Since these non-axisymmetric structures are observed in most disk galaxies today, the evolution along the Hubble sequence, from late to early types, must be very rapid, on time-scales shorter than the Hubble time. A bar can be destroyed through mass concentration, but another one might be formed later, provided that the galaxy has accreted enough fresh gas. The evolution could then be punctuated by several interacting events, triggering the formation of bars. To understand why galaxies have still some gas today, and their disks are not all very thick due to interactions, the accretion of large amounts of gas must be invoked during evolution. A possibility is that dark baryons exist in the form of cold gas in the outer parts of galaxies. This gas will be heated through galaxy collisions in clusters, and account for the huge amounts of hot gas observed there through X-rays.

References

[1] Andredakis Y.C., Peletier R.F., Balcells M. *MNRAS* **275** (1995) 874
[2] Appleton P.N., Struck-Marcell C. *ApJ* **312** (1987) 566& *ApJ* **318** (1987) 103
[3] Appleton P.N., Struck-Marcell C. *Fund. of Cosmic Phys.* **16** (1996) 111
[4] Ashman K.M., Zepf S.E. *ApJ* **384** (1992) 50
[5] Barnes J.E. (1993) in "Physics of Nearby Galaxies: Nature or Nurture?", Rencontres de Moriond, eds. Thuan T.X., Balkowski C., Van, J.T.T., Editions Frontières, p. 301
[6] Barnes J.E., Hernquist L. *ARAA* **30** (1992a) 705

[7] Barnes J.E., Hernquist L. *Nature* **360** (1992b) 715

[8] Barth A.J., Ho L.C., Filippenko A.V., Sargent W.L.W. *AJ* **110** (1995) 1009

[9] Bertin G., Romeo A. *A&A* **195** (1988) 105

[10] Block D.L., Elmegreen B.G., Wainscoat R.J. *Nature* **381** (1996) 675

[11] Boselli A., Gavazzi G., Combes F., Lequeux J., Casoli F. *A&A* **285** (1994) 69

[12] Boselli A., Gavazzi G., Lequeux J. et al *A&A* **300** (1995) L13

[13] Boselli A., Mendes de Oliveira C., Balkowski C.et al A&A, 1996, preprint

[14] Bosma A. *AJ* **86** (1981) 1825

[15] Bottema R. *A&A* **275** (1993) 16

[16] Braine J., Combes F. *A&A* **269** (1993) 7

[17] Broeils A. (1992) PhD thesis, Groningen Univ.

[18] Burstein D. *ApJ* **234** (1979) 829

[19] Carignan C. (1996) in "New Extragalactic Perspectives in the New South Africa", ed. D. Block, Kluwer, p.

[20] Carlberg R.G. *ApJ* **310** (1986) 593

[21] Casertano S., van Gorkom J.H. *101* **1991** (1231.

[22] Cayatte V.) BalkowskiC., van Gorkom J.H., Kotanyi C. *AJ* **100** (1990) 604

[23] Chamaraux P., Balkowski C., Fontanelli P. *A&A* **165** (1986) 15

[24] Combes F., Gerin M. *A&A* **150** (1985) 327

[25] Combes F. (1988) in "Galactic and Extragalactic Star Formation", ed. R.E. Pudritz & M. Fich, Kluwer, p. 475-494.

[26] Combes F., Dupraz C., Casoli F., Pagani L. *A&A* **203** (1988) L9

[27] Combes F., Dupraz C., Gerin M. (1990a), in "Dynamics and Interactions of Galaxies", Heidelberg Conf. ed. R. Wielen, p. 205.

[28] Combes F., Debbash F., Friedli D., Pfenniger D. *A&A* **233** (1990b) 82

[29] Combes F., Prugniel P., Rampazzo R., Sulentic J.W. *A&A* **281** (1994) 725

[30] Combes F. (1994) in "Mass-transfer induced activity in galaxies", ed. I. Shlosman, p. 170

[31] Contopoulos G., Grosbol P. *A&AR* **1** (1989) 261

[32] Couch W.J., Ellis R.S., Sharples R.M., Smail I. *ApJ* **430** (1994) 121

[33] Courteau S., de Jong R.S., Broeils A.H. *ApJ* **457** (1996) L73

[34] Dame T.M., Ungerechts H., Cohen R.S. et al *ApJ* **322** (1987) 706

[35] David L.P., Jones C., Forman W. *ApJ* **445** (1995) 578

[36] Davies R.D., Lewis B.M. *MNRAS* **165** (1973) 231

[37] De Jong R.S. *A&A* **313** (1996) 45

[38] De Paolis F., Ingrosso G.,Jetzer P., Roncadelli M. *A&A* **295** (1995) 567

[39] de Vega H., Sanchez N., Combes F. *Nature* **383** (1996) 56

[40] Dickey J.M., Hanson M.M., Helou G. *ApJ* **352** (1990) 522

[41] Dressler A., Oemler A., Butcher H.R., Gunn J.E. *ApJ* **430** (1994) 107

[42] Dubinski J., Mihos J.C., Hernquist L. *ApJ* **462** (1996) 576

[43] Dupraz C., Casoli F., Combes F., Kazes I. *A&A* **228** (1990) L5

[44] Elmegreen B.G., Kaufman M., Thomasson M. *ApJ* **412** (1993) 90

[45] Fabbiano G., Schweizer F. *ApJ* **447** (1995) 572

[46] Falgarone E., Puget J-L., Perault M. *A&A* **257** (1992) 715

[47] Ferland G.J., Fabian A.C., Johnstone R.M. *MNRAS* **266** (1994) 399

[48] Freeman K.C. *ApJ* **160** (1970) 811

[49] Friedli D., Martinet L. *A&A* **277** (1993) 27

[50] Garcia-Burillo S., Combes F., Gerin M. *A&A* **274** (1993) 148

[51] Gavazzi G., Jaffe W. *A&A* **186** (1987) L1

[52] Gavazzi G., Pierini D., Boselli A. *A&A* **312** (1996) 397

[53] Geisler D., Forte J.C. *ApJ* **350** (1990) L5

[54] Genzel R., Weitzel L., Tacconi-Garman L.E. et al *ApJ* **444** (1995) 129

[55] Gerhard O., Silk J. (1996) preprint

[56] Giavalisco M., Livio M., Bohlin R.C., Macchetto F.D., Stecher T.P. *AJ* **112** (1996) 369

[57] Goldreich, P., Tremaine, S. *ApJ* **233** (1979) 857

[58] Gunn J.E. (1987) in "Nearly Normal Galaxies", ed. S.M. Faber, Springer, New York, p. 459

[59] Harwit M.O., Houck J.R., Soifer B.T., Palumbo G.G.C. *ApJ* **315** (1987) 28

[60] Hasan H., Norman C. *ApJ* **361** (1990) 69

[61] Hasan H., Pfenniger D., Norman C. *ApJ* **409** (1993) 91

[62] Heller C.H., Shlosman I. *ApJ* **424** (1994) 84

[63] Hernquist, L., Weil M.L *MNRAS* **261** (1993) 804.

[64] Hernquist, L., Mihos, J.C. *ApJ* **448** (1995) 41.

[65] Hernquist L., Barnes J.E. (1994) in "Mass-transfer Induced Activity in Galaxies", Lexington Conference, ed. I. Shlosman, Cambridge University Press, p. 323

[66] Hibbard J.E., van Gorkom J.H., Kasow S., Westpfahl D.J. (1993) in "The Evolution of Galaxies and their Environment", III Grand Teton Summer School, eds. J.M. Schull a,d H.A. Thronson, p. 367

[67] Hibbard J.E., Guhathakurta P., van Gorkom J.H., Schweizer F. *AJ* **107** (1994) 67

[68] Hibbard J.E. (1995) PhD thesis, Columbia University

[69] Hibbard J.E., Mihos J.C. *AJ* **110** (1995) 140

[70] Higdon J.L. *ApJ* **455** (1995) 524

[71] Higdon J.L. *ApJ* **467** (1996) 241

[72] Hoffman G.L., Helou G., Salpeter E.E. *ApJ* **324** (1988) 75

[73] Holtzmann J.A., Faber S.M., Shaya E. et al *AJ* **103** (1992) 691

[74] Horellou C., Casoli F., Combes F., Dupraz C. *A&A* **298** (1995) 743

[75] Hubble E. *ApJ* **64** (1926) 321

[76] Impey C., Bothun G. *ApJ* **341** (1989) 89

[77] Ishizuki S., Kawabe R., Ishiguro M. et al *Nature* **344** (1990) 224

[78] Jog C., Solomon P.M. *ApJ* **276** (1984) 114& 127

[79] Jog C., Solomon P.M. *ApJ* **387** (1992) 152

[80] Jog C. *ApJ* **390** (1992) 378

[81] Jog C. *MNRAS* **278** (1996) 209

[82] Junqueira S., Combes F. *A&A* **312** (1996) 703

[83] Kalnajs A.J. *ApJ* **166** (1971) 275

[84] Kenney J.D., Young J.S. *ApJ* **344** (1989) 171

[85] Kenney J.D. (1990) in "The Interstellar Medium in Galaxies", ed. H.A. Thronson & J.M. Shull, Kluwer, p. 151

[86] Kennicutt R.C. *AJ* **88** (1983) 483

[87] Kennicutt R.C. *ApJ* **344** (1989) 685

[88] Kormendy J., Richstone, D. *ARAA* **33** (1995) 581

[89] Lallemand A., Duschene M., Walker M.F. *PASP* **72** (1960) 76.

[90] Larson R.B. *MNRAS* **194** (1981) 809

[91] Larson R.B. (1987) in "Starbursts and galaxy evolution", ed. T.X. Thuan, T. Montmerle, J. T. T. Van, Ed. Frontières, p. 467

[92] Lauer T.R., Faber S.M., Groth E.J. et al *AJ* **106** (1993) 1436

[93] Lavery R., Seitzer P., Walker A.R., Suntzeff N.B., da Costa G.S. *ApJ* **467** (1996) L1

[94] Leon S., Combes F., Menon T.K. (1996) A&A preprint

[95] Lewis J.R., Freeman K.C. *AJ* **97** (1989) 139

[96] Lin D.N.C, Pringle J.E. *MNRAS* **225** (1987a) 607

[97] Lin, D.N.C., Pringle, J.E. *ApJ* **320** (1987b) L87

[98] Lynden-Bell, D., Kalnajs, A.J. *MNRAS* **157** (1972) 1

[99] Lynds B., Toomre A. *ApJ* **209** (1976) 382

[100] Mamon G. *ApJ* **307** (1986) 426

[101] Marston A.P., Appleton P.N. *AJ* **109** (1995) 1002

[102] Mendes de Oliveira C., Hickson P. *ApJ* **427** (1994) 684

[103] Menon T.K. *MNRAS* **274** (1995) 845

[104] Merritt D. *ApJ* **276** (1984) 26

[105] Meurer G.R. (1996) in "The interplay between massive star formation, the ISM and galaxy evolution", ed. D. Kunth, B. Guiderdoni, M. Heydari, T. X. Thuan, Editions Frontières, p. 333

[106] Mihos J.C. *ApJ* **438** (1995) L75

[107] Mihos J.C., Hernquist L. *ApJ* **437** (1994a) 611

[108] Mihos J.C., Hernquist L. *ApJ* **437** (1994b) L47

[109] Mihos, J.C., Hernquist, L. *ApJ* **464** (1996) 641.

[110] Mihos J.C., Richstone D.O., Bothun G.D. *ApJ* **400** (1992) 153

[111] Mirabel I.F., Dottori H., Lutz D. *A&A* **256** (1992) L19

[112]. Moore B., Katz N., Lake G., Dressler A., Oemler A. *Nature* **379** (1996a) 613

[113] Moore B., Katz N., Lake G. *ApJ* **457** (1996b) 455

[114] Mould J.R., Oke J.B., de Zeeuw P.T., Nemec J.M. *AJ* **99** (1990) 1823

[115] Noguchi M., Ishibashi S. *MNRAS* **219** (1986) 305

[116] Parravano A. *ApJ* **462** (1996) 594

[117] Pascarelle S.M., Windhorst R.A., Keel W.C., Odewahn S.C. *Nature* **383** (1996) 45

[118] Peebles P.J.E. (1971) "Physical Cosmology", Princeton Univ. Press

[119] Persic M., Salucci P., Stel F. *MNRAS* **281** (1996) 27

[120] Pfenniger D., Norman C. *ApJ* **363** (1990) 391

[121] Pfenniger D. (1991) in "Dynamics of disc galaxies", ed. B. Sundelius, p. 191

[122] Pfenniger D., Combes F., Martinet L. *A&A* **285** (1994) 79

[123] Pfenniger D., Combes F. *A&A* **285** (1994) 94

[124] Pildis R.A., Bregman J.N., Evrard A.E. *ApJ* **443** (1996) 514

[125] Pildis R.A., Evrard A.E., Bregman J.N.: 1996, AJ preprint

[126] Ponman T., Bourner P., Ebeling H., Bohringer H. (1996) MNRAS preprint

[127] Puche D., Carignan C., Bosma A. *AJ* **100** (1990) 1468

[128] Quinn P.J., Hernquist L., Fullagar D.P. *ApJ* **403** (1993) 74

[129] Rand R.J. *ApJ* **410** (1993) 68

[130] Read A., Ponman T.J., Wolstencroft R.D. *MNRAS* **277** (1995) 397

[131] Reshetnikov V., Combes F. (1996) A&A, preprint

[132] Robin A.C., Haywood M., Creze M., Ojha D.K., Bienaymé O. *A&A* **305** (1996) 125

[133] Romeo A.B. *MNRAS* **256** (1992) 307

[134] Romeo A.B. *A&A* **286** (1994) 799

[135] Rubin V.C., Kenney J.D.P., Young J.S. (1996) AJ in press

[136] Safronov V.S. *Ann. d'Ap.* **23** (1960) 979

[137] Sancisi R., van Albada T.S. (1987) in "Dark Matter in the Universe" IAU Symp. 117, ed. J. Kormendy, G.R. Knapp, Reidel, p. 67

[138] Sandage A. (1961) "Hubble Atlas of Galaxies", Carnegie Institution of Washington

[139] Sanders R.H., Huntley J.M. *ApJ* **209** (1976) 53

[140] Sanders D.B., Soifer B.T., Elias J.H. et al *ApJ* **325** (1988) 74

[141] Sanders D.B., Scoville N.Z., Soifer B.T. *ApJ* **370** (1991) 158

[142] Sargent A.I., Scoville N.Z. *ApJ* **366** (1991) L1

[143] Scalo J.M. (1985) in "Protostars and Planets II", ed. D.C. Black, M.S. Matthews, Tucson, p. 201

[144] Schwarz M.P. *ApJ* **247** (1981) 77

[145] Schweizer F. (1990) in "Dynamics and Interactions of Galaxies" ed. R. Wielen, Springer, p. 60

[146] Schweizer F., Seitzer P. *AJ* **104** (1992) 1039

[147] Schweizer F., Sejtzer P. *ApJ* **417** (1993) L29

[148] Schweizer F (1993) in "Physics of Nearby Galaxies: Nature or Nurture?", Rencontres de Moriond, eds. Thuan T.X., Balkowski C., Van, J.T.T., Editions Frontières, p. 283

[149] Scoville N.Z., Hibbard J.E., Yun M.S., van Gorkom J.H. (1994) in "Mass-transfer Induced Activity in Galaxies", Lexington Conference, ed. I. Shlosman, Cambridge University Press, p. 191

[150] Scoville, N.Z., Sargent, A.I., Sanders, D.B., Soifer, B.T. *ApJ* **366** (1991) L5

[151] Sellwood J.A., Carlberg R.G. *ApJ* **282** (1984) 61

[152] Shlosman I. (1994), editor of "Mass-transfer Induced Activity in Galaxies", Lexington Conference, Cambridge University Press

[153] Shlosman I., Noguchi M. *ApJ* **414** (1993) 474

[154] Shlosman I., Frank J., Begelman M. *Nature* **338** (1989) 45

[155] Shu F. H., Tremaine S., Adams F. C., Ruden S. P. *ApJ* **358** (1990) 495.

[156] Simkin S.M., Su H.J., Schwarz M.P. *ApJ* **237** (1980) 404

[157] Stanford S.A., Sargent A.I, Sanders D.B., Scoville N.Z. *ApJ* **349** (1990) 492

[158] Steidel C.C., Giavalisco M., Dickinson M., Adelberger K.L. *AJ* **112** (1996) 352

[159] Struck-Marcell C., Scalo J.M. *ApJS* **64** (1987) 39

[160] Sulentic J.W., de Mello Rabaça D. *ApJ* **410** (1993) 520

[161] Telles E., Terlevich R. *MNRAS* **275** (1995) 1

[162] Theys J.C., Spiegel E.A. *ApJ* **208** (1976) 650

[163] Toomre A. *ApJ* **139** (1964) 1217

[164] Toth G., Ostriker J.P. *ApJ* **389** (1992) 5

[165] van der Kruit P.C., Searle L. *A&A* **95** (1981) 105

[166] van der Kruit P.C., Searle L. *A&A* **105** (1982) 351

[167] van der Kruit P.C., *A&A* **192** (1988) 117

[168] van den Bergh S., Abraham R.G., Ellis R.S. et al *AJ* **112** (1996) 359

[169] Walker, I.R., Mihos, J.C., Hernquist, L. *ApJ* **460** (1996) 121.

[170] Wang Z., Schweizer F., Scoville N.Z. *ApJ* **396** (1992) 510

[171] White S.D.M., Navarro J.F., Evrard A.E., Frenk C.S. *Nature* **366** (1993) 429

[172] Whitmore .C., Schweizer F. *AJ* **109** (1995) 960

[173] Williams B.A., MacMahon P., van Gorkom J. *AJ* **101** (1991) 1957

[174] Williams B.A., Rood H.J. *ApJS* **63** (1987) 265

[175] Yee H.K.C., Ellingson E. *ApJ* **445** (1995) 37

[176] Young J.S., Scoville N.Z. *ARAA* **29** (1991) 581

[177] Yun M.S., Ho, P.T.P., Lo K.Y. *ApJ* **411** (1993) L17

[178] Yun M.S., Scoville N.Z., Knop R.A. *ApJ* **430** (1994) L109

[179] Zepf S.E. *ApJ* **407** (1993) 448

[180] Zepf S.E., Ashman K.M. *MNRAS* **264** (1993) 611

[181] Zhang X. *ApJ* **457** (1996) 125

From Star To Galaxy Formation

J. Silk

Departments of Astronomy and Physics, and Center for Particle Astrophysics
University of California, Berkeley, CA 94720, USA

----

1. INTRODUCTION

Star formation is the key to understanding galaxy formation. Knowledge of the initial mass function, the rate of star formation, and its efficiency are essential inputs to understanding the large-scale structure of the universe. For galaxies, viewed as ensembles of stars, are the beacons that chart the cosmos. One has to understand how star formation operates, not just in the solar neighborhood, but throughout entire galaxies, and in the past, when conditions differed from those of the present-day environment.

This series of lectures is aimed at reviewing our current understanding of large-scale star formation. It is natural that I commence with local star formation, since the theory of star formation is entirely driven by observations, especially of star-forming regions within a kiloparsec of the Sun. Section 2 is devoted to the initial mass function (IMF). I review observations, and present a summary of the theoretical status of the IMF. In the third section, I develop a semi-phenomenological theory for the star formation rate and history in the Milky Way. These results are applied more generally to star formation in galactic disks (Section 3). I proceed to discuss star formation in spheroidal systems. Here there is no theory, and an empirical model is developed for starbursts (Section 4). Phenomenological arguments are used to compare ultraluminous starbursts with the inferred signatures of elliptical galaxy formation (Section 5). The final section discusses star formation issues in the general context of galaxy formation. I review theories for galaxy properties, including the luminosity function and the rotation and size of galaxies, and galaxy evolution, including the spectrophotometric signatures of distant galaxies and faint galaxy counts. I conclude with a summary of the star formation history of the universe.

2. THE INITIAL MASS FUNCTION

2.1. Observations of the IMF

Direct observations of the IMF at low masses are inevitably local, although the massive star luminosity function can be measured in nearby extragalactic star forming regions. The IMF is defined to be the total number of stars ever born per unit volume in some specified region. Clearly some correction has to be made for the finite stellar lifetime that affects the numbers of massive stars, and the history of the past birthrate must also be specified in order to deduce the IMF.

The principal direct observable is the distribution of stellar luminosities, for stars of known distance. This has been evaluated in nearby open clusters and for the field near the Sun. The Orion association, centered on the Trapezium, provides one of the most complete studies of the luminosity function[1]. The mass-to-luminosity ratio is not well known at low masses, and conversion to the IMF is further complicated by this uncertainty. Nonetheless, a distribution of stellar masses is found that is indistinguishable from the Miller-Scalo IMF. This is a log-normal distribution that fits local field star masses over 0.1 to 100 $M_\odot$. One can crudely approximate the Miller-Scalo function by a power-law of index $dN/dm \propto m^{-1-x}$ at $m > 1\,M_\odot$. A reasonable approximation allows the index to vary, with a systematic and increasing degree of flattening between 1 and 0.1 $M_\odot$, and steepening above $\sim 1 M_\odot$, described by $x = 1 + \log(m/M_\odot)$. The Salpeter IMF is represented by a constant index $x = 1.35$. Modern studies of the field IMF near the sun find that between 0.1 $M_\odot$ and 0.3 $M_\odot$, the slope flattens, and possibly turns over. At very low masses, below $\sim 0.1 M_\odot$, the IMF may resume its rise: at least there are no observational constraints that forbid this, and at least one data set indicates a hint of an upturn[2] in the lowest mass bin near 0.1 $M_\odot$.

The mass function has also been measured in globular star clusters between approximately 0.1 and 1 $M_\odot$. Provided one considers clusters for which the core relaxation time is long, the mass function appears to be similar from cluster to cluster, and is again indistinguishable from the Miller-Scalo function[3]. A Salpeter function of constant power law index to 0.1 $M_\odot$ is excluded, both in open clusters and in globular clusters. The halo field star mass function for nearby high velocity stars has also been determined[4]: again, this is consistent with the IMF in dynamically young (unrelaxed) globular clusters.

Stellar luminosity functions have been determined for several open clusters and associations in our galaxy, and for associations in several nearby galaxies. No significant variations are seen. Some indirect arguments do suggest that the IMF may vary in more extreme environments. These aspects of the IMF will be addressed in a later section.

2.2. The Characteristic Stellar Mass

A fundamental understanding of the origin of stellar masses has proved elusive. Hoyle pioneered the argument that fragmentation in a collapsing cloud would ultimately hit the opacity limit and provide a minimum stellar mass. The instantaneous value of a fragment, acording to the linear theory of fluctuation growth, is given by the Jeans criterion, resulting in a mass scale proportional to $T^2 \rho^{-1/2}$, where T is temperature and ρ is the density in an intially uniform and cold cloud. The gas is approximately isothermal, as it can radiate freely, until the density increases sufficiently that the optical depth across the fragment exceeds unity. During this phase, fragment masses decrease. Subsequently, however, the cloud is optically thick, and individual fragments can only continue to contract adiabatically. One attains a minimum fragment mass of about[5] $10^{-3} \, M_\odot$. Unfortunately, this is far too small to be a star.

It is recognized that the fragmentation process is highly non-linear. Such physical effects as cloud accretion, fragment collisions and coalescence[6], must inevitably occur, and will generally tend to rise the minimum fragment mass. Numerical simulations are on the verge of attaining the dynamical range to follow fragmentation, but have hitherto had limited success. Collapse is found to be highly non-homologous. However, the final evolution of the developing, accreting seeds is uncertain. Hence for the moment, recourse must be had to analytic arguments. These generally suggest that the complex non-linear physics must raise the minimum fragment mass relative to that estimated from opacity-limited fragmentation by an order of magnitude or more. In fact, there is a crucial ingredient that is omitted from the fragmentation *ansatz*, namely feedback from forming low mass stars, as well as evolved and dying massive stars.

One can recognize the need for a new ingredient simply by considering the accretion time-scale. If Δv is the generalized sound velocity, including possible contributions from Alfven waves and turbulence, then the accretion time-scale for a core of mass M to grow is $t_{acc} \approx GM/\Delta v^3$. Hence the core increases without limit, at least until the mass reservoir is exhausted. Now the mass reservoir is the typical mass of a dense core, which in turn is given by the Jeans mass in a typical molecular cloud. Measured core masses are in the range $1 - 100 \, M_\odot$. Hence one would never form stars of mass, say $0.3 \, M_\odot$, corresponding to the typical mass of most nearby stars, were it not for some additional physical effect.

Feedback from stars is the only plausible possibility that can limit accretion. Nuclear energy release provides significant feedback into the interstellar medium, via supernovae as well as winds and H II regions from OB stars. However, this comes too late for local feedback to help: one has already formed a star of $\gtrsim 10 \, M_\odot$. Non-local feedback from massive stars, however, is important in star formation regions. Spectacular evidence of evaporation of molecular clouds, leaving dense globules behind, is seen in Orion and in M16.

However, it may require an earlier source of energy in order to limit accretion and form low mass stars. The logical source is gravitational energy, associated with the protostellar Kelvin-Helmholtz contraction phase. The time scale for Kelvin-Helmholtz contraction is $t_{KH} \propto M^{-1}$, so that it can be less than the accretion time for masses as low as $\sim 1\,M_{\odot}$. Now protostars that are gravity-powered are observed to be sources of jets and bipolar molecular gas outflows. The momentum input observed in bipolar outflows suffices to provide support against gravitational collapse on scales comparable to that of the Orion molecular cloud. Protostellar jets appear to power these outflows, and are generated along the minor axes of accretion disks. The precise mechanism for the jets must involve the ultimate energy source, gravitational contraction, with coupling to some combination of convective energy of the protostar, rotational energy of the accretion disk, and magnetic torquing of the infalling gas.

I will adopt the following simple model. I assume that to halt accretion, one must supply enough energy and momentum from the outflow to balance the accretion inflow. It is necessary to be able to radiate away the accretion energy in order for infall to continue. The condition that the rate of injection of outflow energy, assumed to be of the order of the Kelvin-Helmholtz luminosity, be comparable to the rate of accretion energy release is

$$\frac{64\pi\sigma T_c^3 r_*^2}{3\kappa\rho_c}\frac{dT}{dr} \approx \frac{GM_cM}{r_d}, \tag{4}$$

where r_* is the protostellar radius, ρ_c and T_c are the central density and temperature, M_c is the core mass, and κ is the Rosseland mean opacity appropriate for a mixture of gas and grains. I have assumed that the accretion shock occurs at the outer accretion disk radius, given by the condition for centrifugal balance, as

$$r_d = G^3 M^3 \Omega^2 / 16\Delta v^8.$$

Here M is the interior, *i.e.* disk, mass that forms inside radius r_d surrounding the central core. I remove the cloud rotation dependence by using an observed scaling relation. The rotation rate of molecular cloud cores can be fit by[7,8]

$$\Omega = 10^{-0.3\pm0.2}R_{pc}^{-0.4\pm0.2}\ \mathrm{km\,s^{-1}\,pc^{-1}},$$

where the cloud size is R_{pc} pc.

Comparison of the two heating rates suggests that the accretion luminosity can be balanced and overcome, presumably halting accretion, provided that the protostellar core mass satisfies

$$M > M_{crit} = 1.60(\pm0.25)(\Delta v_{km/s})^{13/5}\kappa_{20}^{1/5}\ M_{\odot},$$

where Δv is expressed in km s^{-1} and κ is in units of 20 cm^2 g^{-1}. Uncertainty in the appropriate value of κ may well dominate the normalization of M_{crit}. Hence only stars with $M < M_{crit}$ can form via accretion. This suggests that one can indeed account for the characteristic mass of a star, and indeed a range of stellar masses, for clouds of sufficient cold, and varying, ΔV.

Obviously, the connection between infall and outflow is a gross oversimplification. The feedback may operate by mechanical energy outflows, for example precessing jets driving bipolar flows, or by radiation, for example via X-rays from the convective protostar that ionize the surrounding gas, recouple both the magnetic field lines and MHD waves, and repressurize the accretion envelope, thereby inhibiting infall. These mechanisms, especially the latter, are important for low mass protostars, whereas radiation pressure and photoevaporation are important for massive protostars.

There are observational indications of a relation between protostars and ΔV. For example, there is an empirical correlation[9] for dark cloud cores between the inferred mass of the most luminous embedded star from $IRAS$ observations, and the linewidth from NH_3 observations:

$$M_*^{max} = 1.82(\pm 0.1)(\Delta v_{km/s})^{2.38 \pm 0.17} \ M_\odot. \tag{7}$$

This relation holds over the range $0.3 \lesssim M_*^{max} \lesssim 30 M_\odot$. The coincidence with the prediction of the characteristic protostellar mass suggests that the potential well depth of the cloud core indeed is the controlling influence on the most massive stars to form. Moreover, cold core linewidths decrease with distance from the nearest embedded groupings of protostars[10]. This evidence supports the general idea of self-regulation between m and Δv, although it is far from being definitive.

Indeed, indications that there must be a more complex connection come from the observations of starless cores, which show similar narrow but suprathermal linewidths as do cores containing stars. Massive cores which contain many embedded stars have broader linewidths than the low mass cores, which typically contain a single young stellar object, often inferred to be a binary. This is suggestive of multiple fragmentation being important on large scales. However, linewidths are essentially constant within dense cores, on scales $\lesssim 0.1$ pc. It is below this scale that star counts in Taurus-Auriga show that there is about one low mass companion per star: the number of companions increases on larger scales[11]. This observational evidence suggests that different processes operate to determine the distribution of stellar masses below and around a solar mass. Binaries provide an additional correction to the observed IMF at low masses[12], and binary formation may dominate on scales comparable to the observed correlation length. This scale is also typical of the scale of the densest dark cores, and comparable to the mean separation of stars in the youngest and densest open clusters, such as the Trapezium cluster.

The line width-spatial scale scaling relation, $\Delta v \propto R^a$, steepens from $a \sim 0$ below 0.1 pc in individual cores to the universal value of about 0.5 found for comparisons of different clouds on larger scales[13]. The $\Delta v \propto R^{0.5}$ scaling has been interpreted as evidence for approximately uniform magnetohydrodynamic pressure support, with $B \sim 30 \, \mu G$. MHD turbulence can support molecular clouds, but must be replenished against dissipation of MHD waves due to ion–neutral friction. The interactions of protostellar outflows with the molecular environment will generate MHD turbulence at a level capable of supporting clouds against collapse. MHD waves provide a means of supporting cloud cores even in the absence of embedded protostars. This may provide the mechanism needed to understand dark clouds that somehow survive and have turbulent linewidths, yet contain no apparent ongoing star formation, at least as monitored at near infrared wavelengths. Such clouds are rare, but exist[14], and have similar clump properties to star-forming clouds[15].

2.3.   Theory of the IMF

Suppose that protostellar outflows preserve the momentum that supports molecular clouds and sustains the linewidths. One can describe a massive, star-forming molecular cloud as being a complex network of interacting molecular flows. Such flows are observed in their most compact, and consequently youngest, configurations to have outflow velocities of several hundred kilometers per second. The older, extended flows are more numerous and are observed at velocities of $\gtrsim 10\,\mathrm{km\,s^{-1}}$ as bipolar flows in CO and other molecular tracers. In the best studied molecular clouds, the momentum in the flows appears to be sufficient to provide support against gravitational collapse. As aging flows sweep up ambient matter, they eventually intersect remnants of other flows. The dense decelerating shells are unstable to Rayleigh-Taylor instabilities, and break up even before running into fragments of neighboring flows. One can therefore depict a molecular cloud as a hierarchy of interacting and colliding shell fragments, mostly at low velocity, but with rare, younger fragments moving at higher velocities. The internal velocity dispersion of a fragment will be comparable to its mean velocity, to the extent that we can regard the outflows as driving molecular cloud turbulent motions.

Since the internal velocity dispersion of a molecular fragment regulates the maximum mass of the protostar that can form within it, one can infer the mass distribution of newly formed stars from the distribution of turbulent velocities $f(> \Delta v)$ above Δv. For a spherically symmetric outflow, the radius of the shell can be written as $R \propto t^\delta$, where $\delta \approx 0.4 - 0.5$ depending on whether the outflow is sporadic or continuous. I suppose that the shell breaks up into fragments of velocity Δv, as it interacts with other shells, that I identify with the instantaneous value of dR/dt. One finds that $f(> \Delta v) \propto \Delta v^{-\frac{3\delta+1}{\delta-1}}$. Adopting the relation $M \propto \Delta v^\epsilon$ that relates protostellar mass to turbulence velocity either observationally or theoretically, I now derive an expression for the IMF:

$$\frac{dN}{dM} = f(> \Delta v)/\frac{dM}{d\Delta v} \propto M^{-1-x},$$

where $x = \frac{3\delta+1}{\epsilon(1-\delta)}$. Remarkably, with the characteristic value of δ and observed value of ϵ, one obtains a value of $1 + x \approx 2.2 - 2.4$ that is close to what is observed for the IMF over $\sim 0.3 - 100\,M_\odot$. This result is surprisingly robust.

One can include effects of non-uniformity in the cloud density modelled as a gradient in density, or non-steady flows, and there is only a minor modification in the predicted index.

A similar argument should apply to the distribution of cloud fragment masses. Now however the direct support is via pressure, as inferred by the observed scaling laws. Indeed, one observes $\epsilon \approx 4$ for cloud cores and clumps. This is appropriate for clouds that are self-gravitating and pressure-supported. Hence one derives a clump mass function that is flatter than the mass function for stars. This indeeed is what is observed: e.g. in the Rosette nebula[16], $dN/dM \propto M^{-1.3}$. The fact that the IMF is steeper than the clump mass function demonstrates that the star formation efficiency increases towards smaller scales, as indeed is observed directly in studies of dark clouds.

2.4. Implications for the IMF

There is little observational evidence on the IMF beyond the solar neighborhood. The gravitational microlensing experiments have measured the distribution of low mass stars towards the galactic bulge, and found consistency with the Miller-Scalo IMF. Otherwise, we only measure massive stars in HII regions, both in our galaxy and in nearby galaxies. In more distant galaxies where one cannot resolve individual massive stars, we have to resort to spectral synthesis diagnostics to infer the number of ionizing photons and thereby yield an integral constraint on the numbers of massive stars. Emission lines such as $[OII]$, $H\alpha$ and $Ly\alpha$ are sensitive to the O star frequency, and Balmer absorption lines are used to infer the contribution of intermediate mass stars to the stellar population. Another approach utilizes Wolf-Rayet star spectral features, but the results are inconclusive[17]. There are no definitive indications of any deviations from the local IMF.

In regions of extremely intense star formation, such as within luminous starbursts, there are controversial arguments, based on the required efficiency of star formation, that may favor an IMF that is deficient in low mass stars[18]. Such a top-heavy IMF can provide up to an order of magnitude more luminosity per unit mass of gas available for forming stars than would be generated by the canonical IMF. A top-heavy IMF also has an enhanced nucleosynthetic yield: the relative frequency of Type II supernovae compared with Type I supernovae is higher than in the Milky Way. In this way, one can increase the generation of α-nuclei (such as O, Mg, Si) relative to carbon and iron by as much as a factor of 3. One sees enhanced α-nuclei abundances in old halo stars, where this is simply a consequence of the fact that Type I supernovae from low mass stars had not been produced in significant numbers when the halo formed. However, similar signatures are seen both in the nuclei of elliptical galaxies[19] and in the intracluster gas in rich clusters[20], suggestive of a possible origin in starbursts with a top-heavy IMF.

One interpretation of the MACHO findings for microlensing events in our galactic halo seen towards the LMC is that the halo dark matter largely consists of white dwarfs. This interpretation also requires a top-heavy IMF for the matter that formed the halo, in order to render it sufficiently dark by excluding any significant number of solar mass stars. If the halo is older than ~ 15 Gyr, the white dwarfs are too dim to be detectable in proper motion surveys[21]. Other signatures of white dwarf halos include an early luminous phase as well as ejection of enriched stellar matter, most notably helium and carbon from intermediate mass stars. However, one can evade all of these constraints with sufficient ingenuity.

The stellar mass is determined by the time for the lensed object to traverse the Einstein cone, the radius of which is proportional to $M^{1/2}/v$. This timescale is directly observed as the duration of the amplification event. In principle, the halo velocity dispersion v is known, at least statistically, but it depends in detail on the dynamical model for the distribution of the halo dark matter. One could envisage the MACHOs being in a compact spheroid or thick disk, and thereby being more slowly moving, and hence less massive, than would be predicted by a halo model with an isotropic velocity dispersion. Hence it is by no means excluded, given the uncertainties in halo models, that the halo MACHOs are very low mass stars, or rather brown dwarfs below the $\sim 0.1\,M_\odot$ threshold for nuclear burning by stars of primordial abundance.

Theory is very flexible on the question of possible systematic variations in the IMF. Imagine a galaxy merger, where cloud collisions occur at a greatly enhanced frequency. The resulting cloud turbulence could plausibly deplete the low Δv tail of the distribution of the clump random motions. The preceeding theory suggests that the slope is invariant, and the net effect would be a top-heavy IMF that is simply deficient in low mass stars.

The converse situation seems less likely, but perhaps not impossible. I next imagine a quiescent molecular cloud, for example in the outer disk or in the halo of a forming galaxy, where collisions with other clouds are infrequent. The temperature is likely to be low, and the initial level of cloud turbulence will also be low. The accretion rate onto cores will be low, and despite the long pre-main sequence contraction time, very low mass stars might then predominate. Possibly, such an isolated cloud might spend a prolonged period in a low Δv phase before a sufficient number of massive stars have formed to generate the high Δv tail of the turbulence distribution. It would seem unlikely that star formation above, say, $\sim 0.1\,M_\odot$, was suppressed, but it might well be inhibited. This situation could result in a bottom-heavy IMF, compared with what might prevail in the inner galaxy clouds. Here there would, at any given time, be injection of turbulence from ongoing massive star formation into nearby clouds. In general, a star-forming galaxy should be a mixture of cold and warm clouds, the relative proportions of which may play an important role in determining the final IMF.

3. GLOBAL STAR FORMATION IN GALACTIC DISKS

Most star formation today occurs in disk galaxies, There is considerable data on star formation rates, gas distributions, and chemical abundances of gas and stars in nearby galaxies, and in particular for our own galaxy. There is also extensive theoretical work on star formation is disk galaxies, and especially on the stability of disks. Star formation lacks any fundamental theory. Hence our best model for disks is semi-phenomenological, a blend of observation and theory.

3.1.   Global Phenomenology for Disks

Two basic ingredients of star formation observed in disks are that the star formation is highly inefficient and that the disk star formation rate has not varied much over the past several Gyr. That the typical global efficiency of star formation is at most a few percent can be seen from the requirement that a disk of $\sim 10^{11} M_\odot$ must convert its gas into stars over $\sim 10^{10}$ yr, for a mean star formation rate of $\sim 10 M_\odot/\text{yr}$. Over a dynamical time ($\sim 10^8$ yr), only ~ 1 percent of the gas can be converted into stars, and if the efficiency is approximately constant, so is the star formation rate. One observes today typical star formation efficiencies of this order, for example by estimating the molecular gas and young star budget for a molecular cloud complex.

The past star formation rate in disks did not differ greatly from the present day star formation rate. One can track the star formation history in the solar neighborhood using cluster isochrone ages or chromospheric ages for nearby stellar samples, and conclude that over the past 10 Gyr the star formation rate may have been at most a factor of ~ 3 higher than that at the present epoch[22]. Deep redshift surveys confirm this result for disk galaxies in general: about 1 magnitude of evolution in star-forming galaxies is seen[23] to a median $z \sim 0.5$.

To construct a recipe for disk star formation, it is necessary to provide another ingredient, namely the dependence of star formation rate on cold gas content. Molecular clouds provide the reservoir for star formation, and their formation in turn is controlled by the reservoir of atomic gas. Molecular clouds appear to form via aggregation driven by differential rotation and spiral density waves. Gravitational instabilities in the cold, self-gravitating disk also drive random motions that lead to cloud aggregation. Global disk instabilities result in formation of a central bar. This in turn drives tidal torques which help transfer angular momentum outward and drive gas inward. Empirically, it is found that the disk star formation rate satisfies[24]

$$SFR \propto \epsilon \Sigma_{gas}^n + \text{threshold criterion.}$$

Here ϵ is the measured star formation efficiency, Σ_{gas} is the total gas surface density, and $1 < n < 2$. A best fit to data on nearby galaxies yields[25] $n \approx 1.4$. The threshold gas surface density corresponds to about $10 M_\odot \, \text{pc}^{-2}$.

The theoretical interpretation of the star formation rate divides into two related issues: star formation efficiency and star formation threshold. The global star formation efficiency is low in disks, inferred to lie between 1 and 10 percent per dynamical time. As one observes star-forming clouds on progressively larger scales, the directly observed instantaneous star formation efficiency declines: from $\gtrsim$ 30 percent in some molecular cores to $\lesssim$ 1 percent in the giant cloud complexes that are more representative of the interstellar medium. However, the lifetime of the cloud complexes is considered to be $\sim 5 \times 10^7$ yr, or about 25 percent of an orbital time at the solar radius, thereby reconciling the local and global efficiencies. The low global efficiency is required in order for disks to remain gas-rich after $\sim 10^{10}$ yr of star formation have elapsed. Approximately one-third of the disk surface density is in gaseous form at the solar galactocentric radius.

A rough estimate of ϵ is as follows. Let v_{cool} be the velocity (≈ 400 km s^{-1}) at which supernova remnants first undergo substantial radiative cooling, and m_{sn} ($\approx 250 M_\odot$) denote the mass formed in stars in order for one supernova to be produced. If E_{sn} is the initial kinetic energy of the supernova ($\approx 10^{51}$ ergs), the specific momentum injected into the interstellar medium by supernovae is

$$v_{sn} = 2E_{sn}/(v_{cool}m_{sn}) \approx 10^3 \, \text{km s}^{-1}.$$

Since the old supernova remnant interactions that accelerate clouds are approximately momentum-conserving, one can write

$$\epsilon \approx \sigma_g/v_{sn},$$

where σ_g is the *rms* velocity dispersion of interstellar clouds, about 11 km s^{-1}. I infer that $\epsilon \approx 0.01$.

The star formation threshold can arise from two distinct mechanisms. Molecular clouds form by coagulation. Their interactions are highly inelastic, and non-circular motions driven by spiral density waves are responsible for the cloud collisions. Hence one expects the rate at which clouds grow to be proportional to[26] $\Omega(r) - \Omega_p$, where $\Omega(r)$ is the differential rotation velocity and Ω_p is the angular pattern speed of the spiral density wave. Gravitational instabilities of a cold disk are controlled by the Toomre parameter, $Q \equiv \Sigma_{cr}/\Sigma$, where the critical surface density $\Sigma_{cr} \approx \kappa\sigma_g/\pi G$ with κ equal to the epicyclic frequency, $\kappa \sim 2^{1/2}\Omega$, for a flat rotation curve. Disks are unstable to both local and (approximately, under certain conditions) to global instabilities if $Q \lesssim 1$. The linear theory of fluctuation growth yields a growth rate[27] $\sim \Omega(1 - Q^2)^{1/2}/Q$. Non-linear effects and ultimately feedback from star formation must limit the growth rate from above.

In fact Q is observed to be close to unity throughout the star-forming disk, implying that disks indeed self-regulate. One could readily imagine that if $Q \ll 1$, massive star formation would result in cloud heating and acceleration by old supernova remnants and massive star winds that would raise the effective value of Q. There is a natural feedback, since quenching star formation removes the heat source. It is also possible that global gravitational instabilities, such as bar formation and spiral density waves, provide sufficient heating to raise Q. In this case, cooling poses a problem if stars are to continue forming, and one possibility is that infall of gas provides the requisite amount of cooling if stars subsequently form out of the cold gas. Infall at the present epoch only amounts to at most a tenth of the star formation rate. However, if there is a means of maintaining a large enough gas fraction, the gas will inevitably cool. The star formation efficiency is therefore the crucial ingredient, and it is this that can be kept suitably small, as is indeed observed, if feedback from massive stars is important.

A specific mechanism for accomplishing self-regulating feedback is described below. The gravitational instability arguments motivate a simple and generic prescription for disk star formation that works remarkably well:

$$SFR = \epsilon \, \Sigma_{gas} \Omega(r) \,,$$

modulated by an appropriate threshold condition. This relation fits the radial distribution of star formation rate and gas density. One can compute the past history of star formation, and derive the metallicity distribution and radial gradient of the stars and the gas. All of this data can be fitted for a single choice of ϵ and initial gas distribution. Infall or outflow of gas are common additions to the generic model, provided that the present epoch inflow or outflow rates are small. In fact, infall has been argued to be necessary to account for the disk star metallicity distribution.

3.2. Implications for Cosmology

Nearby disk star formation rates can be integrated backwards in time to obtain the past star formation history. The star formation rate is highest in the inner galaxy, and especially when combined with infall, this implies that disks were substantially smaller in the past[28]. Deep HST images reveal a trend of smaller galaxy angular sizes at faint magnitudes. The predicted evolution in angular size amounts to as much as a 30 percent decrease at redshift unity. This overwhelms the cosmological signature of geodesic curvature on angular sizes.

Surface brightness is also expected to evolve as disks decrease in size towards higher redshift. It is possible to quantify this only if a model for bulge formation is adopted. Suppose that bulges are already in place when disks form. Then the surface brightness increases with redshift especially strongly for early type, bulge-dominated disks. A brightening of about 1 magnitude relative to the Freeman law is found by $z \approx 0.6$ for both cluster and field disk galaxies[29]. Post star-burst galaxies and blue star-forming galaxies are the dominant population in clusters[30] at $z \gtrsim 0.4$, and their irregular counterparts dominate the deep galaxy counts that probe to even greater depth[31,32].

The models of disk star formation also predict the evolution of metallicity with redshift. An alternative approach uses the damped Lyman alpha clouds at high redshift as a measure of the total gas fraction available for disk star formation to compute star formation rates and metallicity evolution[33]. Both approaches predict that protodisks have mean metallicities that typically are a tenth of the current average disk value, as expected in the most massive closed box model for a gas fraction that is a factor of 10 higher than the present gas fraction. Infall is important primarily for bringing the old disk abundances up to about one-tenth of the stellar value, since most of the mass of the disk is generated from infall after there already has been substantial metal production. The observed metallicities of damped Lyman alpha clouds, however, are typically about one percent of the solar value at $z \approx 3$. At lower redshift, there is a broad dispersion in metallicities.

Can the damped Lyman alpha clouds nevertheless be protodisks, despite the mismatch in metallicity? The apparent conflict has been attributed to a bias arising from the presence of interstellar dust in the absorbing clouds[34]. Interstellar extinction leads to a magnitude-limited survey of quasars systematically deficient in such objects along lines of sight that traverse the more metal-rich absorbing clouds. Alternatively, it may be that the damped Lyman alpha clouds are associated with the early stages of spheroid formation. Kinematic evidence based on the distribution of interstellar heavy element components, such as *Si II* and *Al II*, that are produced by individual clouds within the damped Lyman alpha systems, suggests that there is a systematic rotation, comparable to that of disk galaxies. However, the overall scale height must be at least several kpc, or equivalently the aspect ratio must be about 1:10, in order to account for the inferred inclinations of the absorbing systems to the lines-of-sight. This follows from the measured rotation in damped Lyman alpha clouds. A third, and simplest, option might be that one simply is sampling protodisks at large galactocentric radii, where solar neighborhood-based predictions of metallicity at early epochs are simply irrelevant and the scale heights are naturally expected to be large. Large scale heights H arise because $H \propto \sigma_g^2 / G\Sigma$, and the surface density decreases approximately as r^{-1}, while the gas velocity dispersion in the disk is expected to be nearly independent of galactocentric radius.

3.3.   Disk Self-Regulation

Gravitational instabilities are responsible for disk star formation. The instabilities operate marginally, such that $Q \approx 1$. Evidently there is a negative feedback from star formation that prevents a runaway instability. This must operate via the interstellar medium, which of course provides the raw material for star formation. The interstellar medium is multiphase. The cold, molecular component forms stars, and is fed by the cold atomic component. The hot, highly ionized phase at $T \propto 10^6$ K occupies a substantial fraction of the volume, and one may argue[35] that it is the prevalence of the hot phase that is responsible for driving feedback as manifested, for example, by large scale eruptions of gaseous matter out of the disk. One observes interstellar loops, bubbles, filaments, and chimneys, and in extreme cases, galactic winds.

One can describe the destabilizing tendency of the hot phase by its porosity P. The volume fraction occupied by hot gas is $1 - e^{-P}$, and a galactic wind is driven if $P \gg 1$ and if the specific momentum injected by supernovae exceeds the escape velocity. The porosity is $P = \nu_{sn}\dot{\rho}_* m_{sn}^{-1}$, where $\dot{\rho}_*$ is the star formation rate per unit volume, m_{sn} is the mass in stars formed per Type II supernova (equal to about 250 $M_\odot$ for a Miller-Scalo IMF), and ν_{sn} is the 4-volume occupied by a supernova remnant in a medium of ambient pressure p_g, with the scalings obtained numerically[36]:

$$\nu_{sn} = \text{constant} \times p_g^{-1.36} n^{-0.11}.$$

Now the specific momentum injected by supernovae into the interstellar gas is

$$v_{sn} = 500 n_g^{-1/7} m_{250}^{-1} \text{km s}^{-1},$$

where $m_{250} \equiv m_{sn}/250 M_\odot$. This is comparable to σ (the halo velocity dispersion) for massive (L_*) galaxies. I conclude that blow-out can be avoided only if $P \lesssim 1$.

If interstellar clouds are accelerated by old supernova remnants and decelerated by cloud collapse, their velocity dispersion is $\sigma_g = \epsilon v_{sn} t_{coll}/t_{dyn}$, where ϵ is the star formation efficiency per dynamical time t_{dyn}, or $\epsilon = \dot{\rho} t_{dyn}/\rho_g$, and $t_{coll} = H \mu_{cloud}(\sigma_g \mu_g)^{-1}$ is the mean cloud collision time in a gas disk of column density μ_g. Here H is the disk gas scale height, equal to $\sigma_g (G\rho)^{-1/2}$, and $\mu_{cloud} = (p_g/G)^{1/2}$ is the cloud column density if the clouds are marginally bound and pressure confined. It follows that

$$\sigma_g = 7 P^{-0.6} n_g^{0.1} \text{km s}^{-1}.$$

In other words, the gas velocity dispersion depends primarily on the porosity.

To derive the star formation rate from ϵ, one needs to know the gas density as well as σ_g. The Toomre parameter can be applied to rewrite the gas fraction in the disk, for a flat rotation curve is $f_g = \sigma_g(Q v_{rot})^{-1}$. Hence the gas fraction is $f_g \approx 0.02 P^{-0.6} Q^{-1} v_{200}^{-1}$, where $v_{200} = v_{rot}/200 \text{km s}^{-1}$. The star formation rate is

$$\dot{M}_* \approx 1.4 v_{200}^{5/2} n_g^{0.3} m_{250} P^{-0.9} Q^{-1.5} M_\odot \text{yr}^{-1}.$$

The gas fraction and star formation are in excellent agreement with observed values for the Milky Way, where $P \approx 0.3$, $Q \approx 1$ and $v_{rot} \approx 220 \text{km s}^{-1}$. Moreover the robustness of the star formation rate is explicitly due to the coupling between P and Q: as Q decreases, more massive stars form, generating more supernovae that drive P upwards.

As for the scaling with other disk galaxies, I note that the gas fraction varies approximately inversely with v_{rot}, and that the blue luminosity, dominated by current star formation, varies approximately as $v_{rot}^{2.2}$ according to the observed blue-band Tully-Fisher relation. The Tully-Fisher relation steepens towards longer wavelengths[37], and this is presumably due to the dominance of the

older stellar population. While the global star formation rate is robust, implying that so is the Tully-Fisher relation, the local star formation rate need not
be. For example, if Q is large because the surface density is low, the star formation efficiency is also low. However, P will decline in this case. Low surface
brightness galaxies should therefore satisfy the same Tully-Fisher relation as
do normal galaxies.

4. STARBURSTS

The preceeding discussion shows that there is at least a rudimentary theory
for star formation in disk galaxies. However, there is no theory for star formation in dynamically hot systems. A phenomenological approach is mandatory.
Starbursts provide the nearby analogue of star formation in dynamically hot
systems.

A phenomenological model for starbursts is motivated by numerical simulations of tidal interactions and galaxy mergers. State-of-the-art simulations
cannot resolve star formation scales, but do provide an illuminating model of
gas physics on scales down to ~ 1 kpc. Non-circular cloud motions are generated by tidal torques arising from the formation of a massive central bar.
Such a transient triaxial configuration is common to major mergers. In less extreme situations, infall of a satellite galaxy, constituting a minor merger, also
enhances non-circular cloud motions[38]. The cloud random motions result in
enhanced cloud coalescence. Cloud interactions are inelastic because the interstellar gas is highly dissipative, with an approximately isothermal equation of
state. Self-gravity is important: the presence of a preexisting massive spheroid
enhances the cloud coalescence rate by driving larger random motions. Angular momentum is efficiently transferred via tidal torquing, and kinetic energy
is lost via dissipative cloud build-up. Consequently much of the interstellar gas
is driven to within the central kiloparsec of the final system. All of this is well
represented by numical simulations.

What happens next in this massive gas complex, of $10^9 - 10^{10} M_\odot$ in a typical
merger, is more conjectural, based on the observed intimate connection between
starbursts and mergers. Ultraluminous starbursts are invariably associated
with major mergers. Within a crossing time the gas is concentrated to within
the central kiloparsec[39]. The central gas concentration forms stars with high
efficiency. Rather than the global efficiency of a percent or so associated with
disk star formation, starbursts have an efficiency of 30 percent or more in
converting gas into stars. This high efficiency may be due to the interplay of
several physical effects, ranging from the high masses of the gaseous subunits,
and hence their resistance to disruption, and the high pressures associated with
the onset of an intense phase of massive star formation.

4.1. Interstellar Pressure

One simple result provides some insight into the enhanced star formation efficiency. A major difference between physical conditions in the interstellar gas in a starburst and that of the local interstellar medium is that the interstellar gas pressure is enhanced by about 3 orders of magnitude. The pressure can be estimated by the following arguments. The most direct measurement of interstellar pressure utilizes far infrared lines produced in $H\,II$ regions and in the photodissociation regions around molecular clouds in the central kpc of nearby starburst galaxies[40]. The ionized gas fills upto a few percent of the volume of the starburst region and has a pressure $p/k \sim 3 \times 10^6$ cm-3 K. The short expansion time of the atomic (and inferred molecular) and $H\,II$ clouds implies the existence of a pervasive, hot $(T \sim 10^6 - 10^7$ K) confining medium that fills the starburst volume at a density of ~ 1cm^{-3}.

An independent estimate of the pressure of the hot intracloud medium in the starburst is derived from the inferred supernova rate. The observed ionizing photon flux in the NGC 253 starburst, which extends over ~ 250 pc, for the $[OIII]\,53\mu$ and $[SIII]\,33\mu$ fine structure line intensities requires some 2×10^5 $O7.5$ stars. These stars die at a rate of ~ 0.05 year^{-1}, which is therefore the implied supernova rate R_{sn}. The dynamical pressure generated by the expanding supernova remnants is

$$
\begin{aligned}
p/k &= R_{sn}m_{sn}V_{sn}/4\pi r^2 \\
&\approx 2 \times 10^7 (R_{sn}/0.05\,\mathrm{yr}^{-1})(m_{sn}/250M_\odot)(300\,\mathrm{pc}/r)^2 \mathrm{cm}^{-3}\,\mathrm{K}\,.
\end{aligned}
$$

This is also the pressure in the hot gas within the supernova-driven shells, but does not allow for radiative losses once the shells collide and form a network of interacting fragments. Direct evidence for such a hot diffuse medium comes from the detection of extended soft X-ray emission in starbursts.

Yet another interstellar medium pressure estimate can be obtained from the extended radio emission common to starbursts. If equipartition is assumed between the magnetic field pressure and the cosmic ray pressure, one can deduce from the observed synchrotron emission that the cosmic ray pressure in the starburst region is, in the case of M82,

$$
p/k \approx 10^7 \left[(n_{p,cr}/n_{e,cr})/100\right] \mathrm{cm}^{-3}\,\mathrm{K},
$$

where $n_{p,cr}/n_{e,cr}$ is the ratio of cosmic ray protons to electrons.

A final pressure estimate is derived from thermal balance in the hot intercloud medium, for which $\frac{3}{2}p/t_c = R_{sn}E_{sn}/\frac{4}{3}\pi r^3$, where t_c is the local cooling timescale. This yields

$$
p/k \approx 8 \times 10^6 \left(T/3 \times 10^6\,\mathrm{K}\right)^{1.3} (300\,\mathrm{pc}/r)^{1.5} \left(R_{sn}/0.05\,\mathrm{yr}^{-1}\right)^{1/2} \mathrm{cm}^{-3}\,\mathrm{K},
$$

and is the most reliable estimate for conditions in the hot medium.

All of these pressure estimates confirm that the interstellar pressure in starbursts is enhanced over that in the normal interstellar medium by some 3 orders of magnitude. At the same time, the star formation rate per unit volume is enhanced by about 4 orders of magnitude. The porosity of the hot phase is found to be relatively large,

$$P = 0.2 \left(10^7 \, \text{cm}^{-3} \, \text{K}/p\right)^{1.4} \left(R_{sn}/0.05 \, \text{yr}^{-1}\right) \left(300 \, \text{pc}/r\right)^3.$$

The hot phase is sufficiently pervasive to control the interstellar medium pressure in the starburst. Pressure actually stabilizes the starburst by restricting the extent of the supernova remnant bubbles, since $P \propto p^{-1.4}$ for a given star formation rate per unit volume. However, note that, utilizing the thermal balance estimates for pressure, $P \propto \left(R_{sn}/r^3\right)^{1/3} \propto \dot{M}_*^{1/3}$. This suggests that ultimately as the star formation rate increases, porosity must rise and a wind will be inevitable.

4.2.   The Jeans Mass

The Jeans criterion for gravitational instability gives a mass scale

$$M_{Jeans} = \frac{\pi^{5/2}}{6} \frac{v_s^4}{G^{3/2} p_g^{1/2}}.$$

More generally, this is (to within a factor of order unity) the mass of a self-gravitating isothermal sphere that is marginally stable against collapse and subject to an external pressure p_g. For the local diffuse interstellar medium, with $T \approx 50$ K and $p/k = 3000 \, \text{cm}^{-3}$ K, one obtains $M_{Jeans} \approx 3000 \, M_\odot$. The temperature of the molecular gas, which is the dominant source of interstellar gas in starbursts, is higher by about a factor of 2 over the temperature in nearby H I clouds, whereas the pressure is higher by a factor of perhaps 10^4. Consequently, in a starburst, where one expects that $T \approx 100$ K and $p/k \approx 10^7$ cm^{-3} K, the Jeans mass is of order $100 \, M_\odot$.

In the local interstellar medium, the large value inferred for the Jeans mass is generally considered to represent a barrier to efficient star formation. One has to resort to ambipolar diffusion, on a time-scale about an order of magnitude longer than the gravitational collapse time, in order to allow stellar mass fragments to condense. However, in the starburst environment, where the interstellar pressure is enhanced by a factor of 1000 or more relative to that in the local interstellar medium, the minimum fragmentation scale arising from gravitational instability is reduced to $\sim 100 \, M_\odot$. Hence massive stars can form directly, and therefore efficiently, on a free-fall time-scale.

A crude estimate of the enhanced efficiency achieved by lowering the critical mass is inferred from the argument that formation of the first massive stars can disrupt the surrounding clump of gas and locally limit star formation. With clumps of mass 10^2 – $10^4 \, M_\odot$, and modest star formation efficiency, formation of the most massive stars is a random process. The more clumps there are, the more likely one is to form more stars. Reduction of the critical mass correspondingly increases the number of clumps, and should therefore enhance the star formation efficiency. Hence one might guess that the star formation efficiency should increase as M_{Jeans}^{-1}. One might therefore hope to understand why in starbursts the star formation efficiency is as high as 30 or even 50 percent, whereas in galactic star forming complexes, the star formation efficiency is at most a few percent.

4.3. Starbursts "Explained"

A starburst galaxy is defined to be a galaxy that is temporarily undergoing a greatly elevated burst of star formation. Most of the emission from a starburst galaxy is in the infrared, peaking near 100μm, with L_{IR}/L_B ranging from 10 to 100 for extremely luminous starbursts. The star formation is mostly shrouded in dust and confined to within the central kiloparsec of the galaxy. Ultraluminous starbursts, defined to have $L_{IR} \gtrsim 10^{12}L_{\odot}$, or about 100 times the luminosity of the Milky Way ($L_{IR} \sim L_B \sim 10^{10}L_{\odot}$), are invariably associated with mergers or strong tidal interactions with neighboring galaxies. Large concentrations of molecular gas are found, with surface density in excess of $10^4 M_{\odot}$ pc^{-2} for objects such as Arp 220. The efficiency of star formation is remarkably high, compared to the locally measured star formation efficiency.

One can try to understand the high star formation efficiency of a starburst in terms of the ineffectiveness of feedback, as follows. I take the pressure to be turbulent, $p_g = \rho_g\sigma_g^2$. I note first that

$$P \propto \rho_g^{-1.47}\sigma_g^{-2.72}\dot{\rho}_* \propto \rho_g^{-0.47}\sigma_g^{-2.72}\epsilon/t_{dyn},$$

where as before $\epsilon/t_{dyn} \equiv \dot{\rho}_*/\rho_g$. Now I use the collision ansatz from above, $v_{sn}\epsilon/t_d = \sigma_g t_{coll}^{-1}$, to infer that

$$P \propto \rho_g^{-0.47}\sigma_g^{-1.72}v_{sn}^{-1}t_{coll}^{-1}.$$

The cloud-cloud collision time is, now without any assumption about being in a disk,

$$t_{coll} = \mu_{cloud}\rho_g^{-1}\sigma_g^{-1},$$

where $\pi G\mu_{cloud}^2 = \frac{4}{3}\rho_g\sigma_g^2$. Hence $\sigma_g \propto P^{-0.6}\rho_g^{0.1}$ regardless of whether or not a disk geometry is assumed.

Now rewrite the expression for porosity using this derivation of gas velocity dispersion to obtain $\dot{\rho}_* \propto \rho_g^{1.74}P^{-0.6}$. I assume that the star formation rate is given by

$$\dot{\rho}_* \propto \rho_g^n,$$

where star formation models for cold disks based on gravitational instability growth rates, as well as observations of star-forming disks, suggest that $n \approx 1.5$. However, in a merger of gas-rich galaxies, one might expect cloud-cloud collisions and coalescence to be important triggers of star formation. This would suggest that $n \approx 2$ might be more appropriate for a starburst involving a major merger. Regardless of the precise value of the index, the porosity is inferred to vary as

$$P \propto \rho_g^{1.74-n}.$$

I infer that porosity, which is a measure of feedback, is important in disks which have $n \approx 1.5$. Hence one may solve the overcooling problem associated with disk formation via a two-phase interstellar medium: as gas density increases during disk formation, feedback rises to compensate and may inhibit excessive contraction. This should help resolve a major difficulty in disk formation, where over-efficient angular momentum transfer in a clumpy dissipative collapse results in a disk scale-length that is too small by about a factor of 5 for standard cold dark matter initial conditions (see below). Ensuing star formation will naturally be inefficient.

However, in a starburst, provided that $n \approx 2$, feedback is suppressed. As the density increases, the porosity, and hence feedback, is reduced. This plausibly leads to runaway, high efficiency star formation. The star formation will at least initially be stable and therefore continue unimpeded, since the starburst-induced increase in pressure and/or gas turbulence at specified gas density also results in a reduction of porosity.

5. STARBURSTS AND ELLIPTICAL GALAXY FORMATION

Compare these properties of a starburst with our expectations for a forming elliptical. Population synthesis based on a template of stellar spectra for stars of differing masses and evolutionary stages shows that the bulk of the stars in nearby ellipticals must have formed at least 6 Gyr ago. In fact, to obtain colors similar to those of globular cluster stars, an age of $\gtrsim 10$ Gyr is inferred for luminous ellipticals in most analyses. The colors of luminous ellipticals in galaxy clusters at $z \sim 0.4$ demonstrate that elliptical formation occurred at[41] $z \gtrsim 1-2$. Considerable star formation efficiency is also required, to avoid delayed infall of gas and continuing star formation that otherwise would form a disk. This is best achieved if the bulk of star formation occurs within a dynamical timescale, about 1 Gyr for the initial collapse of a protogalaxy. In this limiting case, the newly formed stars form a spheroid that is supported by the anisotropic velocity dispersion of the stellar orbits. If star formation continues over much longer than a dynamical time-scale, the intervening time allows gas pressure to maintain the figure of the forming galaxy, and the resulting spheroid would be flattened and rotationally supported. However, observations of rotational velocities show that luminous ellipticals are not rotationally supported[42]. Hence a high protoelliptical star formation rate is inferred, amounting to several hundred solar masses per year.

Galaxies form by gravitational instability of small energy density fluctuations during the matter-dominated phase of the expanding universe. The fluctuation spectrum is such that the amplitudes of density fluctuations are larger on smaller scales. Such a spectrum is inferred on cluster and supercluster scales, extending up to the horizon scale, from deep redshift surveys and studies of cosmic microwave background anisotropies. The spectrum naturally leads to hierarchical structure formation, with the smallest structures forming first on subgalactic scales, and progressively merging to form larger and larger objects, culminating at the present epoch in the formation of superclusters. Mergers of gas-rich protogalaxies are an inevitable phenomenon in hierarchical structure formation, and are likely to provoke starbursts. Mergers effectively concentrate gas into the inner kiloparsec of the final system, via the inelasticity of the enhanced rate of gas cloud encounters that occur during the merging process. It is the major mergers, between systems of mass ratio no more than 10 to 1, that are destined to be the precursors of ellipticals. Only these will be capable of efficient star formation, and of effective angular momentum transfer as the dense stellar substructures orbitally decay into the centers of the merged systems[43]. Any preexisting disks are likely to be destroyed in major mergers and creation of new disks is avoided by virtue of the exhaustion of the gas supply.

The central surface brightness and surface brightness profiles provide further clues. The efficiency of star formation measured for ultraluminous starbursts suffices to convert the observed gas surface density into the observed surface brightness of evolved ellipticals, $\sim 10^3 \, L_\odot \, \mathrm{pc}^{-2}$. Moreover, the surface brightness profiles measured in light from the older stars for post-starburst galaxies, and even for starbursts when measured at near infrared wavelengths, is fit by a de Vaucouleurs profile[44]. Dynamical relaxation to a spheroid-like profile evidently occurs by violent relaxation within the $\sim 10^8$ yr duration of the starburst, and certainly within the ~ 1 Gyr elapsed for the post-starburst systems where light profiles are more readily measured.

Ultraluminous starbursts and mergers are rare today. Only ~ 1 % of galaxies show evidence of strong tidal distortions. However, the situation is quite different at $z \sim 1$. Irregular galaxies, mergers, and galaxies with tidal features dominate the deep galaxy counts, and the measured redshift distributions show that such systems are the dominant galaxy population by $z \sim 1$. Luminosity evolution is measured for the disk galaxies, consistent with a formation epoch at $z \sim 1$. In contrast, however, little evolution is found for ellipticals out to this redshift. The fate of the distant irregulars must be to form present epoch S0 and dwarf galaxies.

This begs the question of when elliptical galaxies formed. Perhaps, if they formed via a dust-shrouded starburst, elliptical formation could only be visible at far infrared wavelengths, a wavelength regime that has yet to be adequately surveyed at high sensitivity and angular resolution. Although there is no direct evidence of elliptical formation, there are several indirect hints. These involve both morphological and chemical merger signatures in characteristic elliptical galaxy properties, intracluster gas enrichment, the tentative detection of a diffuse far infrared background, and the first ISO detections of distant ultraluminous infrared galaxies.

5.1. Galaxies as Fossils

Various properties of galaxies must have been generated at or soon after birth. These include their velocity dispersions, light profiles, central surface brightnesses, luminosities and metallicities. The fundamental plane for ellipticals and SO galaxies is the relation between half-light radius, central surface brightness and velocity dispersion. One finds that combination of the fundamental plane relation with the virial theorem results in the single parameter relation[45]:

$$M/L \propto L^{1/5}.$$

Correction for the non-homologous nature of elliptical galaxy profiles modifies this relation to[46]

$$M/L \propto \sigma.$$

Whichever relation is adopted, there is undoubtedly more to the fundamental plane than the virial theorem.

One may imagine that there are several classes of explanation for the tilt of the fundamental plane. For example, the origin of this relation may lie in the efficiency of star formation decreasing systematically with increasing mass, followed by mass loss of the gas that has not formed stars. However, this is the opposite trend to what one would anticipate. It could also arise from the differential concentration of baryons that subsequently form stars relative to dark matter, due to enhanced dissipation in lower mass systems: again, consideration of cooling timescales favors the opposite trend.

Simple models for dwarf galaxies predict that M/L should increase as M decreases, as indeed is observed for dE galaxies. A clue to the likely origin of the M/L variation comes from the observed variation of magnesium abundance with central velocity dispersion. In fact, the tightest correlation involving the magnesium line strength index is with escape velocity, both global and local, for luminous ellipticals. Metallicity, as inferred from color, correlates with escape velocity over a wide range of luminosity. The natural inference is that galactic winds have limited the metallicities of ellipticals. The correlation of Fe abundance with global escape velocity applies to dynamically hot systems, from dwarfs to giants.

However, there are also significant differences between sub–L_* ellipticals and luminous ellipticals. The lower luminosity ellipticals are rotationally supported whereas luminous ellipticals are supported by anisotropic velocity dispersion. This is strongly suggestive of major mergers with accompanying efficient star formation producing the luminous ellipticals, whereas minor mergers, with accompanying inefficient star formation, dominate the formation of the lower luminosity ellipticals. There are theoretical reasons for believing that major mergers result in an efficient burst of star formation, as the debris from supernovae is effectively trapped: this would not be the case for the smaller systems with shallow potential wells that are responsible for minor mergers.

There are observational indications that lower luminosity ellipticals have a broad range of ages. They often show evidence of past starbursts, as a younger component, and display disky isophotes, indicative of an early dissipative merger. The fundamental plane becomes poorly defined below $\sim 0.03L_*$, and the surface brightness of ellipticals peaks at about $0.1L_*$. All of this is suggestive of a distinct history for low luminosity elliptical formation.

Luminous ellipticals are observed to have high Mg/Fe, by a factor of ~ 2 relative to solar abundances, reflecting a chemical enrichment that is generated either by early mass loss or by a top-heavy IMF. Either possibility allows Type II SN yields to dominate the stellar abundances. The more uniform stellar populations, in terms of age, imply an efficient star formation history that is characteristic of major mergers. Major mergers provide an attractive environment for stimulating ultraluminous starbursts that would have provoked an early wind, and possibly the extreme conditions that might have biased the initial stellar mass function towards higher masses.

In order to connect metal production in ellipticals with the enrichment of the intracluster medium, the gas must be ejected. Luminous ellipticals are the most likely source, since low luminosity ellipticals have synthesized relatively few metals (although one could presumably invent a population of disrupted dwarfs that are no longer visible). However, the M/L dependence on M and dependence of metallicity on escape velocity argue for luminous ellipticals as the source of intracluster Fe, as is indirectly inferred from the correlation of intracluster Fe mass with luminosity of the early-type galaxies. Such a correlation suggests that the Fe source is dominated by L_* galaxies. Other arguments suggest that ellipticals have lost a substantial amount of metallicity, and hence gas, in an early wind. Relatively little gas is measured in the halos of isolated elliptical galaxies in studies of ROSAT X-ray images, and measurements of metallicity show that the Fe/H abundance in the elliptical halos is typically $\sim 1/3$ solar. Most of the metals produced by Type I supernovae must have been ejected from the elliptical galaxy environment, and in particular from the halo. Ram pressure stripping is unlikely to have played much of a role, since these isolated ellipticals are not in rich cluster environments.

Many ellipticals outside rich clusters show evidence of past major mergers, as inferred from the occurrence of fine structure in the form of shells, twisted isophotes and dust lanes[47]. Typical timescales inferred imply that the merger occurred a few Gyr ago, from dynamical modelling. The more dramatic examples of past mergers like the Antennae, with 100 kpc long tidal tails, merged ~ 1 Gyr ago, as confirmed by detection of A-type stellar features. Yet these systems have already developed elliptical galaxy-like light profiles for the older stars, and hence the merger and ensuing starburst may have resulted in the acquisition of the central surface density characteristic of ellipticals.

Could major mergers a few Gyr ago have triggered star-burst driven winds that cleared the halos of debris from prior supernovae, so that current epoch observations of elliptical halos would only measure the accumulation of stellar ejecta over a few Gyr? In clusters, such mergers occurred rather earlier in cosmic history, during the epoch of cluster collapse at $z \sim 1 - 3$, and could have supplied the enriched debris to account for the intracluster Fe.

5.2.   The Intracluster Medium

Much as globular star clusters provide a controlled environment for studying
stellar evolution via the Hertzsprung-Russell diagram, galaxy clusters can be
used to probe galaxy evolution for a sample of galaxies that formed more or less
contemporaneously. It is very likely that the cluster environment determines
the morphology of galaxies. One observes the early morphological types of
galaxies to be strongly concentrated towards rich cluster centers, ellipticals
dominating the central dense cores of rich clusters. Simulations demonstrate
that prolific merging occurred during the initial collapse phase of a rich cluster.
Hence it is a plausible deduction that galaxy morphology is determined by
mergers in this environment.

The mergers must have provoked starbursts. While starbursts are short-
lived, occur early in the universe, and are hence difficult to detect, the post-
starburst phase has spectroscopic signatures that persist for up to several Gyr.
These include Balmer absorption lines, the 4000 Å break, and blue light from
intermediate mass stars. Spectroscopic data for several clusters[48] reveals a
spread in various age diagnostics that is probably due to addition of new stars
in an earlier starburst. Since the post-starburst fraction of cluster galaxies
increases with redshift, one could infer that the merged fraction added to ellipticals and S0s increases systematically with redshift, to an inferred redshift
$\gtrsim 1$.

There are some galaxies with colors too red and 4000 Å breaks too large to
be fitted by this model. Dust cannot account for both of these effects, and it
has been suggested[49] that the culprit may be an IMF that is dominated by
massive stars, above 2 or 3 $M_\odot$. Within a Gyr or so after the starburst, the
light from the newly added component is dominated by giants. Such a top-
heavy IMF may be required to account for the observed dispersion in stellar
properties. Observations of the luminosity, supernova rate and radio emission
of the nearby starburst galaxy M82 have also been interpreted in terms of a
top-heavy IMF[50]. Complementary observations of abundances actually may
require such an IMF to have prevailed in past starbursts.

The intracluster medium provides a reservoir that traps any enriched ejecta
from massive stars that may well have escaped the host galaxies in a sufficiently
vigorous starburst or else been driven out by ram pressure stripping. That such
ejection occurred in the past is suggested by the amount of iron measured in
the intracluster gas via iron K-line emission. The average iron abundance is
1/3 of the solar value for a large sample of rich clusters. The mass in iron is
found to satisfy

$$M_{\mathrm{Fe}}^{ICM} \approx 0.02 L_V (\mathrm{E} + \mathrm{S0}),$$

where M_{Fe} and L_V are measured in $M_\odot$ and $L_\odot$, respectively.

The iron mass is proportional to the luminosity in early-type galaxies. For
comparison, the mass of iron in the old stellar component is

$$M_{\mathrm{Fe}}^* \approx 0.01 L_V (\mathrm{E} + \mathrm{S0}).$$

Now the iron in the Milky Way disk, with a mean metallicity of about 1/2 solar, amounts to $(M_{\mathrm{Fe}}/M_*)^{MW} \approx 1 \times 10^{-3}$, adopting $M/L_V \approx 6$ as characteristic of old stellar populations, whereas the iron mass in clusters is $(M_{Fe}/M_*)^{cluster} \approx 5 \times 10^{-3}$. The high cluster yield can be understood if the early IMF was top-heavy and led to gas ejection from ellipticals. The predominance of massive stars allowed a substantial contribution to the metallicity without affecting the present-day light, which is from lower mass stars.

5.3.   Protoellipticals as Top-Heavy Starbursts

By analogy with star formation locally, we may infer that star formation is bimodal, with low mass stars forming by ambipolar diffusion in cold clouds of subcritical mass, and stars of all masses forming in the collapse of supercritical mass warm clouds. The critical mass corresponds to the Bonner-Ebert scale, suitably generalized to include magnetic as well as thermal support. If this bimodality is responsible for the predominance of low mass stars in the local IMF, due in turn to the predominance of cold molecular gas over warm molecular gas, then one may reasonably expect that in the starburst environment where the gas temperature is elevated on account of the high star formation rate, the massive star-forming mode should be the dominant mode. Hence a top-heavy IMF seems an inevitable consequence once a starburst is initiated, and the ensuing wind will be capable of driving the enriched debris out even from massive galaxies.

One may now construct a simple model for proto-ellipticals[51]. Minor mergers result in star formation with a normal IMF. Typical masses involved are initially small, of dwarf galaxy scale, and the induced star formation is inefficient in the shallow potential wells of the infalling clouds and of the growing protogalaxy. The onset of galaxy clustering dramatically enhances the merging rate: eventually a major merger occurs between what by now is a massive protogalaxy and a gas-rich object amounting to a tenth or more of its mass.

Star formation is efficient in a major merger event: the ejecta from massive stars and supernovae are effectively trapped and recycled into continuing star formation. The associated starburst is likely to generate a top-heavy IMF, as a consequence of the high interstellar medium pressure and molecular cloud turbulence that are a consequence of the concentration of gas via inelastic cloud encounters into the inner kpc of the protogalaxy. The gas pressure and turbulence are sustained by the continuing supernovae as massive stars form and die, until a galactic wind, perhaps inevitably especially if the IMF is top-heavy, sweeps out the residual interstellar gas. An elliptical or S0 galaxy is left behind.

The top-heavy IMF in the starburst has two notable consequences. The mass-to-light ratio of the resulting stellar population is enhanced because of the increased number of white dwarfs and neutron stars relative to a normal IMF. Moreover the nucleosynthetic yield is enhanced, both in absolute terms, and in the ratio of alpha nuclei to iron. This is reflected partly in the abundances of the surviving stars, since even with a top-heavy IMF, there must be an appreciable mass fraction of solar mass stars, and most dramatically in the enriched wind. The wind escape velocity from a giant galaxy, comparable to the stellar velocity dispersion of $\sim 300\,\mathrm{km\,s^{-1}}$, exceeds the relative velocities of galaxies in loose groups. Hence one would expect the enriched gas to accumulate within galaxy clusters, and be detectable via iron line emission from the intracluster medium. This enhanced yield is precisely what is needed to account for the iron abundance, of about one-third the solar value, observed in the intracluster gas of rich clusters.

The starburst hypothesis, of a top-heavy IMF, seems capable of accounting both for the mass-to-light ratios of giant ellipticals, which are systematically larger than for lower mass ellipticals approximately according to $M/L \propto M^{1/6}$, and for the ratio of iron mass to visible light in galaxy clusters. At the same time, because elliptical halos will generally not retain the top-heavy IMF induced wind ejecta, one avoids over-enriching elliptical galaxy halos.

To quantitatively account for both of these properties, one has to use the observed M/L ratio as a measure of the compact remnant population and hence as a monitor of the mass in newly added massive stars, and therefore of the cumulative intracluster gas enrichment, in the major starburst that heralded the end of the protoelliptical phase. Concordance is found by adopting a Schechter luminosity function provided that the enriching starbursts generally occur for ellipticals of luminosity that exceed $L_{cr} = (0.01 - 0.1)L_*$. While there is no precise explanation of this critical luminosity, it is of interest to note that L_{cr} demarcates a transition in several, possibly related, structural properties of early-type galaxies. Below L_{cr}, ellipticals are oblate and rotationally supported, whereas most ellipticals more luminous then L_{cr} have anisotropic velocity dispersions and are prolate or triaxial systems. The fundamental plane thickens appreciably below L_{cr}, and the surface brightness of early-type galaxies reaches a maximum near L_{cr}, declining approximately as $\Sigma \propto L^{3/4}$ towards small L at $L \lesssim L_{cr}$, and $\Sigma \propto L^{-3/2}$ towards large L at $L \gtrsim L_{cr}$.

5.4. Predictions

There are a number of testable predictions that result if a top-heavy IMF was generated in protogalactic starbursts. A unique signature is provided by the ratios of alpha element abundances to iron. These ratios are enhanced by a factor of 2 – 3, as is empirically demonstrated in the observed abundance ratios of old, metal-poor population II stars, in which only the debris from massive star precursors has had time to be incorporated. One expects the stellar abundances, especially in the inner kpc, to systematically show enhanced $[\alpha/Fe]$ for the more luminous ellipticals. Magnesium and iron line indices support this prediction[53,52]. The intracluster gas should also contain $[\alpha/Fe]$ enhancements, visible as K and L X-ray line emission from Si, O, Mg etc. Results from the ASCA satellite show such enhancements, by a factor of ~ 2, in several rich clusters of galaxies[54]. The intergalactic medium outside rich clusters should also be enriched to a fraction of the ICM level if early-type galaxies account for a fraction ϵ of all galaxies in the field. With $\epsilon \sim 0.1$, one might expect enrichment to a level of $\sim 0.03 Z_\odot$ by the present epoch.

It is of interest to note that Lyman alpha forest clouds, inferred from their lack of clustering, deficiency of heavy elements and narrow line widths to be the most primitive intergalactic gas clouds probed via quasar observation line studies, have metallicities of about $10^{-2.5} Z_\odot$ at $z \sim 3 - 4$. These may also possess enhanced [Si/C], which if confirmed would support a starburst-enriched origin. The quasar phenomenon is itself often considered to be an event that characterizes fueling by a major merger in a protogalaxy of a massive central black hole. A starburst would presumably be an external accompaniment of such a merger. There are indications that the element ratios in the broad emission line regions of quasars have anomalously high N/O ratios, and this has been interpreted as evidence for enrichment by a top-heavy IMF[55]. One presumably is witnessing a transient stage where WN stars are dominating the enrichment.

For a host galaxy of a given mass or luminosity, one would expect a dispersion in the masses of objects participating in the final protogalactic major merger. This could result in a thickening of the fundamental plane. However, the mass distribution of captured objects will be steeply declining, so that the resulting dispersion need not be excessive. The distance uncertainties from fundamental plane analyses are less than about 20 percent, so that one requires the mass spread to not exceed

$$\frac{\Delta M}{M} = \frac{6}{5} \frac{\Delta L}{L} = \frac{12}{5} \frac{\Delta r}{r} \lesssim 0.5 \,.$$

Typically, about half of the mass can be attributed to the merger-generated remnants, so that one could tolerate a dispersion of up to a factor of 2 in merged masses. One would expect the residuals from the fundamental plane and the Mg or Fe abundance versus velocity dispersion trend to be correlated. Moreover the Mg/Fe ratio should be correlated with galaxy mass as well as with fundamental plane and (Mg, σ) residuals. The environmental dependence is weak, since mergers are expected to occur mostly in groups rather than in rich clusters or in the field.

The diffuse extragalactic background light inevitably incorporates the contribution from the enhanced nucleosynthetic activity driven by a top-heavy IMF in the early universe, even though individual objects may not be detectable. Production of a heavy element abundance fraction Z of the cosmic density ρ_0 results in a cosmic background light flux

$$\nu i_\nu = \epsilon_{nucl} Z \rho_0 (c/4\pi)(1 + z_e)^{-1},$$

where z_e denotes the redshift at which the metals are generated in protoellipticals and $\epsilon_{nucl} \approx 0.004 c^2 \, \mathrm{ergs}\,\mathrm{g}^{-1}$ is the nuclear energy released to form iron, oxygen etc. One cannot predict the frequency distribution of this cosmic background signal without a detailed model for protogalactic starbursts, as well as for the other contributing sources such as disk galaxy star formation. However, the starburst hypothesis for protogalaxies suggests that at least half of the starburst luminosity will be generated in the far infrared, peaking near $100\mu m$ in the rest frame by analogy with nearby dust-shrouded starbursts.

Searches for the diffuse background light at optical wavelengths have hitherto been unsuccessful. Current upper limits amount to $\nu i_\nu \approx 20\,nw\,sr^{-1}$, or about a factor of 10 above predictions for the diffuse light produced by the observed stellar populations in galaxies, evolved backwards in time to an appropriate formation epoch. There is a tentative claim of detection of a correlated component of the diffuse background light near 2000 Å at about the predicted level and with approximately the expected correlation strength of remote protogalaxies[56]. The expected starburst contribution may be normalized to the inferred iron mass-to-stellar mass ratio, enhanced by about 5 relative to the Milky Way, so that

$$\frac{(\nu i_\nu)_{starburst}}{(\nu i_\nu)_{disks}} = 5\left(\frac{1+z_d}{1+z_e}\right),$$

where z_d and z_e are the respective formation redshifts of disks and of protoelliptical starbursts. Now $z_d \sim 1$, as is inferred from deep HST studies of faint galaxy counts and from deep redshift surveys, where the disk/blue galaxy population is found to be undergoing strong evolution. One might expect ellipticals to form at about twice the redshift of disks of the same mass, if these objects correspond to respective 2σ and 1σ peaks in a primordial Gaussian density fluctuation field[57]. Cluster ellipticals would presumably form at a slightly higher redshift, because of the larger scale overdensity that characterizes their local environment. I take $z_e \sim 3$ and infer that one might plausibly expect $(\nu i_\nu)_{starburst} \approx 5\,nw\,sr^{-1}$, and the resulting diffuse radiation would peak at $\sim 100\,(1+z_e)\mu m$, or $\sim 400\,\mu m$. It is of interest to note that Puget et al[58] have reported a tentative detection of a cosmic far infrared background signal from analysis of FIRAS data, at a level of $3 - 10\,nw\,sr^{-1}$ and peaking near $400\,\mu m$.

6.   HOW TO REALLY FORM GALAXIES

There are two distinct approaches to galaxy formation that may be dubbed the "forwards" and "backwards" approaches. The forwards approach assumes plausible initial conditions that are interpolated from models for the large-scale structure of the universe. These conditions specify the evolution of the dark matter distribution until the mass of a galaxy halo separates out. *Ad hoc* prescriptions are adopted for star formation that seem plausible but are hardly unique. Numerical simulations are preferred using state-of-the-art smoothed particle or grid-based hydrodynamics combined with gravity for a multi-component fluid consisting of baryons and weakly interacting dark matter particles. For an L_* galaxy, the best resolution achievable in three dimensions utilizes $\sim 10^7$ particles, and corresponds to a mass resolution of about $10^5\,M_\odot$.

One can try to improve on this with variable grids, tree-based codes, or hybrid Eulerian/Lagrangian schemes, but additional refinements in computing power are also needed to cope with the complex non-gravitational physics of star formation. This includes following the evolution and radiative transfer of ionizing photons (and energetic particles) produced by the first massive stars to form, and the associated evolution in ionization and chemical abundances that regulates gas cooling and dissipation, which in turn controls the mass reservoir available for star formation. All of this must be incorporated into a code, along with assumptions about the IMF of the first stars, production of dust and yields of heavy elements. Needless to say, such issues as magnetic fields and protostellar outflows or supernova remnant-driven turbulence are usually ignored. Clearly, the forwards approach to galaxy formation has inevitable limitations with regard to its predictive power.

The backwards approach commences with nearby galaxies and in particular, with the Milky Way. The semi-phenomenological model for disk star formation accounts for the observed star formation rate, gas abundances, gas and stellar metallicities as a function of galactocentric radius, the stellar metallicity distribution, and the stellar metallicity as a function of stellar age. The model provides an adequate description of current star formation in nearby disk galaxies. These successes inspire enough confidence in the model to extrapolate it to early times. The primary missing ingredient is that of the spheroidal component of galaxies: this is added as an empirical starburst model. Implicit in backwards modelling of the star formation history is the assumption that the parameters of the model do not change with time. Specifically, the star formation efficiency and the IMF in the past are assumed to be identical to the model ingredients that work so well for nearby systems. This empirical approach nevertheless allows the possibility of a high star formation efficiency, as in nearby starbursts, and even of a top-heavy IMF, as inferred from tentative interpretations of abundance data for intracluster gas and elliptical galaxies.

6.1. Forwards Modelling

Gravitational instability in the expanding universe accounts for large-scale structure. The growth rate is sensitive to the adopted cosmological model as well as to a possible admixture of hot dark matter. It is especially sensitive to initial conditions. Specification of the amplitude and spectrum of the initial density fluctuations is a critical ingredient for simulating structure formation.

6.1.1. Linear Evolution

Effective growth of density fluctuations only occurs during the matter-dominated expansion phase, at $z \lesssim 3 \times 10^4\,\Omega h^2$. The associated temperature fluctuations $\delta T/T \sim 10^{-5}$ have been measured by the COBE satellite. $\delta T/T$ reflects the primordial amplitude of the density fluctuations, when appropriately coupled via a transfer function obtained from time-integrating the Boltzmann equation for the photons through the epoch of last scattering at $z \approx 10^3$. A cosmological model is necessary to extrapolate from the range probed by the COBE DMR ($> 300h^{-1}$ Mpc in comoving scale) to the relevant scale ($\lesssim 1h^{-1}$ Mpc) for galaxy formation. All viable models match the COBE spectrum with constant (scale-independent) gravitational potential fluctuations (if the primordial fluctuations are adiabatic) but the small-scale extrapolations differ by factors of up to ~ 10 in power. This reflects the uncertainty in current epoch large-scale structure probes on scales $\lesssim 100h^{-1}$ Mpc. The model yields the linear theory predictions of density fluctuations on comoving scales appropriate to those of galaxies.

Nonlinearity first occurs at $z \approx 10 - 30$. For cold dark matter, the density fluctuation spectrum is almost flat, diverging only logarithmically to small masses, on subgalactic scales. All primordial galactic scale fluctuations are erased if the dark matter is hot, and the ensuing failure to form galaxies at early epochs eliminates a hot dark matter-dominated universe as a possible cosmological model. A small admixture ($\lesssim 30$ percent) of hot dark matter is allowed, and provides a means of reconciling the $\Omega = 1$ universe with most observational constraints. A cold dark matter universe overproduces power in density fluctuations on large scales ($\lesssim 100h^{-1}$ Mpc) if $\Omega = 1$.

The allowed suite of CDM models includes an open model with $\Omega \approx 0.3$, a flat vacuum-dominated model with $\Omega_{CDM} \approx 0.3, \Omega_0 = 1$, and a flat ($\Omega_0 = 1$) mixed model with $\Omega_{CDM} = 0.7$ and $\Omega_{HDM} \approx 0.25$.

The associated initial density fluctuation spectrum, given by $\nabla^2 \delta\phi = 4\pi G \delta\rho$, is $\delta\rho/\rho \propto M^{-(n+3)/6} D(t)$, where the growth factor $D(t)$ is $t^{2/3}$ for $\Omega = 1$. Normalization is usually effected at a scale of $8h^{-1}$Mpc, where the variance of $\delta\rho$ as measured in the luminous galaxy counts has unit amplitude. Tilting the spectrum helps reconcile CDM at $8h^{-1}$Mpc, with the COBE fluctuations. For example, a fourth model with $\Omega_{CDM} = 0.95$ and $\Omega_0 = 1$ is allowed by slightly tilting the primordial fluctuation spectrum to reduce the large-scale power. A change in spectral index, defined by gravitational potential fluctuations $\delta\phi \propto M^{(1-n)/6}$, from the scale invariant value $n = 1$ to $n \approx 0.8 - 0.9$ suffices to reconcile a flat CDM model with observational constraints.

6.1.2. Nonlinear Evolution

A spherical shell model can be patched onto linear theory to give the nonlinear evolution following shell collapse. This is effected at the epoch of maximum shell size. If the shell collapse occurs to a radius that is half the maximum (in comoving coordinates), the density contrast at this stage is $18\pi^2$. This is the epoch when collisionless matter would have virialized. The virial velocity scales as $M^{(1-n)12}$, and the epoch of virialization scales as $M^{-(n+3)/4}$. Since $n > -3$, and $n \approx -2$, on galaxy scales, there is a clear hierarchical sequence of bottom-up structure formation.

If the primordial density fluctuations are Gaussian-distributed, as inflationary models suggest, one can infer the mass function of virialized lumps. It is necessary to filter the density field, since in cold dark matter, fluctuations are present on all scales, to define a mean density fluctuation over some filter scale R. The mass density fluctuations can be written as $\sigma^2(R,t) = \langle (\delta\rho/\rho(x)^2) \rangle = \sigma^2(R)D(t)^2$, where $\sigma^2(R) = 1$ for galaxy counts over spheres of radius $8h^{-1}$ Mpc. For the underlying mass fluctuations, one can define $\sigma_8 \equiv \sigma(8h^{-1}\,\text{Mpc})$, and a corresponding bias parameter by $b = \sigma_{galaxies}/\sigma_{matter} = 1/\sigma_8$. In fact, the scale at which $\sigma_{galaxies}$ is unity depends on the adopted galaxy catalog: $\sigma_{IRAS} \approx 0.7$ and $\sigma_{CfA} \approx 1$. Armed with $\sigma(R,t)$ one can write the fraction of mass points in spheres with $\delta\rho/\rho > \delta_c$, where δ_c is the linear theory amplitude at collapse to half the maximum radius, as

$$F(R,t) = \int_{\delta_c}^{\infty} \frac{d\ell}{\sqrt{2\pi}\sigma} e^{-\delta^2/2\sigma^2} \, .$$

If this mass fraction can be identified with the fraction of mass points in lumps of mass in excess of $\frac{4}{3}\pi\rho R^3$, where ρ is the background density, one can write the mass function of lumps as

$$\frac{dN}{dM} = -2\frac{\rho}{M}\frac{dF}{dR}\frac{dR}{dM} = -\sqrt{\frac{2}{\pi}}\frac{\rho}{M^2}\frac{\delta_c}{\sigma}\frac{d\ln\sigma}{d\ln M} e^{-(M/M_{nl})^{\frac{3}{n+3}}} \, ,$$

the factor of 2 accounting for the fact that half of the mass is in undersize regions and is accreted by the overdense regions.

6.1.3. Galaxy Mass Function

To compare the predicted mass function with data is not straightforward. One observes the cluster mass function and the galaxy luminosity function. Consider first the cluster mass function. The predicted mass function has only one characteristic mass scale, that of non-linearity, which today corresponds to $\sim 10^{14} M_{\odot}$, far from the mass of a galaxy but about equal to the mass of a modest cluster. The proximity of M_{nl} to the cluster mass means that one can reliably normalize the models by using the abundance of clusters. Studies of the cluster mass function suggest that[59] $\sigma_8 \approx 0.6 \pm 0.1$ in order to result in the observed number density of Abell clusters if $\Omega \approx 1$. There is a weak dependence on Ω: $\Omega^{0.6}\sigma_8 \approx 0.6$, that arises from the dependence of the fluctuation growth rate on Ω. Comparison with the COBE DMR fluctuations in the CMB sets an orthogonal constraint on the combination σ_8/Ω, leading to a narrow range in the allowed (Ω, σ_8) parameter space.

For galaxies, comparison of observations and theory is far more difficult. For example, there is a strong bias against galaxies of low surface brightness. These galaxies may contribute significantly to the mass budget of the universe. If one makes the comparison by assuming a constant mass-to-light ratio, one infers that theory overpredicts the number of dwarf galaxies. Yet the observed luminosity function has a characteristic luminosity $L_* \approx 10^{10} h^{-2} L_{\odot}$. The functional form is approximated by $dN/dL = \phi_0 L^{-\alpha_L} \exp(-L/L_*)$, where $\alpha_L \approx 1$.

Consider first the reconciliation of the theoretical mass function with slope $\alpha_M \approx 2$ at the low mass end with the observed slope $\alpha_L \approx 1$. Possible solutions involve either destruction of dwarfs or avoiding their formation. Early winds driven by massive stars can disrupt the gas supply, leaving a low surface brightness remnant. Mergers would incorporate the dwarfs into more massive systems. An IMF deficient in stars of mass above 1 $M_\odot$ would leave behind dim dwarfs that would perhaps only be detected in wide area HI surveys. Preheating of the intergalactic medium provides perhaps the most attractive solution[60]. This would, after a first generation of dwarfs, raise the Jeans mass sufficiently that early massive galaxy formation occured subsequently until the IGM has adiabatically cooled. Dwarf formation can occur as early as $z \sim 30$, when H_2 cooling suffices to allow baryons to condense into objects as small as $\sim 10^6$ $M_\odot$.

The origin of L_* is also not well understood. Cooling constraints do not impose a sharp limit on the cooled baryonic mass within a dark matter potential well. The mass that can cool within a dynamical time can be estimated from the ratio of cooling to dynamical time scales for a baryon distribution modelled by an isothermal sphere:

$$\frac{t_{cool}}{t_{dyn}} \approx \frac{G^2 m_r^2 M}{\Lambda \sigma} \equiv \frac{M}{M_{cool}},$$

where Λ is the cooling rate and is a weak function of velocity dispersion σ over the relevant range, and $M_{cool} = \Lambda \sigma / G^2 m_p^2$. Since Λ is not a decreasing function of σ at $T \gtrsim 10^6$ K (or $\sigma \gtrsim 100 \, \text{km s}^{-1}$), I infer that the cooled mass increases without limit as the potential well depth (σ^2) increases.

Feedback from star formation must be invoked to limit baryonic accretion. A similar argument has to be invoked in order to account for disk galaxy sizes. Early estimates of disk sizes asserted that in a quasi-uniform collapse of baryons in a dark matter potential well, the radial collapse factor of the baryons is λ_i, where the dimensionless initial angular momentum $\lambda_i = |E|^{1/2} J / G M^{1/2} \approx 0.06$. In the absence of a dark halo, the self-gravitating baryon collapse factor is only $\lambda_i^{1/2}$. For an initial gas extent of ~ 100 kpc, the inferred halo size, one could form a self-gravitating disk in centrifugal balance of about the right size only in the presence of a preexisting dark halo against which the infalling baryons could torque and conserve specific angular momentum. However, numerical simulations of clumpy halo collapse find that angular momentum transfer is highly effective as a consequence of dynamical friction and gas cooling. A disk within a dark halo is now a factor of 5 too small.

One solution might be to resort to a halo with high initial baryon content. Another is to appeal to feedback from star formation to prevent excessive infall. In disk formation, porosity was found to decrease as the gas density increases, suggesting that feedback is increasingly important as the baryons accumulate. With a two-phase interstellar medium, the disk overcooling problem can be resolved.

6.1.4. Understanding Galaxy Luminosities

What determines the characteristic luminosity of a galaxy? The energetics of supernova feedback might work as follows in imposing a limit on the least and on the most massive galaxies that can form. The rate of binding energy radiated by the protogalactic gas certainly does not exceed the rate of injected supernova energy. In fact, balance between these rates must occur in order for the gas to continue to accrete and contract. This is what one might expect in a cooling flow at the center of which a galaxy develops. Equating the cooling and heating rates yields

$$\sigma^2 = 2\frac{t_{cool}}{t_{dyn}}\frac{E_{sn}}{m_{sn}}\,\epsilon\,,$$

where the star formation efficiency ϵ is the ratio of stars formed per dynamical time relative to gas mass, m_{sn} is the mass in stars formed per supernova, and E_{sn} is the initial kinetic energy of a supernova remnant. Now in order to form stars, the cooling time t_{cool} must be less than the dynamical time t_{dyn}. I infer that there is an upper limit on gas velocity dispersion

$$\sigma < \sigma_* = 300 \left(\frac{\epsilon}{0.2}\right)^{1/2} \left(\frac{E_{sn}}{10^{51}\mathrm{ergs}}\right)^{1/2} \left(\frac{250M_\odot}{m_{sn}}\right)^{1/2} \mathrm{km\ s}^{-1}.$$

The need to simultaneously have gas accretion and star formation in the presence of supernova feedback implies that there is an upper limit on σ. For star formation efficiencies that are inferred for forming ellipticals, the expression for σ_* can help provide an explanation of L_*.

Now I examine the condition imposed by the two-phase interstellar medium. There is little doubt that low mass galaxies are effectively shredded when they first form stars by the onset of a vigorous supernova-driven wind[61]. Combination of the expression for porosity with the condition that cooling balances heating and that the cooling time is less than the dynamical time implies that

$$P > 0.6(\sigma/100\mathrm{km\ s}^{-1})^{-0.7}(0.1\rho/\rho_{gas})^{1/2}.$$

Strong winds are inevitable if $P \gg 1$. Hence only relatively deep potential wells survive as luminous galaxies, efficient protogalactic star formation requiring the central velocity dispersion to exceed about 100 km/sec.

However, the star formation efficiency in spheroids is the key to understanding the origin of galaxy luminosities. Its origin must be sought in dynamical modelling of the interactions of massive gas clouds in a spheroidal potential well. This is likely to be very different from star formation in disks. In disks, the global star formation efficiency is a few percent. In forming spheroids, ϵ is an order of magnitude larger. A similar conclusion about ϵ was inferred for starbursts. Perhaps the resolution is in the nature of the merging process that triggers star formation in ultraluminous starbursts and is an essential ingredient in forming ellipticals.

Simulations show that preexisting massive and dense substructures undergo strong dynamical friction in the merger process and efficiently transfer angular momentum to end up in the central bulge of the merged galaxy. Gas dynamical simulations show that the presence of an initial bulge in a merger between disk galaxies strongly enhances the resulting gas concentration in the merger product, in which the gas inflow is driven by tidal torques exerted by a transient bar. These results suggest that deep potential wells that result from major mergers and massive substructures are likely to guarantee efficient star formation. Such mergers should have been most frequent in the densest regions of the universe when galaxy clusters are forming. It is perhaps no coincidence that ellipticals and S0s are the prevalent morphological type in dense regions. In contrast, minor mergers will tend to develop lower mass substructures. Disruption by massive stars will be important and result in globally inefficient star formation. No doubt this could account for the prevalance of disk galaxies in the field, where major mergers would be relatively rare.

6.2. Backwards Modelling of Galaxy Formation

Given our abysmal ignorance of star formation in the extreme environment of the early universe, an alternative approach is to begin with present day star formation. The semiphenomenological expression for the disk rate of star formation accounts for many observed aspects of star formation. The Milky Way provides an ideal template. With few adjustable parameters, most notably star formation efficiency ϵ and infall rate, one can account for the observed radial distributions of gas, star formation rate, gas metallicity and stellar metallicity, as well as the age-metallicity relation in the solar neighborhood. One can predict the history of the gas, star and metallicity distributions in our galaxy. Varying the star formation efficiency allows one to reproduce the differing mix of old and young stars that distinguishes between Hubble types.

One can now predict the angular size distribution of galaxies as a function of redshift. Disk galaxies are expected to be very compact at $z > 1$. Sab's, in particular, are even smaller than ellipticals viewed at high redshift. This is because as the disk shrinks, the galaxy light is dominated by the bulge. Observations of high redshift galaxies at $z > 3$, as identified by the Lyman break, are compact systems, as are a substantial fraction of the galaxies in the Hubble Deep Field. These galaxies are typically inferred to be at $z > 1$, generally confirming the expected compactness of high redshift galaxies.

The luminosity function of field galaxies for different morphological types can be extrapolated to past epochs if mergers are neglected. One can compute the deep galaxy counts, as functions of redshift or of apparent magnitude. This approach, that of pure luminosity evolution, is successful in explaining the numbers of faint blue galaxies provided that a low density ($\Omega \sim 0.3$) cosmological model is adopted. Disk size evolution is mandatory in order to account for the observed distribution of half-light radii in the HDF. Backwards evolution of disks fails in one respect, however, The Canada-France Redshift Survey finds a peak in the measured UV/blue light density for the survey galaxies at $z \approx 1 - 2$. Such a peak is confirmed by the Lyman break galaxies at $z \cdot \sim 3$, which measure a reduced intensity per comoving volume of young star-forming galaxies. If one neglects any potentially large contribution from dusty systems at $z > 1$, these results are simply interpreted as a peak in the star formation rate. The slow evolution of disks cannot generate such a peak.

The peak is too recent to be identified with elliptical starbursts. Luminous ellipticals are too red to have formed so recently. Disk and spheroid comparisons suggest thet spheroids are at least formed with disks, and indeed the Lyman break objects at $z > 3$ are plausibly forming spheroids. Given that the evolution to $z \sim 1$ is inferred to be due to disk luminosity, a new class of disks must be sought that shows stronger evolution than ordinary spiral galaxies. These may in fact be irregulars that are in the process of becoming normal disks.

One possibility is to tinker with early infall. Provided the metal-poor gas infall lasted about 3-4 Gyr, it may simultaneously resolve the local G dwarf problem (the paucity of metal-poor disk stars) and give a rapidly declining early star formation rate as seems to be required by the cosmological observations. The early star formation would have produced the initial disk enrichment. Another candidate is formation of S0 galaxies. These evolve initially as disks, but with higher efficiency, and higher star formation rate, than spirals, until their gas supply is disrupted. Perhaps if this disruption is triggered by a satellite merger and ensuing starburst one might naturally have an enhanced late source of formation that could account for the star formation peak near $z \approx 1$. Alternatively, dwarf ellipticals, many of which may be the precursor field counterparts of the low redshift dE's that rise steeply in number at the low luminosity end of nearby cluster luminosity functions, could be the culprits.

Where are the young counterparts of luminous ellipticals? This is one of the most urgent issues that remains to be resolved. High star formation rates of sevcral hundred solar masses per year are inferred from population synthesis models, yet narrow band optical and near infrared searches have failed to find such objects at $z < 5$. There are two possible resolutions. These objects may be dust-shrouded. Only deep far-infrared surveys would be able to detect them, and such surveys are presently being planned. Alternatively, it is possible that the bulk of the star formation in protoellipticals preceded the last major merger. In this case the star formation would be spread out over several Gyr, and star formation rates of tens rather than of hundreds of solar masses per year would be predicted from individual clumps. Our knowledge of star formation in protospheroids is too meager to give much theoretical guidance: observations from such experiments as ISO, SCUBA, and WIRE may provide more empirical clues in the not too distant future.

Acknowledgments

I am indebted to my collaborators Rychard Bouwens, Stephane Charlot, Colin Norman, Mathias Steinmetz, Rosemary Wyse and Steve Zepf for many discussions of the topics covered in these lectures. My research at Berkeley has also been supported in part by grants from NASA and NSF.

References

[1] Hillenbrand, L. 1996, *ApJ*, submitted
[2] Tinney, C. G. 1993, *ApJ*, **414**, 279
[3] King, I. R. *et al.* 1996,   Formation of the Galactic Halo...Inside and Out, ASP
Conf. Ser., Vol. 92, H. Morrison and A. Sarajedini, eds., p. 277
[4] Monet, D. G. *et al.* 1992, *AJ*, **103**, 630
[5] Low, C. and Lynden-Bell, D. 1976, *MNRAS*, **176**, 367
[6] Silk, J. and Takahashi, T. 1979, *ApJ*, **229**, 242
[8] Goodman, A. A., Benson, P. J., Fuller, G. A. and Myers, P. C. 1993, *ApJ*, **406**, 5
[7] Arquilla, R. and Goldsmith, P. F. 1986, *ApJ*, **303**, 356
[9] Myers, P. C. and Fuller, G. A. 1993, *ApJ*, **402**, 635
[10] Caselli, P. and Myers, P. C. 1995, *ApJ*, **446**, 665
[12] Kroupa, P. 1995, *ApJ*, **453**, 358
[13] Goodman, A. *et al.* 1997, *ApJ*, submitted
[11] Larson, R. B. 1995,, *MNRAS*, **272**, 213L
[14] Maddalena, R. J. and Thaddeus, P. 1985, *ApJ*, **294**, 231
[15] Williams, J., De Geus, E. and Blitz, 1994, *ApJ*, **428**, 693
[16] Williams, J. P. , Blitz, L. and Stark, A. A. 1995, *ApJ*, **451**, 252
[17] Schaerer, D. 1996, *ApJ*, **467L**, 17
[18] Doyon, R., Joseph, R. D. and Wright, G. R. 1994, *ApJ*, **421**, 101
[19] Fisher, D., Illingworth, G. and Franx, M. 1995, *ApJ*, **448**, 119
[20] Loewenstein, M. and Mushotzky, R. F. 1996, *ApJ*, **466**, 695
[21] Tamanaha, C. M. *et al.* 1990, *ApJ*, **358**, 164
[22] Noh, H. and Scalo, J. 1990, *ApJ*, **352**, 605
[23] Lilly, S. J. *et al.* 1995, *ApJ*, **455**, 1081
[24] Kennicutt, R. C. 1989, *ApJ*, **344**, 685
[25] Martin, C. and Kennicutt, R. C. 1997, in preparation
[26] Wyse, R. F. G. and Silk, J. 1989, *ApJ*, **339**, 700
[27] Wang, B. and Silk, J. 1993, *ApJ*, **427**, 759
[28] Cayon, L., Silk, J. and Charlot, S. 1996, *ApJ*, **467L**, 53
[29] Schade, D. *et al.* 1996, *ApJ*, **465L**, 103
[30] Oemler, A., Dressler, A. and Butcher, H. R. 1997, *ApJ*, **474**, 561
[31] Glazebrook, K. *et al.* 1995, *MNRAS*, **275**, 157
[32] Driver, S. P. *et al.* 1995, *ApJ*, **449**, L23
[33] Pei, Y. C. and Fall, S. M. 1995, *ApJ*, **454**, 69
[34] Fall, S. M. and Pei, Y. C.1993, *ApJ*, **402**, 479
[35] Silk, J. 1997, *ApJ*, in press
[36] Cioffi, D., Mckee, C. F. and Bertschinger, E. 1988, *ApJ*, **334**, 252
[37] Strauss, M. A. and Willick, J. A. 1995, *Phys. Rev.*, **261**, 272
[38] Mihos, C. J. and Hernquist, L. 1994, *ApJ*, **425L**, 13

[39] Mihos, C. J. and Hernquist, L. 1996, *ApJ*, **464**, 641
[40] Carral, P. *et al.* 1994, *ApJ*, **423**, 223
[41] Bender, R., Ziegler, B. and Bruzual, G. 1996, *ApJ*, **463L**, 51
[42] Fisher, D., Illingworth, G. and Franx, M. 1995, *ApJ*, **438**, 539
[43] Zurek, W. H., Quinn, P. J. and Salmon, J. K. 1988, *ApJ*, **330**, 519
[44] Wright, G. S. *et al.* 1990, *Nature*, **344**, 417
[46] Graham, A. and Colless, M. 1997, *MNRAS*, in press
[45] Bender, R., Burstein, D. and Faber, S. M. 1992, *ApJ*, **399**, 462
[47] Schweizer, F. and Seitzer, P. 1992, *AJ*, **104**, 1039
[48] Charlot, S. & Silk, J. 1994, *ApJ*, **432**, 453
[49] Charlot, S., Ferrari, F., Mathews, G.J., and Silk, J. 1993, *ApJ*, **419**, L57
[50] Doane, J.S., & Mathews, W.G. 1993, *ApJ*, **419**, 573
[51] Zepf, S. and Silk, J. 1995, *ApJ*, **466**, 114
[52] Davies, R. L., Sadler, E. M. and Peletier, R. F. 1993, *MNRAS*, **262**, 650
[53] Worthey, G, Faber, S. M. and Gonzalez, J. J. 1992, *ApJ*, **398**, 69
[54] Mushotzky, R. *et al.* 1996, *ApJ*, **466**, 686
[55] Hamann, F. and Ferland, G. 1993, *ApJ*, **418**, 11
[56] Martin, C., Hurwitz, G. and Bowyer, S. 1991, *ApJ*, **379**, 549
[57] Blumenthal, G., Faber, S., Primack, J., & Rees, M.J. 1984, *Nature*, **311**, 517
[58] Puget, J.-L. *et al.* 1996, *A&A*, **308L**, 5P
[59] Cole, C. *et al.* 1997, *MNRAS*, in press
[60] Blanchard, A., Valls-Gabaud, D. and Mamon, G. A. 1993, *A&A*, **264**, 365
[61] Dekel, A. and Silk, J. 1983, *ApJ*, **303**, 39

Figures

KENNICUTT paper
fig. 1 (ApJ), fig. 2.(ApJ), fig. 3 (ApJ), fig. 4 (ApJ), fig. 6 (A&A), fig. 7 (MNRAS), fig. 8 (AJ), fig. 9 (ApJ), fig. 10 (ApJ), fig. 12 (ApJ), fig. 17 (ApJ)

FALGARONE paper
fig. 1 (A&A), fig. 2 (A&A), fig. 4 (ApJ), fig. 11 (Nat)

PALLA paper
fig. 1 (MNRAS), fig. 2 (ASP series), fig. 3 (ARAA), fig. 4 (A&A), fig. 5 (ApJ), fig. 6 (ApJ), fig. 7 (ApJ), fig. 8 (MNRAS)

MEYNET paper
fig. 1 (A&A), fig. 2 (A&A)

COMBES paper
fig. 1 (MNRAS), fig. 2 (MNRAS), fig. 8 (ApJ), fig. 9 (AJ), fig. 11 (MNRAS), fig. 12 (ApJ)